D1382946

Auger Electron Spectroscopy

CHEMICAL ANALYSIS

A SERIES OF MONOGRAPHS ON
ANALYTICAL CHEMISTRY AND ITS APPLICATIONS

VOLUME 74

A WILEY-INTERSCIENCE PUBLICATION

JOHN WILEY & SONS

New York / Chichester / Brisbane / Toronto / Singapore

Auger Electron Spectroscopy

MICHAEL THOMPSON
MARK D. BAKER

University of Toronto
Lash Miller Chemical Laboratories
Toronto, Ontario
Canada

ALEC CHRISTIE

Vacuum Generators Scientific Ltd.
East Grinstead
United Kingdom

JULIAN F. TYSON

University of Technology
Loughborough, Leicestershire
United Kingdom

A WILEY-INTERSCIENCE PUBLICATION

JOHN WILEY & SONS

New York / Chichester / Brisbane / Toronto / Singapore

Library of Congress Cataloging in Publication Data:
Main entry under title:

Auger electron spectroscopy

  (Chemical analysis; v. 74)
  "A Wiley-Interscience publication."
  Includes index.
    1.   Photoelectron spectroscopy.   2.   Auger effect.
I. Thompson, Michael, 1942-   .   II. Series.

QD96. P5A9   1984   543'.0858   84-21922
ISBN   0-471-04377-X

Printed in the United States of America
10 9 8 7 6 5 4 3 2 1

PREFACE

Historically, Auger electron spectroscopy appears to be one of a number of techniques that are characterized by a long induction period. Some 50 years elapsed from the time of the brilliant observations of Pierre Auger to the establishment of the technique as a particularly valuable tool in the range of methods available to the surface analyst. The tremendous importance of surface science in practical interfacial technology continues to grow rapidly, and there is no doubt that analysis by AES is regarded as indispensable in certain areas—for example, surface metallurgy. Accordingly, we felt that the many facets of AES deserved the preparation of a comprehensive book directed at the analytical chemistry community.

Chapter 1 presents a brief introduction to the place of AES in the family of electron spectroscopic methods and a short description of its history. This is followed, in Chapter 2, by a look at Auger nomenclature and a concise review, with a historical flavor, of the important theoretical aspects, such as Auger energies and transition rates, Coster–Kronig transitions, and fluorescence yields. The next chapter constitutes an introduction to the instrumentation required to obtain an Auger electron energy spectrum and the various modes in which the technique is used, that is, direct versus differential recording, scanning Auger spectroscopy, and concentration-depth profiling. Application of these method is treated in subsequent sections of the book.

Gas-phase Auger spectroscopy is discussed in the following chapter. This area is often dealt with in a rather cursory fashion because emphasis is placed very much on the surface technological aspect. However, in this volume we have attempted to provide a comprehensive review of the field in view of the importance of the technique as a probe of atomic and molecular electronic structure and as an adjunct to surface analysis with respect to molecular adsorption to materials. Chapter 5 concentrates on a discussion of the origins and applications of the Auger chemical shift. Although it is fair to say that correlation of energy shifts with chemical environment is not as straightforward as in the case of x-ray photoelectron spectroscopy (which is not always facile either), it can be very useful in selected cases. In this section we also include a summary of some applications with a strong chemical bias, such as electrochemistry and catalysis.

The enormous literature on the use of AES in materials science and surface analysis is condensed to form Chapter 6. Here we discuss three-dimensional analysis, metallic surfaces, oxide surfaces, films and coatings, internal interfacial analysis (which includes topics such as adhesion and tribology), and semiconductor analysis. The final chapter gives a brief comparison of the properties of AES as a tool for the examination of surfaces with other common techniques.

The content of the book deliberately reflects the background and interests of the authors in an attempt to be as comprehensive as space allows. However, we recognize that even a group of individuals possesses a limited range of collective expertise and, therefore, someone else's pet topic may be omitted. We can only hope that the text still represents a contribution of some merit to the understanding and practice of AES.

Finally, we wish to express our sincere thanks to Susanne McClelland of the Department of Chemistry of the University of Toronto not only for expertly typing the manuscript, but more importantly, for her constant and highly good-natured attempts to prod us in to action on the next chapter.

MICHAEL THOMPSON
MARK D. BAKER

ALEC CHRISTIE
JULIAN F. TYSON

Toronto, Ontario, Canada

East Grinstead, United Kingdom
Loughborough, Leicestershire, United Kingdom
January 1985

CONTENTS

CHAPTER

1

INTRODUCTION

1. ELECTRON SPECTROSCOPY

Auger electron spectroscopy (AES) occupies a place in the family of techniques that is now known as electron spectroscopy. This is the generic
title given to the various methods that are based on the study of the kinetic
energy of electrons produced by a collision between the sample species and
an impacting photon or particle (Fig. 1.1). Although the various techniques
were developed somewhat independently by groups working in such diverse
areas as surface physics and molecular spectroscopy, the consensus of opinion
among interested parties is that the field has now taken its place as a distinct
branch of the "tree" of spectroscopy. The enormous growth in recent years
in fundamental and practical studies using the techniques of electron spectroscopy has undoubtedly been fueled by significant advances in instrument
technology such as the construction of high-resolution electron-energy
analyzers and equipment for production of ultra-high vacuums. A second
important factor is the tremendous growth in interest in such specific areas
as the physics and chemistry of surfaces, where application of such techniques
as x-ray photoelectron spectroscopy (XPS) and AES have shed new light on
on the surface properties of materials. Finally, the natural if belated development of commercial multipurpose equipment has occurred in recent times.
This is clearly an important factor considering the price tag that is generally
applied to electron spectroscopic instrumentation.

The individual techniques encompassed by electron spectroscopy are
generally classified according either to the method of inducing ionization or
excitation or to the nature of the process that accompanies the emission of
electrons. In order to illuminate the place of AES in electron spectroscopy a
brief summary of basic concepts underlying the major techniques and an
outline of their applications are given subsequently. In addition, the salient
features are included for comparison in Table 1.1 together with a number of
references to appropriate texts and review papers. This discussion is followed
by a short description of the history of the development of AES and an introduction to its advantages as a tool for analysis, particularly in the field of
surface study.

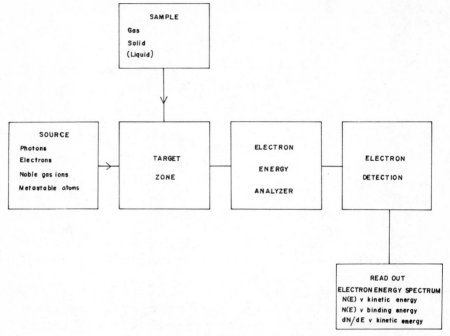

Figure 1.1.   The electron spectroscopy experiment.

1.1.   Vacuum Ultraviolet Photoelectron Spectroscopy (UPS)

In UPS the atom or molecule is exposed to a source of radiation in the vacuum ultraviolet region of the electromagnetic spectrum. The photon energy used for most conventional work is the He(I) line at 21.22 eV produced from an electrical discharge in helium gas. However, other energies are available, for example, the He(II) and H Lyman-α lines at 40.81 and 10.20 eV, respectively. The resulting effect of the photoionization of a large number of like atoms or molecules with photons of fixed energy hv is to yield a poly-energetic emission of electrons described by the adjusted Einstein relation,

$$E_n = hv - I_n \qquad (1.1)$$

where I_n is the ionization energy (binding energy) of the nth species of electron and E_n is the kinetic energy of the bunch of electrons ejected by a photon of energy hv. In the case of molecules a certain amount of the photon energy will cause concomitant vibrational and rotational excitation, hence a more correct form of the Einstein law is

$$E_n = hv - I_n - E_{\text{vib}} - E_{\text{rot}} \qquad (1.2)$$

Table 1.1. Some Techniques in Electron Spectroscopy

Technique	Abbreviation	Mechanism Schematic	Principal Features	References
Vacuum ultraviolet photoelectron spectroscopy	UPS	$h\nu + M \rightarrow M^+ + e^-$	Valence ionization potentials, spin-orbit coupling, ion-vibrational levels, exchange structure	1–8
x-ray photoelectron spectroscopy	XPS	$h\nu + M \rightarrow M^+ + e^-$	Binding energies, spin-orbit coupling, chemical shift, shake-up and multiplet structure	5–10
Auger electron spectrometry	AES	$(h\nu \text{ or } e^-) + M$ $\rightarrow M^+ + e^-$ $M^+ \rightarrow M^{++} + e^-_{\text{Auger}}$	Auger electron peaks, chemical shift, shake-up and shake-off satellites, autoionization, Coster–Kronig electrons	Text references
Penning ionization spectroscopy	PIS	$\text{He}^* + M \rightarrow \text{He} + M^+ + e^-$	Compare UPS	11–13
Electron impact energy loss spectroscopy	EELS	$e^- + M \rightarrow M^* + e^-_{\text{loss}}$	Electronic excitation levels; compare optical absorption	14–17
Ion neutralization spectroscopy	INS	$M + \text{He}^+ \rightarrow M^+ + \text{He} + e^-$	Compare Auger spectroscopy	18

3

It is an analysis of the kinetic energy distribution, $N(E)$, that is performed in UPS, and a plot of this parameter versus E is called a photoelectron spectrum (ultraviolet excitation). Equation (1.2) can only be regarded as approximate since it does not take into account the fraction of photon energy imparted to the ion or the possible occurrence of other processes such as two-photon–single-molecule collisions. However, the contributions of these phenomena to the total energy picture are clearly expected to be very small and, therefore, are generally ignored.

UPS measurements can be conducted on gases and solids, but by far the greater amount of work has been carried out on the former. For an excellent review of photoemission studies on solids the reader should consult Reference 19. The principal features exhibited by typical gas-phase UP spectra are those of electronic detail and vibrational fine structure. There are two basic kinds of electronic substructure, namely, that generated by a neutral system containing a partly filled electronic subshell and secondly the effect of spin-orbit coupling. A good example of the former is exhibited by the UP spectrum of NO.[20] Ionization of a $\sigma2p$ electron results in a triplet state, $^3\Pi$, in which the σ electron has its spin opposed to that of the π^* electron, and a $^1\Pi$ state with σ electron with parallel spin (exchange structure). Furthermore, electrostatic coupling between a $\pi2p$ hole resulting from filled $\pi2p$ photoionization and the π^* electron generates the six states, $^{3,1}\Delta$, $^{3,1}\Sigma^+$, and $^{3,1}\Sigma^-$. A summary of the possible states and suggested spectral assignments are given in Table 1.2. Spin-orbit coupling refers to the spectral band splitting that occurs as a result of the production of a molecular ion in an orbitally degenerate state. The hydrogen halide molecules show the expected

Table 1.2. Electrostatic Coupling Effects in Nitric Oxide Resulting from Photoionization[a]

Orbital Ionized	States	Vertical Ionization Potential (eV)
π^*	$^1\Sigma^+$	9.4
$\sigma2p$	$^3\Pi$	16.5
	$^1\Pi$	18.3
$\pi2p$	$^{3,1}\Delta$, $^{3,1}\Sigma^-$, $^3\Sigma^+$	16–20
	$^1\Sigma^+$	~23
σ^*2s	$^3\Pi$, $^1\Pi$	21.7

[a]Configuration: $(\sigma^*2s)^2(\sigma2p)^2(\pi2p)^4(\pi^*2p)^1$.

progressive increase in splitting of the orbitals of halogen *np* character and are, therefore, a good example of the phenomenon of spin–orbit coupling (HF, 0.033 eV; HCl, 0.073 eV; HBr, 0.32 eV; and HI, 0.66 eV).

A photoelectron band obtained from a molecule contains a representation of vibrational energy quanta of excitation that occurs with photoionization. The elements of the band are often collectively termed the Franck–Condon envelope owing to the relationship between potential energy curves for a molecule and one of its ionized states. Ionization of a nonbonding electron results in little change in the internuclear distance and a brief vibrational progression, whereas removal of a bonding or bond-contributing electron causes significant changes in the bond length and a distribution of possible

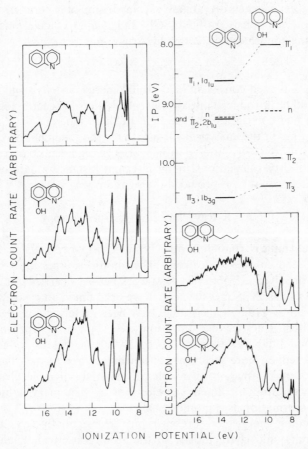

Figure 1.2.   He(I) photoelectron spectra of 8-hydroxyquinoline and derivatives.[21] (Reprinted with permission of Elsevier Scientific Publishing Company, New York.)

vibrational transitions. Vibrational structure has been used to study vibrational frequencies in ions; to distinguish bonding, nonbonding, and antibonding electrons; and to examine the consequences of both vibronic coupling and the Jahn–Teller theorem.

Examples of typical He(I) UP spectra are those of 8-hydroxyquinoline and some of its derivatives (Fig. 1.2).[21] In the case of the parent molecule, the four narrower bands at vertical ionization potentials of 8.0, 8.2, 9.1, and 10.0 eV correspond to ionizations from the π_1, $n(N)$, π_2, and π_3 orbitals, respectively, and the structure between 10.5 and 20.0 eV relates to remaining π and σ levels. UPS has been applied to problems in analytical chemistry and studies of molecular conformation, geometry, free radicals, transient species, and tautomeric equilibria. It should be noted that the relevance of UPS in theoretical chemistry, namely, comparision of experimental bonding energies with self-consistent-field (SCF) and *ab initio* calculations, is largely based on the application of Koopmans' rule,

$$I_n = -\varepsilon_n^{SCF} \qquad (1.3)$$

where ε_{SCF} is the self-consistent field orbital energy. Strictly speaking corrections should be introduced into orbital calculations to account for electronic relaxation and correlation. In the light of these inherent limitations, it is heartwarming for the analytical chemist to see the statement by Orchard:[22] "The literature of UPS is littered with incorrect assignments based exclusively on unreliable MO calculations, many of them in instances where an empirical line of attack would have provided the right answer."

1.2. X-Ray Photoelectron Spectroscopy

In XPS the sample is irradiated with a beam of soft x-rays causing the ejection of valence and core-level electrons. The photon sources available for the technique are given in Table 1.3; most work has been carried out with Al $K\alpha$ and Mg $K\alpha$ lines. XPS measurments have been made on gases, solids, and liquids. With the latter, some studies have been performed on high-pressure streams directed across the ionization zone, but, not surprisingly, many experimental difficulties are experienced. By far the majority of results have been reported on core electrons, although valence studies have proven fruitful in certain areas, for example, in the study of molecules adsorbed onto surfaces. Study of inner-shell electrons by XPS yields a hierarchy of chemical information as expounded below.

The binding energies of core electrons are essentially constant, within narrow limits, and characteristic of the atom concerned. Such measurements can be obtained for all elements of the periodic table except H and He. Thus, from the point of view of qualitative analysis, a compre-

Table 1.3. Photon Sources for XPS

x-radiation	Energy (eV)	FWHM[a] (eV)
Y $M\zeta$	132.3	0.44
Zr $M\zeta$	151.4	0.84
Na $K\alpha$	1041.0	0.6
Mg $K\alpha$	1253.6	0.8
Al $K\alpha$	1486.6	0.9
Cu $K\alpha$	8047.8	~ 3

[a]Full width at half maximum.

hensive identification of elements can be achieved, although in a multi-element sample peak assignment problems may be encountered. To some degree this problem can be solved by consideration of spin-orbit splitting and relative intensities. For solids the sensitivity of XPS expressed in bulk terms is only modest ($\sim 0.1\%$), but, in contrast, its high surface sensitivity puts the technique in a different league as a qualitative tool for surface analysis ($\sim 0.1\%$ monolayer).

The exact binding energy of a core electron varies with the oxidation state of chemical environment of an element. This chemical-shift effect (compare PMR and Mössbauer spectroscopy) tends to be rather small (generally within 10 eV), however, it has attracted considerable attention from those interested in structural analysis and from workers who are more concerned with theoretical treatments of photoionization. Notably, the question of the magnitude of the chemical shift from compound to compound relative to possible changes in relaxation energy has occupied the theoreticians for some time. Several models have been prepared to describe the shift effect. In the charge potential model of Siegbahn and coworkers,[10] the ionization energy of a core subshell (i_A) is given approximately by

$$I(i_A) = K_i^A q_A + eV_A \tag{1.4}$$

where q_A is the partial charge on the atom A, K_i^A is an adjustable parameter, and V_A is the potential of atom A due to other atoms. Other models include the valence potential model and the equivalent-cores approach.

Additional features of core-electron spectra are shake-up and multiplet structure. Shake-up satellites occurring on the low-kinetic-energy side of the main photoline are, effectively, a manifestation of relaxation phenomena. Ejection of a core electron is followed by a contraction toward the positive hole, creating the possibility of simultaneous excitation of a valence electron. With regard to the spectra of transition-metal complexes, interpretation of shake-up satellites has great promise in the elucidation of the nature of metal–ligand bonding. However, such potential is not likely to be fulfilled until the questions concerning the mechanism of shake-up effects have been settled. The charge-transfer transition has been implicated as the mode of electron promotion,[23] whereas the opposing view has been expressed that the $3d \rightarrow$ conduction-band transition is responsible for shake-up effects.[24] Multiplet splitting is the result of spin interaction between an unpaired electron resulting from the ionization process and other unpaired electron(s) present in the system.[25] An example of this effect is the removal of a $3s$ electron from Mn(II) leading to ^5S or ^7S states.

Some of the principal features described above are shown in the S $2p$ and

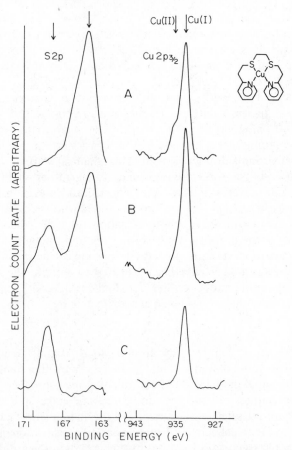

Figure 1.3.   Sulfur $2p$ and Cu $2p_{2/3}$ XP spectra of the Cu(II) complex of 1,8-bis-(2'-pyridyl)-3, 6-dithiaoctane:[26] (*a*) 20 min x-ray exposure; (*b*) 12 h exposure; and (*c*) 10 s exposure to 10-μA argon-ion beam. (Reprinted with permission of American Chemical Society, Washington, D.C.)

Cu $2p_{3/2}$ spectra of the Cu(II) complex of 1, 8-bis- (2′-pyridyl)-3, 6-dithia-octane obtained in our studies (Fig. 1.3).[26] The chelate exhibits an x-ray-induced surface redox reaction. Note the chemical shift of the "normal" S $2p$ (164 eV) signal to one originating from an oxidized sulfur moeity at 168 eV with increased beam exposure. Furthermore, both Cu(II) and Cu(I) peaks are present in the Cu $2p_{3/2}$ region. Finally, the S $2p$ signal is slightly skewed due to the presence of an unresolved S $2p_{1/2}$ signal on the high-binding-energy side of S $2p_{3/2}$.

With regard to the applications of XPS, even a cursory examination of the literature leads one to the view that the number of possibilities are quite large, with a few examples being studies of polymers, catalysts, textiles, corrosion, metalloproteins, particulate pollution, geochemistry, and archaeological chemistry. The reader is advised to consult the references in Table 1.1 for papers in the applied field.

1.3. Penning Ionization Spectroscopy (PIS)

In PIS (references in Table 1.1), excited-source species, such as atoms in the 2^3S and 2^1S metastable states, are allowed to impinge on the target molecules in the gas phase. The kinetic energy of the electron emitted as a result of the break up of a short-lived collision complex between the sample molecule and source atom is determined. Since ionization of the sample molecule occurs, the information gained from this technique resembles that derived from UPS. Not surprisingly, there has been nowhere near the interest shown in this field that has been demonstrated for ultraviolet photo-ionization. However, the technique has proved useful in the study of the physical chemistry of collision complexes.

1.4. Electron Impact Energy Loss Spectroscopy (EELS)

In EILS (references in Table 1.1), the electron-energy spectrum obtained from the impact of low-energy electrons on the gas or solid sample is record-ed. Unlike the other techniques mentioned to this point, EELS does not involve ionization. Transfer of energy from the source particle occurs, resulting in electron promotion in the sample molecule (interband transition). Hence, the energies of electronically excited states of molecules can be derived from electron-energy-loss spectra in a manner similar to the methods used in conventional optical absorption spectroscopy, although the selection rules for the two kinds of spectroscopy are not the same. Much of the work with gases has been of a theoretical nature, but there has been a suggestion that the techniques could be used profitably in gas analysis. Because of the complex nature of EEL spectra, the cost and lack of mobility of the equipment, and the availability of excellent alternatives, the technique does not appear to be seriously attractive to the contemporary analytical chemist. With solids, EELS has been applied to surface chemistry, plasmon spectroscopy, and semiconductor technology.

1.5. Auger Electron Spectroscopy (AES)

In order to introduce the subject of this text we need to consider the fate of a "hole" produced in a core subshell by an appropriate means, for example, x-ray photoionization or electron-impact ionization. The possibilities for decay of such a hole are depicted in Fig. 1.4. Note that the energy levels are shown to be stabilized, which is a direct result of the relaxation mechanism mentioned previously. Deexcitation of the system can occur via the emission of x-ray fluorescence or the emission of a secondary electron (leaving the system doubly positive charged) in a radiationless manner. The mechanism of the former transition is, of course, well known to the inorganic analytical chemist. The secondary electron emitted via the latter process is named after its discoverer, Auger, and can be thought of as an internal x-ray photo-effect. In Auger spectrometry, we study the kinetic energy of these secondary electrons emitted from gas or solid matrices. Radiative decay and the Auger transition are, in fact, competing processes that are dependent on the element being studied. The nature of fluorescence yield versus Auger yield will be discussed in more detail later. A further important point concerns the nature of core-ion production. Auger peaks observed in XP spectra can be easily distinguished from primary photolines by changing the x-ray energy, since the Auger process is clearly source-energy independent. However, most of the work presented in this text has been performed with the inherently more-sensitive electron-beam source; therefore, this will be understood unless stated otherwise.

In addition to the "normal" Auger transition mentioned above there are several other possible processes that can occur which will result in ejection of an electron whose energy will be measured in the electron spectrometer.[27] These are the so-called "satellite" transitions. (Some of the basic types of satellite processes are depicted and discussed in Figure 4.1.) There are two categories of monopole transitions; these are monopole excitation or shake-up similar to that described for XPS and monopole ionization or shake-off. Monopole ionization can also occur at the same time as the Auger process itself, resulting in electrons with lower energy than the Auger transition. Additional high-energy satellites can also occur as a result of the autoionization process. This involves an initial excitation of an electron to an unfilled orbital via, for example, an inelastic collision of an incident electron beam with the sample species. If the energy of excitation

V, X, Y atomic levels in atom M

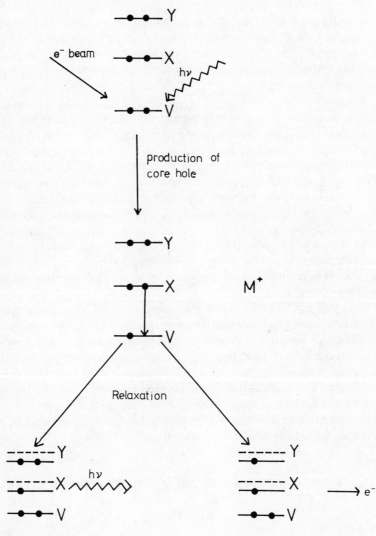

X-ray fluorescence

Figure 1.4. Possibilities for decay of a core vacancy.

11

is greater than that required to ionize any one of the other electrons present in the system, then the Auger-type process can occur to produce an ejected "autoionization" electron. If x-ray excitation is used, autoionization peaks are generally absent in photoelectron spectra, since excitation to a discrete state would result in the loss of only part of the x-ray energy. This is forbidden by the photoelectric effect which requires that all the energy of the photon be annihilated. Autoionization will only be detected in photoelectron spectroscopy if the photon energy used is in "resonance" with the appropriate energy of excitation.

A further satellite process to be introduced here is that of the Coster–Kronig transition. It is generally agreed that electron impact will produce core holes in the various subshells of a particular shell in rough proportion to their occupation number, for example, with ionization in the L-shell the initial distribution of core holes should be in the ratio $1:1:2$ for $2s:2p_{1/2}:$ $2p_{3/2}$. The Coster–Kronig process involves migration of the core hole from one subshell to another higher-lying subshell, for example, $2s$ hole$\rightarrow 2p_{1/2}$ or $2p_{3/2}$ hole and $2p_{1/2}$ hole$\rightarrow 2p_{3/2}$ hole, thus changing the core-hole distribution. Clearly, the relative number of Auger transitions originating from each subshell will be altered with the obvious consequences for relative AES peak intensities. When the Coster–Kronig process occurs, energy is released and a weakly bound electron is ejected, in much the same manner as with the Auger process.

Although not strictly an Auger process, we shall mention at this point the technique of ion-neutralization spectroscopy (INS). In this method a slow beam of He^+ ions is allowed to impinge on a surface. An electron can fall from the valence level (sample) to the vacant level of He^+ resulting in ionization "outside" the surface by means of an electron-tunneling mechanism. The study of the ejected electron is termed INS, and the method does appear to resemble the deexcitation mechanism that occurs in the Auger effect. The main application of INS to date has been in comparative studies of species adsorbed onto surfaces with data from UPS.

2. DEVELOPMENT OF AES AND SCOPE OF ITS APPLICATIONS

2.1. History

The radiationless deexcitation of atoms was predicted from theoretical considerations by Rosseland[28] in 1923. At about the same time Auger[29-31] was carrying out experiments with x-ray irradiation of noble gases in a Wilson expansion chamber. In a study of the photoelectric effect he observed a short fat track originating from the same point as the long photoelectron track. The length of these short tracks was independent of the energy of the incident x-ray but depended on the atomic species being excited. The tracks were correctly ascribed to a secondary electron ejected from the same atom as the photoelectron. The nature of the Auger process could then be used to explain "extra" lines obtained previously by Robinson[32] in his experiments in photoelectron spectroscopy. In the 1920s and 1930s interest in the Auger effect generally lay in explaining variations in x-ray fluorescence yields, and it was not until the 1950s that experimental and theoretical work increased in pace. Two reviews to their respective dates were published by Burhop[33] in 1952 and Bergström[34] in 1955. In 1953 Lander[35] first used electron-excited Auger spectroscopy and discussed the possibility of using the Auger effect in surface analysis. Peaks in the spectra of backscattered electrons due to various metals and the effects of excitation and other features of the Auger transition were all identified. Further experimental work was carried out by Harrower[36] and Powell et al.[37] During this period and into the 1960s several theoretical predictions of the K Auger spectrum were carried out to explain relativistic effects, intermediate coupling, and configuration interaction. Further review articles also appeared during the early 1960s.

Probably as a result of developments in the design and technology of electron spectrometers, the discovery of satellite lines in Auger spectra was reported by Körber and Mehlhorn[38] and others in the late 1960s. The development of electron-stimulated AES as a powerful tool for the analysis of surfaces began with the work of Harris, Weber, and Peria in 1967. Harris[39] demonstrated the potential of differentiation of the Auger-electron-energy distribution and used the technique in studies of surface segregation and carbon evaporation from a cathode. The important contribution of Weber and Peria[40] was to show that the profusion of low-energy-electron-diffraction (LEED) instruments available at that time could be modified for use in Auger spectrometry in the differential mode. (The considerable quantity of LEED work carried out in the 1960s is catalogued in the bibliography of Haas et al.[41]) In 1969 Palmberg, Bohn, and Tracey[42] showed that the cylindrical-mirror analyzer could be used to achieve high sensitivity in AES with a resultant increase in the speed of an analysis.

Since approximately 1970 the growth in the number of applications of AES reported in the literature has been very marked. The advances that have occurred have taken place in a climate of significant improvements in instrument design and ultra-high-vacuum technology. A good example is the development of a scanning technique that makes it possible to produce a two-dimensional image of the surface distribution of a chosen element which can be compared with a "physical" image from electron microscopy.[43] Recently, the considerable potential of AES for gas-phase analysis was discussed.[44] Both elemental and molecular information on a sample molecule can be obtained. Over the last decade a number of review papers in AES have appeared and these are listed in Appendix I.

2.2. Advantages and Scope of Applications of AES

There is no doubt that electron-excitation AES has become one of the premier tools for the analysis of surfaces. As we shall see in later chapters the technique has not only served as a standard method of qualitative surface analysis, but has also contributed significantly to our understanding of the chemistry and physics associated with the surface properties of materials. Briefly the importance of AES in surface work stems from the following points:

1. The technique is truly surface sensitive and effectively governed by the electron escape depth of the Auger electron.
2. A multielement identification can be achieved.
3. High-intensity electron beams for primary excitation are easily attainable and lead to high sensitivity.
4. Cross sections for ionization are quite large and, within a range of three to six times threshold ionization potential, are reasonably independent of energy.
5. The electron-beam energy is continuously variable and, therefore, can be adjusted to operate under maximum Auger current conditions.
6. The sensitivity of the technique is further increased due to multiple excitation from the primary beam and to ionization caused by back-scattered electrons.
7. Because of high sensitivity, the speed of achieving an analysis is quite fast.
8. The occurrence of the Auger-chemical-shift effect offers the possibility of examining the chemical environment of surface species.
9. Electron beams can be focused readily to spot areas of about 1 μm leading to high spatial resolution (compare XPS) and the possibility of scanning AES.
10. In conjunction with item 9 the technique can achieve "three-dimensional" resolution via concentration depth profiling.
11. The method can easily be combined with other electron spectroscopic techniques in one unified instrument.

In addition to these positive features there are also a number of disadvantages:

1. High-intensity electron beams may degrade the surface.
2. Multiple excitation by the primary beams and ionization caused by backscattered electrons result in the Auger current being not simply related to cross sections. Therefore, use of the technique for quantitative analysis requires considerable care.
3. The electron beam can also cause surface effects such as stimulated desorption.
4. The complexity of the Auger process leads to difficulties in interpreting the Auger chemical shift.
5. Equipment for AES is expensive.

In view of the above comments it is not surprising that the greatest number of applications of AES have appeared in the field of metallurgy. The method has been used in studies of oxidation, corrosion, tribology, embrittlement, segregation, alloy formation, diffusion, and other areas where the sensitivity of AES can be usefully employed. Recently, there has also been some work carried out on insulating materials such as glasses with some success. Although strictly chemical applications are not in as great a number as those mentioned above, the method has been used with great success in areas such as electrochemistry.

As we shall see later gas-phase AES offers the opportunity to achieve both an elemental and molecular identification as well as possibilities in understanding the Auger-chemical-shift and cross-section effects.

REFERENCES

1. D. W. Turner, C. Baker, A. D. Baker, and C. R. Brundle, *Molecular Photoelectron Spectroscopy*, Wiley, London, 1970.
2. A. D. Baker and D. Betteridge, *Photoelectron Spectroscopy—Chemical and Analytical Aspects*, Pergamon, Oxford, 1972.
3. J. H. D. Eland, *Photoelectron Spectroscopy*, Butterworth, London, 1974.
4. J. W. Rabalais, *Principles of Ultraviolet Photoelectron Spectroscopy*, Wiley, New York, 1977.
5. A. D. Baker, C. R. Brundle, and M. Thompson, *Chem. Soc. Rev.*, 355 (1972).
6. A. D. Baker, M. Brisk, and D. Liotta, *Anal. Chem.*, **50**, 328R (1978).
7. A. D. Baker and C. R. Brundle (Eds.), *Electron Spectroscopy*, Academic Press, New York, 1977 Vol. 1; and 1978, Vol. 2.
8. D. Briggs (Ed.), *Handbook of x-ray and Ultraviolet Photoelectron Spectroscopy*, Heyden, London, 1977.

9. K. Siegbahn, C. Nordling, R. Fahlman, R. Nordberg, K. Hamrin, J. Hedman, G. Johansson, T. Bergmark, S. -E. Karlsson, I. Lindgren, and B. Lindberg, *ESCA: Atomic, Molecular and Solid State Structure Studied by Means of Electron Spectroscopy*, Nova Acta Regiae Soc. Sci., Upsaliensis, 1967, Ser. IV, Vol. 20.

10. K. Siegbahn, C. Nordling, G. Johansson, P. F. Heden, K. Hamrin, U. Gelius, T. Bergmark, L. O. Werme, R. Manne, and Y. Baer, *ESCA Applied to Free Molecules*, North-Holland, Amsterdam, 1969.

11. V. Cermak, *J. Chem. Phys.*, **44**, 3781 (1966).

12. H. Hotop, *Advances in Mass Spectrometry*, Institute of Petroleum, London, 1971, Vol. 5, p. 116.

13. M. J. Shaw, *Contemp. Phys.*, **15**, 445 (1974).

14. S. Trajmar, J. K. Rice, and A. Kupperman, *Adv. Chem. Phys.*, **18**, 15 (1970).

15. C. E. Brion, in *MTP International Review of Science*, Physical Chemistry, Ser. 1, Butterworth, London, 1972, Vol. 5, p. 55.

16. G. R. Wright and C. E. Brion, *J. Elec. Spectrosc. Rel. Phenom.*, **4**, 313, 327, 335, 347 (1974).

17. H. Ibach and D. L. Mills, *Electron Energy Loss Spectroscopy and Surface Vibrations*, Academic Press, New York, 1982.

18. H. D. Hagstrum, *J. Vac. Sci. Technol.*, **12**, 1 (1975).

19. P. M. Williams, in D. Briggs (Ed.), *Handbook of x-ray and Ultraviolet Photoelectron Spectroscopy*, Heyden, London, 1977, p. 313.

20. O. Edqvist, L. Åsbrink, and E. Lindholm, *Z. Natursforsch.*, **26a**, 1407 (1971).

21. M. Thompson and E. A. Stubley, *Anal. Chim. Acta*, **119**, 179 (1980).

22. A. F. Orchard, in D. Briggs (Ed.), *Handbook of Photoelectron Spectroscopy*, Heyden, London, 1971, p. 1.

23. K. S. Kim, *J. Elec. Spectrosc. Rel. Phenom.*, **3**, 217 (1974).

24. J. A. Tossel, *J. Elec. Spectrosc. Rel. Phenom.*, **10**, 169 (1977).

25. C. S. Fadley and D. A. Shirley, *Phys. Rev. A*, **2**, 1109 (1970).

26. M. Thompson, R. B. Lennox, and D. J. Zemon, *Anal. Chem.*, **51**, 2260 (1979).

27. W. E. Moddeman, T. A. Carlson, M. O. Krause, D. P. Pullen, W. E. Bull, and G. K. Schweitzer, *J. Chem. Phys.*, **55**, 2317 (1971).

28. S. Rosseland, *Z. Phys.*, **14**, 172 (1923).

29. P. Auger, *Comp. Rend.*, **177**, 169 (1923).

30. P. Auger, *Compt. Rend.*, **180**, 65 (1925).

31. P. Auger, *Ann. Phys. (Paris)*, **6**, 183 (1926).

32. H. R. Robinson, *Proc. Roy. Soc. (London)A*, **104**, 455 (1923).

33. E. H. S. Burhop, *The Auger Effect and Other Radiationless Transitions*, Cambridge University Press, Cambridge, 1952.

34. I. Bergström, in *Beta- and Gamma-Ray Spectroscopy*, K. Siegbahn (Ed.), North-Holland, Amsterdam, 1955.

35. J. J. Lander, *Phys. Rev.*, **91**, 1382 (1953).

36. G. A. Harrower, *Phys. Rev.*, **102**, 340 (1956).

37. C. J. Powell, J. L. Robins, and J. B. Swan, *Phys. Rev.*, **110**, 657 (1958).

38. H. Körber and W. Mehlhorn, *Z. Phys.*, **191**, 217 (1966).

39. L. A. Harris, *J. Appl. Phys.*, **39**, 1419 (1968).

40.   R. E. Weber and W. T. Peria, *J. Appl. Phys.*, **38**, 4355 (1967).

41.   T. W. Haas, G. J. Dooley, J. T. Grant, A. G. Jackson, and M. P. Hooker, *Progr. Surf. Sci.*, **1**, 155 (1971).

42.   P. W. Palmberg, G. K. Bohn, and J. C. Tracey, *Appl. Phys. Lett.*, **15**, 254 (1969).

43.   N. C. MacDonald, *Appl. Phys. Lett.*, **16**, 76 (1970).

44.   M. Thompson, P. A. Hewitt, and D. S. Wooliscroft, *Anal. Chem.*, **48**, 1336 (1976).

THE AUGER PROCESS AND SPECTRUM

In this section we deal with several of the theoretical aspects that are at the heart of Auger spectroscopy. After an introduction to the categories of Auger transitions and nomenclature, we consider the calculation of Auger energies. This discussion includes theoretical work and computation by empirical means. The intensity of Auger signals is considered through sections on transition rates, Auger and fluorescence yields, and Coster–Kronig transitions. For a more-detailed discussion of many of the theoretical parameters included in this chapter, the reader should refer to the excellent reviews by Burhop and Asaad,[1] Sevier,[2] and Bambynek et al.[3] (and references therein).

1. CLASSIFICATION AND NOMENCLATURE OF AUGER TRANSITIONS

1.1. Notation

In the Auger process an initial hole created in a shell or subshell is transferred to an outer shell or orbital, and the energy involved in this transition results in the emission of a secondary electron. The kinetic energy of the Auger electron will be related to the final energy state of the atom or molecule, which is governed by the positioning of the double-hole formation and the interaction between remaining electrons. The double-vacancy configuration can be used to characterize particular Auger transitions in atoms and molecules. Unfortunately, there is no uniform notation for such descriptions because x-ray nomenclature tends to be employed by the physicist for application to atomic transitions, whereas the physical chemist tends to use the usual letters—s, p, d, etc.—in representing subshells and point-group symbolism for molecular orbitals.

In general terms, we can consider the radiationless process to consist of transitions involving holes. Hence, an initial vacancy characterized by the principal, angular momentum, magnetic, and spin quantum numbers $(nlm_lm_s)^i$ resulting in final vacancies (nlm_lm_s) and $(nlm_lm_s)'$ can be represented as follows:

$$(nlm_lm_s)^i \rightarrow (nlm_lm_s)(nlm_lm_s)'$$

This situation is normally written as three consecutive symbols with the arrow depicting the transfer of the initial vacancy being omitted as follows:

$$V_pX_qY_r$$

where p, q, and r are subshell indices or J values. If the initial vacancy is in the K shell, then the Auger electrons derived from the filling of the K level from outer electrons are of the type KX_qY_r and are often termed K-series electrons. Alternatively, the chemist might use the expression

$$1sX_qY_r$$

Although not strictly applicable to the lighter elements, the hole symbols X_qY_r can also be written in two different formats. For example, in K-series transitions with both holes occurring for $n = 2$, the following equivalent expressions could be used:

$$KL_1L_2 \equiv 1s2s2p_{1/2}$$

$$KL_2L_3 \equiv 1s2p_{1/2}2p_{3/2}$$

In a similar manner equivalent expressions could be derived for L-series electrons, where the initial hole is created in one of the L subshells:

$$L_1M_4M_5 \equiv 2s3d_{3/2}3d_{5/2}$$

Table 2.1. *KLL* Auger Energies in the Multiconfiguration Self-Consistent-Field Approximation for Double-Hole States of H_2O

Final State	Symmetry	Computed Auger Energy (eV)
$(1b_1)^{-2}$	$^1A_1$	500.77
$(3a_1)^{-1}(1b_1)^{-1}$	$^1B_1$	499.57
$(3a_1)^{-2}$	$^1A_1$	496.40
$(1b_2)^{-1}(1b_1)^{-1}$	$^1A_2$	495.91
$(1b_2)^{-1}(3a_1)^{-1}$	$^1B_2$	493.84
$(1b_2)^{-2}$	$^1A_2$	488.98
$(2a_1)^{-1}(1b_1)^{-1}$	$^1B_1$	474.64
$(2a_1)^{-1}(3a_1)^{-1}$	$^1A_1$	473.80
$(2a_1)^{-1}(1b_2)^{-1}$	$^1B_2$	468.11
$(2a_1)^{-2}$	$^1A_1$	453.26

Occasionally, in the literature, the following types of expression are also used: $KL_{2,3}L_{2,3}$ and KVV. The former notation refers to an inability to distinguish the subshells, with indices 2 and 3, whereas the latter implies that the double-hole configuration resides in valence levels (usually used for solid-state experiments).

Finally, those concerned with AES of molecular systems in the gas-phase usually use the x-ray notation for a general description of the Auger transitions being considered, together with orbital and final-state symmetry representations. Table 2.1 illustrates an example of this type of notation applied to KLL-Auger energies for H_2O.[4]

In the following section we shall consider the notation described above in the light of coupling, as a function of atomic number, between final atomic states following a KLL transition.

1.2. Russell–Saunders, Intermediate, and j-j Couplings

As a result of a KLL transition the L-shell electrons may be in one of the following three configurations

$$(2s)^0(2p)^6, \qquad (2s)^1(2p)^5, \qquad (2s)^2(2p)^4$$

For the lighter elements ($10 \leqslant Z \leqslant 25$), relativistic effects are small, and we must consider the terms that arise from electrostatic interaction between the electrons in the nonclosed shells. This means that the description of a particular configuration such as $(2s)^2(2p)^4$ is not unique because the p orbitals differ in their magnetic quantum numbers m_l and the electrons differ in their spins $m_s = \pm\frac{1}{2}$. Addition of the m_l numbers results in $2L+1$ possible values of M_L associated with a given L. The spin quantum numbers m_s are coupled to produce the quantum number M_s; there are $2s+1$ values of M_s. Atomic terms are characterized by specific values L and S and have different energies. Furthermore, the conserved quantum numbers L and S may couple to give an overall atomic quantum number J. If $L < S$, $2L+1$ J values are possible; if $L > S$, $2S+1$ J values can be obtained. The Russell–Saunders symbolism for the states is $^{2S+1}L_J$. The spin-orbit coupling results in only small energy differences between levels for the lighter atoms, therefore, the J subscript is often omitted. The L-shell configurations expressed above translate into the following six terms:

$$
\begin{array}{ll}
(2s)^0(2p)^6 & {}^1S \\
(2s)^1(2p)^5 & {}^1P, {}^3P \\
(2s)^2(2p)^4 & {}^1S, {}^1D, [{}^3P]
\end{array}
$$

In the case of pure L-S coupling, Auger transitions resulting in the ^3P term of the $(2s)^2(2p)^4$ configuration are forbidden from parity-conservation considerations.

The electron-orbital coupling mentioned above is applicable to light atoms, but with increasing atomic number larger and larger differences occur between states with the same L and S values, but different J numbers. In the pure j-j coupling scheme no account is taken of interelectron interactions, whereas relativistic effects are included. In this type of coupling j is given for each individual electron and can have values $l+s, \ldots, l-s$ to give J. Where a complete breakdown of L-S coupling is involved, the Auger final

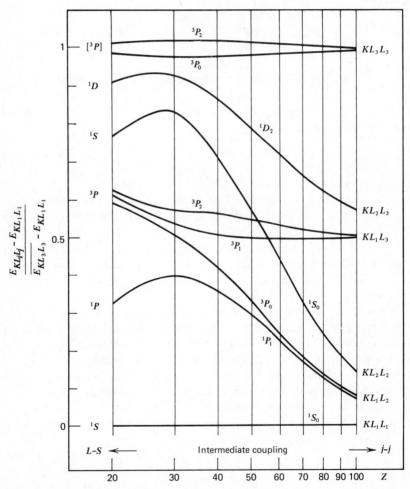

Figure 2.1. Coupling between final atomic vacancy states following a KLL transition for different Z. (Reprinted with permission of Wiley-Interscience, New York.)

states can be expressed in terms of the electrons (l and J values) taking part in the transition. Thus, we can now correlate configurations with transition labels as follows:

$(2s)^0(2p)^6$	KL_1L_1	$1s2s2s$
$(2s)^1(2p)^5$	KL_1L_2	$1s2s2p_{1/2}$
	KL_1L_3	$1s2s2p_{3/2}$
$(2s)^2(2p)^4$	KL_2L_2	$1s2p_{1/2}2p_{1/2}$
	KL_2L_3	$1s2p_{1/2}2p_{3/2}$
	KL_3L_3	$1s2p_{3/2}2p_{3/2}$

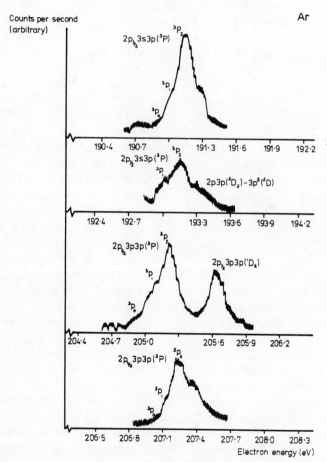

Figure 2.2. Auger ^3P peaks of argon in the gas phase by electron impact.[5] (Reprinted with permission of American Chemical Society, Washington.)

In between the limits of extreme L-S and j-j couplings, both electron interaction and spin-orbit coupling have to be involved to describe the final energy levels of Auger transitions. In this region of the periodic table, where the coupling is often labeled as intermediate, each of the Russel–Saunders terms is effectively split into a multiplet of components. Of the 10 possible resulting energy states, transitions to one of these—the $L_2L_3(^3P_1)$ state—is strictly forbidden under any coupling scheme. This leads to nine possible states. Auger transitions in this intermediate region are symbolized by both L-S and j-j designations as shown in Fig. 2.1. In general terms, the Auger spectroscopist who is interested in low-resolution surface work is not able to distinguish the J splittings even for atoms with a quite low Z number, therefore, the Russell–Saunders terms are often dropped. This is not the case for relatively high-resolution work. In this regard, Fig. 2.2 shows an interesting example where ^3P components of certain Auger transitions of argon $(Z = 18)$ are just resolved in high-resolution gas-phase work.[5]

2. CALCULATION OF AUGER ENERGIES

The interpretation of an Auger spectrum is simplified by working from an empirical formula and comparing calculated with observed values. Two basic methods have been employed in energy calculations. These are the theoretical approaches involving nonrelativistic, relativistic, self-consistent-field (SCF), and parent-orbital-configuration-interaction (POCI) calculations, and, secondly, the totally empirical techniques based on simplified equations. (The more-rigorous calculations have recently been extended into a semi-empirical approach.) In studies of atomic transitions the relativistic model assumes that the transition of one electron from its initial state to the positive energy continuum is caused by the interaction of that electron with the electromagnetic radiation emitted by the other electron when it fills the atomic vacancy. The Auger transition is thereby considered to be one of internal absorption of radiation. The nonrelativistic approximation considers two electrons moving in their initial states in both the field due to the nucleus and the average field due to the other atomic electrons. The coulombic interaction between the two electrons involved in the Auger transition is considered to be the perturbation that causes one electron to undergo a transition from its initial bound state to a final state in the positive energy continuum, whereas the second electron fills the initial atomic vacancy.

Before outlining the theoretical and semiempirical approaches, we shall first introduce the simplified concepts and equations used in the calculation of Auger energies.

2.1. Empirical Methods

One of the most well-known simplified equations is that of Burhop:[6]

$$E^Z(VXY) = E^Z(V) - E^Z(X) - E^Z(Y') \qquad (2.1)$$

where $E^Z(VXY)$ = energy of the ejected Auger electrons from element
with atomic number Z

$E^Z(V)$ = energy of level in which initial core hole is created

$E^Z(X)$ = energy of level from which an electron falls to fill the
initial vacancy

$E^Z(Y')$ = energy appropriate to an atom already singly ionized in
an inner shell (Y is the level from which the Auger
electron is excited)

Note that $E^Z(V)$ and $E^Z(X)$ can certainly be approximated as the atomic
binding energies of electrons in the V and X levels.

Bergström and Hill[7] modified Eq. (2.1) by suggesting that $E^Z(Y')$ probably
lies between $E^Z(Y)$ and $E^{Z+1}(Y)$ (binding energies of level Y in the atom of
interest, Z, and in atom of $Z+1$, respectively). Thus, Eq. (2.1) becomes

$$E^Z(VXY) = E^Z(V) - E^Z(X) - E^Z(Y) - \Delta E^Z(Y) \qquad (2.2)$$

where $\Delta E^Z(Y) = \Delta Z[E^{Z+1}(Y) - E^Z(Y)]$.

The empirically determined quantity ΔZ is called the effective incremental
charge and accounts for the change in binding energy of an electron in shell
Y of an atom Z ionized in shell X. Thus,

$$\Delta Z = \frac{\Delta E^Z(Y)}{E^{(Z+1)}(Y) - E^Z(Y)} \qquad (2.3)$$

The parameter ΔZ only varies slowly with Z and, hence, can be used in quite
a wide range of elements, although different values of ΔZ may be expected
for different shells.

Equation (2.2) implies that after ionization of electron V an electron in
level X is then trapped by the vacancy in V and the excess energy causes the
ejection of an electron in Y. If we consider that the electron Y is trapped by
vacancy V and the electron X is ejected, then a question arises concerning
the equivalent expression

$$E^Z(VXY) = E^Z(V) - E^Z(Y) - E^Z(X) - \Delta Z[E^{Z+1}(X) - E^Z(X)] \qquad (2.4)$$

In both cases the experimental kinetic energies must be the same, but expressions (2.2) and (2.4) do not predict the same values if X and Y are different
(unless ΔZ varies drastically[8,9]). One can make the kinetic energy of the
Auger electron independent of the manner of the transition (since the transi-

tion is a symmetrical process with respect to both X and Y) by taking the average of expressions (2.2) and (2.4):[10]

$$E^Z(VXY) = E^Z(V) - E^Z(X) - E^Z(Y) - \Delta Z'[E^{Z+1}(X) - E^Z(X)]$$
$$- \Delta Z'[E^{Z+1}(Y) + E^Z(Y)] \qquad (2.5)$$

$\Delta Z'$ is a new parameter adjusted to fit experimental data and has a value of 0.5 when X and Y are equivalent.

Another approximate formula for calculating Auger energies, again making use of levels in the next higher atom in the periodic table, was proposed by Geffrion and Nadeau:[11]

$$\left[\frac{1}{\Delta E^Z(XY)}\right]^n = \left[\frac{1}{E^{Z+1}(X) - E^Z(X)}\right]^n - \left[\frac{1}{E^{Z+1}(Y) - E^Z(Y)}\right]^n \qquad (2.6)$$

Note that in this case the expression $\Delta E^Z(XY)$ infers that the actual origin of the Auger electron with respect to the levels X or Y is not distinguished. The parameter n is an empirically determined integer and is usually between the values of 1 and 2.

It is interesting to note that several workers have attempted to predict Auger energies based on modifications of the expressions mentioned above by computer techniques. The reader should consult the papers by Yasko and Whitmoyer[12] and Coghlan and Clausing[13] for examples of this type of work.

The difference between calculated energies based simply on the binding energies of levels V, X, and Y and the experimental energies is now generally considered, at least from a semiempirical point of view, to involve electron coupling and relaxation energy components as a result of the coulombic redistribution of electrons that occurs on ionization. In the next section, we shall review the theoretical work carried out on atomic transitions, mention semiempirical data, and outline the calculations carried out to date on molecular systems.

2.2. Theoretical and Semiempirical Methods

The KLL group of the K series of atomic transitions has attracted by far the most attention from a theoretical standpoint. This is hardly surprising in view of the relative simplicity of the K-series group compared to the L and other series (despite this, the KLL group itself provides a formidable problem). Therefore, we shall begin this section with a summary of the work that has been accomplished in the KLL area.

The energies and relative transition probabilities of the KLL group computed according to nonrelativistic theory and assuming intermediate coupling were published in 1958 in a classic paper by Asaad and Burhop.[14] The gist of this treatment follows.

After a *KLL* transition the energy of the electron in one of the three configurations—$(2s)^0(2p)^6$, $(2s)^1(2p)^5$, and $(2s)^2(2p)^4$—can be expressed according to the Hamiltonian

$$H = -\frac{\hbar^2}{2m_e} \sum_{i=1}^{6} \nabla_i^2 + \sum_{i=1}^{6} V(r_i) + \sum_{i<j} \frac{e^2}{r_{ij}} + \sum_i \zeta(r_i)\rho_i \cdot S_i \qquad (2.7)$$

where the first two terms describe the central-field Hamiltonian, the third is the perturbation arising from electrostatic interaction of pairs of electrons, and the fourth represents spin-orbit coupling of electrons. Also,

$$\zeta(r) = -\frac{e}{2m_e^2 c^2} \frac{1}{r} \frac{\partial V}{\partial r} \qquad (2.8)$$

Expressing the above configurations in terms of equivalent two-electron systems—$(2s)^2$, $2s^1 2p^1$, and $(2p)^2$—leads to the calculation of the *KLL* energies through consideration of the electrostatic interaction between the two electrons. Hence, the Slater–Condon integrals for such interactions in (nl) and $(n'l')$ shells are

$$F^\nu(nl, n'l') = e^2 \int\!\!\!\int_{r_1, r_2 = 0}^{\infty} \gamma_\nu R_{nl}^2(r_1) R_{n'l'}^2(r_2) r_1^2 r_2^2 \, dr_1 \, dr_2$$

$$G^\nu(nl, n'l') = e^2 \int\!\!\!\int_{r_1, r_2 = 0}^{\infty} \gamma_\nu R_{nl}(r_1) R_{n'l'}(r_2) R_{nl}(r_2) R_{n'l'}(r_1^2) r_1 r_2^2 \, dr_1 \, dr_2 \qquad (2.9)$$

where $R_{nl}(r)$ and $R_{n'l'}(r)$ are the radial parts of the wave functions of electrons in the nl and $n'l'$ bound states, respectively, and

$$\gamma_\nu = \begin{cases} \dfrac{r_1^\nu}{r_2^{\nu+1}} & \text{for } r_1 < r_2 \\[2ex] \dfrac{r_2^\nu}{r_1^{\nu+1}} & \text{for } r_2 < r_1 \end{cases}$$

The energies of the nine *KLL* Auger lines as outlined in Section 1.2 can then be specified in terms of both extreme *L-S* and *j-j* coupling notation as follows:

$K - L_1L_1(^1S_0)$: $E(K) - 2E(L_1) - F^0(2s, 2s)$

$K - L_2L_2(^1S_0)$: $E(K) - 2E(L_2) - F^0(2p, 2p) - 0.1F^2(2p, 2p) + 1.5\zeta_{2p}$
$$- \{[0.3F^2(2p, 2p) - 0.5\zeta_{2p}]^2 + 2\zeta_{2p}^2\}^{1/2}$$

$K - L_3L_3(^3P_0)$: $E(K) - 2E(L_3) - F^0(2p, 2p) - 0.1F^2(2p, 2p) - 1.5\zeta_{2p}$
$$+ \{[0.3F^2(2p, 2p) - 0.5\zeta_{2p}]^2 + 2\zeta_{2p}^2\}^{1/2}$$

$K - L_1L_2(^1P_1)$: $E(K) - E(L_1) - E(L_2) - F^0(2s, 2p) + 0.75\zeta_{2p}$
$$- \{[\tfrac{1}{3}G^1(2s, 2p) - 0.25\zeta_{2p}^2]^2 + 0.5\zeta_{2p}^2\}^{1/2}$$

$K - L_1L_3(^3P_1)$: $E(K) - E(L_1) - E(L_3) - F^0(2s, 2p) - 0.75\zeta_{2p}$
$$+ \{[\tfrac{1}{3}G^1(2s, 2p) - 0.25\zeta_{2p}^2]^2 + 0.5\zeta_{2p}^2\}^{1/2}$$

$K - L_1L_2(^3P_0)$: $E(K) - E(L_1) - E(L_2) - F^0(2s, 2p) + \tfrac{1}{3}G_1(2s, 2p)$

$K - L_1L_3(^3P_2)$: $E(K) - E(L_1) - E(L_3) - F^0(2s, 2p) + \tfrac{1}{3}G^1(2s, 2p)$

$K - L_2L_3(^1D_2)$: $E(K) - E(L_2) - E(L_3) - F^0(2p, 2p) + 0.08F^2(2p, 2p)$
$$+ 0.75\zeta_{2p} - \{[0.12F^2(2p, 2p) + 0.25\zeta_{2p}]^2 + 0.5\zeta_{2p}^2\}^{1/2}$$

$K - L_3L_3(^3P_2)$: $E(K) - 2E(L_3) - F^0(2p, 2p) + 0.08F^2(2p, 2p)$
$$- 0.75\zeta_{2p} + \{[0.12F^2(2p, 2p) + 0.25\zeta_{2p}]^2 + 0.5\zeta_{2p}^2\}^{1/2} \tag{2.10}$$

where $\zeta_{2p} = \tfrac{2}{3}[E(L_2) - E(L_3)]$ and is the matrix element of the spin-orbit interaction.

If we consider a purely electrostatic model, that is, $\zeta_{2p} = 0$, as would be appropriate for the lighter elements, then Eq. (2.10) reduces to the six energies given by the "pure" Russell–Sanders scheme:

^1S: $E(K) - 2E(L_1) - F^0(2s, 2s)$

^1P: $E(K) - E(L_1) - E(2p) - F^0(2s, 2p) - \tfrac{1}{3}G^1(2s, 2p)$

^3P: $E(K) - E(L_1) - E(2p) - F^0(2s, 2p) + \tfrac{1}{3}G^1(2s, 2p)$

^1S: $E(K) - 2E(2p) - F^0(2p, 2p) - \tfrac{2}{5}F^2(2p, 2p)$

^1D: $E(K) - 2E(2p) - F^0(2p, 2p) - \tfrac{1}{25}F^2(2p, 2p)$

$[^3P]$: $E(K) - 2E(2p) - F^0(2p, 2p) + \tfrac{1}{5}F^2(2p, 2p)$ (forbidden)   (2.11)

The reader should note that Eqs. (2.10) and (2.11) take the general form

$$E_{\text{Auger}} = E(K) - E(L_p) - E(L_q) + \Delta \tag{2.12}$$

where $E(K)$, $E(L_1)$, and $E(L_2)$ are, for example, appropriate electron binding energies. The nature of this equation is fundamental to practical Auger spectroscopy in that it forms that basis of semiempirical calculations of energies for comparison with experimental data (compare Section 2.1).

The G and F terms were deduced from the expression

$$A(Z - Z_s)(1 - \alpha Z^2)$$

where Z_s is the screening constant, and from experimental work on bismuth carried out by Mladjenovič and Slätis.[15] At that time it was difficult to perform a full "test" of the theoretical work, because of the dearth of experimental data. Hörnfeldt[16] demonstrated that better agreement could be obtained with experimental results if $(1 - \alpha Z^2)$ were replaced by $(1 + \beta Z^3)$; KLL energies from calcium to fermium were computed using A and β values obtained from experimental work. Subsequently, Siegbahn and co-workers[17] showed that for $Z < 40$ a Z^2 dependence was more appropriate for the lighter elements.

The first relativistic calculations, assuming j-j coupling, were carried out by Asaad[18] in 1959 on KLL transitions in mercury. Further calculations using a relativistic self-consistent approach were performed for the elements magnesium, potassium, and copper,[17] and for the series sodium to copper.[19] These and other calculations have confirmed that it is preferable to use relativistic wave functions in the study of Auger energies (and transition rates).[20−22] In 1976 Asaad and Petrini[56] published their relativistic calculations for the KLL Auger spectrum of a number of elements of significantly different Z. This particularly important paper describes the application of a relativistic analog of intermediate coupling with configuration interaction. In this treatment a relativistic Hamiltonian is described as follows [compare that given in Eq. (2.7)]:

$$H = \sum_{i=1}^{6} H^D(i) + \sum_{i<j=1}^{6} \frac{e^2}{r_{ij}} \tag{2.13}$$

where in the first term, the central-field Hamiltonian,

$$H^D(i) = c\boldsymbol{\alpha}(i)\cdot \mathbf{p}(i) + mc^2\beta(i) + V(r_i) \tag{2.14}$$

In the calculations use was made of the binding energies of K and L electrons compiled by Sevier,[2] the relativistic Slater integrals and relaxation term $K(L_1L_1)$ introduced by Larkins,[23] and the matrix elements for the KLL transitions calculated by Bhalla and Ramsdale.[22] Table 2.2 shows the results of the calculation of Asaad and Petrini[56] compared to the work of Hörnfeldt[16] and Siegbahn et al.[17] and experimental values taken from the paper by Burhop and Asaad.[1]

Several workers have attempted to translate the classical theoretical approaches into a semiempirical method for predicting atomic Auger energies for "free" species and for the solid phase based on the equation[24−26]

Table 2.2. Comparison of Calculated *KLL* Energies (keV) According to Asaad and Petrini with those of Hörnfeldt and Siegbahn et al. and Experimental Vales[a]

Element	L_1L_1 $^1S_0$	L_1L_2 $^1P_1$	L_1L_2 $^3P_0$	L_1L_3 $^3P_1$	L_1L_3 $^3P_2$	L_2L_2 $^1S_0$	L_2L_3 $^1D_2$	L_3L_3 $^3P_0$	L_3L_3 $^3P_2$
$_{21}$Sc	3.444	3.526	3.553	3.554	3.557	3.620	3.631	3.638	3.642 AP
	3.456	3.533	3.563	3.564	3.567	3.622	3.638	3.647	3.651 S
$_{30}$Zn	7.202	7.342	7.374	7.385	7.397	7.490	7.517	7.535	7.548 AP
	7.214	7.348	7.384	7.394	7.407	7.493	7.526	7.543	7.558 S
	7.220	7.358		7.399			7.535	7.550	7.566 E
$_{35}$Br	9.827	10.005	10.037	10.062	10.083	10.183	10.232	10.269	10.286 AP
	9.840	10.014	10.049	10.074	10.096	10.189	10.244	10.279	10.300 S
	9.860	10.05	10.05	11.01		10.200	10.26	10.33	10.33 E
$_{55}$Cs	24.407	24.764	24.798	25.092	25.093	25.145	25.442	25.766	25.800 AP
	24.409	24.775	24.812	25.104	25.160	25.109	25.464	25.782	25.825 H
	24.39	24.75		25.08	25.14		25.43		25.79 E
$_{70}$Yb	40.153	40.668	40.709	41.670	41.744	41.140	42.177	43.179	43.225 AP
	40.150	40.673	40.7161	41.675	41.752	41.148	42.191	43.186	43.242 H
	40.149	40.672		41.681	41.762	41.120	42.184	43.234	43.234 E
$_{80}$Hg	53.173	53.815	53.862	55.698	55.787	54.404	56.333	58.218	58.272 AP
	53.143	53.799	53.847	55.682	55.771	54.398	56.331	58.206	58.272 H
	53.118	53.761	53.761	55.672	55.672		56.328	58.223	58.223 E

[a]AP: Asaad and Petrini[56]; S: Siegbahn et al.[17]; H: Hörnfeldt[16]; E: experimental.[1]

$$E^Z(VXY) = E^Z(V) - E^Z(X) - E^Z(Y) - S(XY) + R(XY) \qquad (2.15)$$

where $S(XY)$ represents the sum of final-state coupling terms and $R(XY)$ accounts for relaxation processes. Recently Larkins[27] commented on this work and discussed a basis for the accurate prediction of the relaxation-energy term in an intermediate-coupling model. The predicted energies for transitions of elements 10–100 are within 1–3 eV of experimental values, for the most part.

We now turn to a discussion of calculations, of which there are relatively few, that have been performed on molecules. The first system to be studied[28] was CH_4 for which an experimental spectrum was already available for comparative purposes.[29] In this analysis, extended-basis-set Hartree–Fock SCF wave functions were calculated separately for an initial carbon $1s$-hole and all double-hole final configurations. The computed total energy of the ground state is -40.2148 hartrees, and the K-shell binding energy is 290.88 eV (in good agreement with the experimental value of 290.7 eV). The calculated Auger energies for the various double-hole states are compared with experimental values in Table 2.3. The $^1T_2 - ^3T_2$ splitting accounts for the second and third lines of the experimental spectrum, and the broad fourth band of the spectrum is made up of four unresolved lines assigned to $^1A_1$, $^1T_2$, ^1E, and $^3T_1$ states.

The Auger energies of sulfur $2pMM$ transitions of H_2S and sulfur $2pMM$ and oxygen $1sLL$ transitions of SO_2 have been calculated by multiconfiguration SCF and by frozen orbital (POCI) methods[4,30] and compared to the experimental assignments of Thompson et al.[5] (Note that MM and LL in this case both refer to the valence orbitals of SO_2 and H_2S.) The appropriate results for H_2S are collected in Table 2.4. To achieve an empirical assignment of the experimental spectrum, binding energies from UPS and XPS were used to calculate an "approximate" value for the highest-energy Auger transition. Subtraction of the highest-energy peak in the experimental spectrum from this value yielded the "relaxation energy," which was then used to predict energies for all remaining double-hole possibilities. No attempt was made to assign triplet states to the expected weaker peaks. For the SCF calculations, a spin-orbit correction using the experimental $2p_{1/2} - 2p_{3/2}$ split of 1.3 eV was applied to the theoretical values. The difference between the SCF and POCI values represents a relaxation energy component because of the theoretical models used for the two calculations. Despite the crude nature of the empirical assignment, it is in remarkably good agreement with that obtained from the theoretical work. Any discrepancy between the two approaches results from the patently crude assumption with the empirical assignment that the same relaxation-energy correction can be applied to all double-hole states. The calculations predict that 81 transitions are possible for sulfur and oxygen in SO_2. Clearly, nothing like this number is observed in practice, and this fact underlines the inherent difficulties in the empirical assignment of molecular Auger spectra.

Other calculations have been carried out for H_2O[31,32] and NH_3.[33] The experimental gas-phase spectrum of the latter has been compared with the

Table 2.3. Calculated and Experimental Auger Energies (eV) for Methane: CH_4, T_d, $(1a_1)^2(2a_1)^2(1t_2)^6$, $^1A_1$

Double-Hole State	Symmetry	Calculated	Experimental
$(2a_1)^{-2}$	$^1A_1$	230.43	229.4
$(2a_1)^{-1}(1t_2)^{-1}$	$^1T_2$	238.64	237.0
	$^1T_3$	244.96	243.3
$(1t_2)^{-2}$	$^1A_1$	249.26	
	$^1T_2$	251.72	250.0
	^1E	253.66	
	$^3T_1$	254.70	

Table 2.4. Calculated and Experimental Auger Energies (eV) for Hydrogen Sulfide: H_2S, C_{2v}, $(4a_1)^2(2b_2)^2(5a_1)^2(2b_1)^2$

Double-Hole State	Symmetry	Symmetry SCF	SCF with Spin-Orbit Correction $(J=\frac{1}{2}, J=\frac{3}{2})$	Calculated POCI	Experimental $(J=\frac{1}{2}, J=\frac{3}{2})$	Assignment
$(2b_1)^{-2}$	$^1A_1$	139.22	(140.09, 138.79)	136.13	(139.78, 138.55)	$(2b_1)^{-2}$
$(5a_1)^{-1}(2b_1)^{-1}$	$^3B_1$	138.72	(139.59, 138.29)	135.48		
$(5a_1)^{-1}(2b_1)^{-1}$	$^1B_1$	136.88	(137.75, 136.45)	133.83	(137.6, 136.0)	$(5a_1)^{-1}(2b_1)^{-1}$
$(2b_2)^{-1}(2b_1)^{-1}$	$^3A_2$	136.03	(136.90, 135.60)	133.53		
$(2b_2)^{-1}(2b_1)^{-1}$	$^1A_2$	135.05	(135.92, 134.62)	132.60	(135.3, 133.9)	$(2b_2)^{-1}(2b_1)^{-1}$
$(2b_2)^{-1}(5a_1)^{-1}$	$^3B_2$	133.48	(134.35, 133.05)	131.17		
$(5a_1)^{-2}$	$^1A_1$	133.48	(133.35, 133.05)	131.02	(—, 132.7)	$(5a_1)^{-2}$
$(2b_2)^{-1}(5a_1)^{-1}$	$^1B_2$	131.71	(132.58, 131.28)	129.16	(—, 131.1)	$(5a_1)^{-1}(2b_2)^{-1}$
$(2b_2)^{-2}$	$^1A_1$	128.23	(129.10, 127.80)	122.25		
$(4a_1)^{-1}(2b_1)^{-1}$	$^3B_1$	126.17	(127.04, 125.74)	122.25	(127.4, 125.8)	$(4a_1)^{-1}(5a_1)^{-1}$
$(4a_1)^{-1}(5a_1)^{-1}$	$^3A_1$	123.11	(123.98, 122.68)	119.95		
$(4a_1)^{-1}(2b_1)^{-1}$	$^1B_1$	122.81	(123.68, 122.38)	117.58		
$(4a_1)^{-1}(2b_2)^{-1}$	$^3B_2$	121.26	(122.13, 120.83)	117.95		
$(4a_1)^{-1}(5a_1)^{-1}$	$^1A_1$	118.42	(119.29, 117.99)	114.79		
$(4a_1)^{-1}(2b_2)^{-1}$	$^1B_2$	115.00	(115.87, 114.57)	111.57		
$(4a_1)^{-2}$	$^1A_1$	105.10	(105.97, 104.67)	101.35		

Table 2.5. Calculated and Experimental Auger Energies (eV) for Ammonia: NH_3, C_{3v}, $(1a_1)^2(2a_1)^2(1e)^4(3a_1)^2$

Double-Hole State	Symmetry	Calculated	Experimental
$(3a_1)^{-2}$	$^1A_1$	371.0	371.5
$(1e)^{-1}(3a_1)^{-1}$	^3E	368.8	
$(1e)^{-1}(3a_1)^{-1}$	^1E	366.9	366.6
$(1e)^{-2}$	$^3A_2$	362.7	
$(1e)^{-2}$	^1E	360.4	360.2
$(1e)^{-2}$	$^1A_1$	358.2	356.5
$(2a_1)^{-1}(3a_1)^{-1}$	$^3A_1$	356.1	
$(2a_1)^{-1}(3a_1)^{-1}$	$^1A_1$	350.2	352.4
$(2a_1)^{-1}(1e)^{-1}$	^3E	350.1	
$(2a_1)^{-1}(1e)^{-1}$	^1E	342.4	345.1
$(2a_1)^{-1}(1e)^{-1}$	$^1A_1$	331.1	337.0

theoretical work;[34] the results are given in Table 2.5. Further details of calculations carried out in relation to gas-phase spectra are considered in in Chapter 4.

3. CALCULATION OF AUGER INTENSITIES

The proportion of vacancies in a given inner subshell, i, filled by radiative transitions is termed the fluorescence yield, ω_i. For the K shell, the vacancy can be filled by the emission of K fluorescence (yield ω_K) or of K-series Auger electrons (Auger yield a_k). For the L and higher shells, account must be taken of the possibility of Coster–Kronig transitions. Thus, for transitions involving vacancy transfer to another inner shell only, we describe the yields as a_1, a_2, and a_3 for the three L subshells, and the yields for Coster–Kronig vacancy transfers $L_1 \rightarrow L_2$, $L_1 \rightarrow L_3$, and $L_2 \rightarrow L_3$ are denoted by f_{12}, f_{13}, and f_{23}, respectively. Hence, the following fundamental expressions can be given for initial vacancies in the K and L shells:

$$\omega_K + a_K = 1$$
$$\omega_3 + a_3 = 1$$
$$\omega_2 + a_2 + f_{23} = 1$$
$$\omega_1 + a_1 + f_{12} + f_{13} = 1 \qquad (2.16)$$

Furthermore, the fluorescence yield ω_i can be expressed in terms of transition rates for radiative and Auger processes. If the sum of the transition rates for all possible radiative transitions to fill the vacancy is P_R^i and for all Auger transitions is P_A^i, then we can write

$$\omega_i = P_R^i/(P_R^i + P_A^i) \tag{2.17}$$

Finally, in certain theoretical papers on Auger spectroscopy the reader may also encounter the width of various levels. In general terms the total width Γ_{T_i} for the ith subshell will be the sum of partial widths,

$$\Gamma_{T_i} = \Gamma_{\omega_i} + \Gamma_{a_i} + \sum_{k>i} \Gamma_{ik} \tag{2.18}$$

As with Auger energies theoretical work on calculation of Auger transition rates and probabilities has largely been concerned with the K-series processes. The basic theory behind relativistic and nonrelativistic calculations has been reviewed by Burhop and Asaad[1] as follows. In the non-relativistic model the Hamiltonian for the two electrons involved in the Auger transitions (labeled 1 and 2) can be written

$$H = -(\hbar^2/2m_e)[\nabla_{(1)}^2 + \nabla_{(2)}^2][V(1) + V(2)] + (e^2/r_{12}) \tag{2.19}$$

As mentioned earlier the Coulomb interaction between the two electrons [the last term in Eq. (2.19)] is regarded as the perturbation behind the transition. If $\psi_i(r_1)$ and $\psi_i(r_2)$ represent the initial single-electron wave functions of the two electrons in the atom having an inner shell vacancy and $\psi_f(r_1)$ and $\psi_f(r_2)$ are the final states of these two electrons, then the Auger transition probability can be written according to first-order perturbation theory:

$$W_a = 2\pi h^{-1} \left| \iint \psi_f^*(r_1)\psi_f(r_2) \left(\frac{e^2}{r_{12}}\right) \psi_i(r_1)\psi_i(r_2)dr_1 dr_2 \right|^2 \tag{2.20}$$

Table 2.6.   Some Calculations of KLL-Series Line Intensities Up to 1971

Authors (Year)	Element(s), Transition Type	Method[a]	Reference
Wentzel (1927)	KLL various Z	NR	35
Burhop (1935)	K series, Ag	NR, j-j coupling, H-like wave function, Slater screening constants	36
Pincherle (1935)	Up to $4p$	Unscreened H wave functions	37
Massey and Burhop (1936)	KL_1L_1, KL_1L_2, KL_1L_3, Au	R, Dirac wave functions	38
Ramberg and	KLL, Au	NR, j-j coupling, screening by	39

Table 2.6 Contd.

Authors (Year)	Element(s), Transition Type	Method[a]	Reference
Richtmeyer (1937)		Thomas–Fermi model	
Rubenstein and Snyder (1955)	KL_1L_1, $KL_1L_{2,3}$ $KL_{2,3}L_{2,3}$, Ar, Kr, Ag	LS coupling, SCF	40
Burhop, Asaad (1958)	KLL, KLM, Mn, Zn, Rb, Ag, Hg	NR, intermediate coupling	14
Asaad (1959)	KLL, Hg	R	41
Callan (1961)	KLL, 40 elements $Z = 12$ to 80	NR, LS coupling, H wave functions	42
Listengarten (1961)	KLL, Tl	R, j-j coupling, screening Thomas–Fermi–Dirac model	43
Listengarten (1962)	KLL, Tb, U	R, j-j coupling, screening Thomas–Fermi–Dirac model	44
Asaad (1963)	KLL, several elements	Intermediate coupling	45
Asaad (1963)	General Auger transitions	j-j coupling	46
Krause, Vestal, Johnston, and Carlson (1964)	KL_1L_2, $KL_1L_{2,3}$ $KL_{2,3}L_{2,3}$, Ne	LS coupling, Hartree–Fock	47
Asaad (1965)	KLL, several elements	Includes configuration interaction	48
Mehlhorn and Asaad (1966)	KLL, $Z = 10$ to 36	Includes configuration interaction	49
Chattarji and Talukdar (1968)	KL_1L_1 $Z = 72$ to 80	R and NR	50
McGuire (1969)	KLL, KLM, Be–Ar	Hartree, Fock, Slater equations	51
McGuire (1970)	KLL, KLM, KLN, KMM, Ar–Xe	Hartree, Fock, Slater equations	52
Walters and Bhalla (1971)	KLL $Z = 4$ to 54	NR, Hartree–Fock–Slater	53
Kostroun, Chen, and Crasemann (1971)	KLL, KLM, KLN $A = 10$ to 70	Screened H wave functions	54

[a]R = relativistic; NR = nonrelativistic.

Table 2.7. *KLL* Auger Transition Probabilities ($\times 10^{-3}$ au) Calculated in *j-j* Coupling, Intermediate Coupling (IC), and Intermediate Coupling with Configuration Interaction (CI) According to Chen and Crasemann (Ref. 55)[a]

Element	KL_1L_1 $^1S_0$	KL_1L_2 $^1P_1$	$^3P_0$	KL_1L_3 $^3P_1$	$^3P_2$	KL_2L_2 $^1S_0$	KL_2L_3 $^1D_2$	KL_3L_3 $^3P_0$	$^3P_2$	
$_{13}$Al	1.111	1.156		2.311		0.198	5.146	2.965		*jj*
	1.112	2.707	0.084	0.251	0.422	0.591	7.701	0.001	0.019	IC
	0.826	2.707	0.084	0.251	0.422	0.877	7.701	0.001	0.019	CI
$_{15}$P	1.237	1.319		2.637		0.234	6.024	3.477		*jj*
	1.237	3.115	0.093	0.281	0.466	0.697	8.977	0.003	0.056	IC
	0.864	3.115	0.093	0.281	0.466	1.068	8.977	0.005	0.056	CI
$_{18}$Ar	1.360	1.490		2.977		0.275	6.960	4.025		*jj*
	1.360	3.522	0.104	0.319	0.521	0.810	10.251	0.014	0.191	IC
	0.955	3.522	0.104	0.319	0.521	1.212	10.251	0.017	0.191	CI
$_{20}$Ca	1.435	1.586		3.168		0.297	7.480	4.328		*jj*
	1.435	3.765	0.109	0.340	0.543	0.859	10.863	0.031	0.360	IC
	0.992	3.765	0.109	0.340	0.543	1.287	10.863	0.046	0.360	CI
$_{23}$V	1.513	1.696		3.387		0.322	8.114	4.690		*jj*
	1.513	4.031	0.112	0.378	0.562	0.880	11.414	0.084	0.759	IC
	1.073	4.031	0.112	0.378	0.562	1.269	11.414	0.136	0.759	CI
$_{25}$Mn	1.559	1.760		3.515		0.338	8.487	4.904		*jj*
	1.559	4.166	0.115	0.421	0.575	0.868	11.623	0.142	1.110	IC
	1.120	4.166	0.115	0.421	0.575	1.227	11.623	0.222	1.110	CI
$_{28}$Ni	1.621	1.847		3.687		0.360	8.975	5.187		*jj*
	1.621	4.297	0.118	0.529	0.592	0.815	11.739	0.259	1.706	IC
	1.250	4.297	0.118	0.529	0.592	1.117	11.739	0.327	1.706	CI

Element										Scheme
$_{30}$Zn	1.660	1.903		3.798		0.375	9.306	5.378		j-j
	1.660	4.326	0.122	0.646	0.608	0.772	11.814	0.345	2.124	IC
	1.286	4.326	0.122	0.646	0.608	1.064	11.814	0.428	2.124	CI
$_{33}$As	1.711	1.968		3.925		0.390	9.638	5.566		j-j
	1.711	4.265	0.123	0.887	0.617	0.701	11.747	0.458	2.683	IC
	1.340	4.265	0.123	0.887	0.617	0.955	11.747	0.575	2.683	CI
$_{35}$Br	1.742	2.007		4.002		0.399	9.839	5.680		j-j
	1.742	4.163	0.125	1.098	0.625	0.661	11.706	0.523	3.021	IC
	1.382	4.163	0.125	1.098	0.625	0.893	11.706	0.652	3.021	CI
$_{40}$Zr	1.803	2.084		4.150		0.415	10.201	5.878		j-j
	1.803	3.766	0.126	1.711	0.630	0.587	11.558	0.641	3.697	IC
$_{47}$Ag	1.889	2.195		4.365		0.938	10.790	6.191		j-j
	1.889	3.257	0.129	2.526	0.645	0.539	11.674	0.751	4.440	IC

[a]Note that KL_1L_1 is independent of coupling scheme and KL_1L_2 ($^1P_1$ and $^3P_0$), KL_1L_3 ($^3P_1$ and $^3P_2$), KL_2L_3 ($^1D_2$), and KL_3L_3 ($^3P_2$) probabilities are unaffected by configuration interaction.

Taking the Pauli exclusion principle into account, the initial and final wave functions must be antisymmetrical in their coordinates. Thus the product $\psi_i(r_1)\psi_i(r_2)$ is replaced by $2^{-1/2}[\psi_i(r_1)\psi_i(r_2)-\psi_i(r_2)\psi_i(r_1)]$ and a similar replacement for the final wave function is carried out. The wave functions for the bound states are normalized to unity, and the continuum wave function is normalized to reflect one ejected electron per unit time.

In the relativistic model the calculation is modified to represent the ejection of the Auger electron as being caused by radiation generated by the filling of the inner-level vacancy with the other electron.

The theoretical work on K-series transition probabilities carried out before 1971 using the various coupling considerations is summarized in Table 2.6. As with the Auger energy calculations these efforts suffered somewhat due to a dearth of experimental data. The results are specified in terms of relative intensities, usually to the KL_1L_1 transition, or transition probabilities in atomic units of time (1 au$=4.134 \times 10^{16}$ sec^{-1}). The gradual improvement in the various calculations over the years demonstrated that for intermediate atomic numbers the relative intensities of different lines agreed reasonably well with experimental data. However, discrepancies with the heaviest elements were caused by the neglect of relativity in the nonrelativistic calculations. The graph presented by Chattarji and Talukdar[50] exemplifies this factor for $Z=72$ to 80. The errors for lighter elements were increasingly being recognized as due to the neglect of configuration interaction (which is independent of the coupling scheme used). Thus, it was found that the KL_1L_1 line intensity varied significantly with Z due to interaction between the $(2s)^0(2p)^6$ and $(2s)^2(2p)^4$ configurations.

In 1973 Chen and Crasemann[55] recalculated KLL Auger transition probabilities for elements with atomic number $13\leqslant Z\leqslant 47$ nonrelativistically in j-j coupling and in intermediate coupling, with and without configuration interaction. The system was treated as a coupled two-hole configuration, and the single-particle radial wave functions required in the calculation of radial matrix elements and in the calculation of mixing coefficients in the intermediate-coupling scheme were obtained from Green's atomic independent-particle model. The results of this work are summarized in Table 2.7. In keeping with previous workers Chen and Crasemann deduce from these results that relativity becomes important for elements heavier than bromine ($Z=55$) in KL_1L_1 transitions and zirconium ($Z=40$) in KL_1L_2 transitions, and for KL_1L_3, KL_2L_2, and KL_2L_3 transitions relativistic effects are small for atomic numbers below $Z=50$. Furthermore, KLL Auger spectra for atomic numbers $13\leqslant Z\leqslant 60$ should be computed in intermediate coupling with configuration interaction.

The first relativistic KLL calculations in intermediate coupling with configuration interaction were carried out by Asaad and Petrini as mentioned

Table 2.8. KLL Auger Transition Probabilities ($\times 10^{-3}$ au) According to Asaad and Petrini (Ref. 56)[a]

Element	KL_1L_1 $^1S_0$	KL_1L_2 $^1P_1$	KL_1L_2 $^3P_0$	KL_1L_3 $^3P_1$	KL_1L_3 $^3P_2$	KL_2L_2 $^1S_0$	KL_2L_3 $^1D_2$	KL_2L_3 $^3P_0$	KL_2L_3 $^3P_2$	
$_{21}$Sc	1.19	3.54	0.160	0.418	0.563	1.09	10.02	0.060	0.355	
$_{30}$Zn	1.61	4.13	0.282	0.864	0.717	0.994	11.54	0.438	1.79	
	1.29	4.33	0.122	0.646	0.608	1.06	11.81	0.428	2.12	CC
$_{35}$Br	1.86	4.00	0.380	1.41	0.795	0.874	11.36	0.637	2.62	
$_{21}$Sc	1.38	4.16	0.125	1.10	0.63	0.893	11.71	0.652	3.02	CC
$_{41}$Nb	2.23	3.72	0.547	2.25	0.961	0.800	11.24	0.775	3.37	
$_{48}$Cd	2.73	3.47	0.835	3.02	1.04	0.782	11.34	0.850	3.91	
$_{55}$Cs	3.36	3.62	1.27	3.57	1.22	0.799	11.49	0.882	4.22	
$_{63}$Eu	4.30	4.19	2.02	4.03	1.47	0.846	11.71	0.896	4.41	
$_{70}$Yb	5.38	5.06	3.02	4.38	1.74	0.888	11.90	0.900	4.51	
$_{80}$Hg	7.54	7.13	5.31	4.83	2.25	1.01	12.19	0.901	4.58	
$_{81}$Tl	7.81	7.41	5.61	4.87	2.30	1.02	12.21	0.900	4.58	
$_{93}$Np	12.18	12.41	11.01	5.36	3.35	1.19	12.59	0.897	4.65	

[a]CC = Chen and Crasemann.[55]

39

previously.[56] The results of this work are shown in Table 2.8; the nonrelativistic calculations (in intermediate coupling with configuration interaction) of Chen and Crasemann for zinc and bromine are included for comparison. It is confirmed again that relativistic values for the intensities of KL_1L_1 ($^1S_0$), KL_1L_2 ($^3P_0$), and KL_1L_3 ($^3P_1$ and $^3P_2$) lines are increasingly greater with increasing Z than nonrelativistic ones. On the other hand, relativistic values for the transition rates for KL_2L_3 ($^1D_2$), KL_3L_3 ($^3P_0$), and KL_3L_3 ($^3P_2$) are less than the nonrelativistic values.

During the early 1970s several papers appeared concerning calculations of transition rates of L-, M-, and N-shell Auger and Coster–Kronig transitions.[57−63]. As with the KLL spectral lines, the intensities of the L and higher series vary significantly with atomic number, and attempts were made to correlate the results with experimental data. It should be noted, however, that this is often a difficult task owing to overlapping of lines associated with spectral complexity. With regard to L-shell initial vacancies, Auger transitions to the L_1 level are strongly influenced by Coster–Kronig processes, that is, filling of the vacancy from $L_{2,3}$. The total Auger transition rates for a particular final-state configuration do not depend on the coupling scheme. In fact most studies of L Auger spectra have assumed j-j coupling for both the

Table 2.9. Comparison of Some of the Theoretical Relative Intensities for Different Final-State Vacancy Configurations with the Experimental Data for an Initial Vacancy in the $2p$ Shell

Z	Final Vacancies	Experiment (%)	Theory (%)
18	M_1M_1	7	1.6
	$M_1M_{2,3}$	20	32
	$M_{2,3}M_{2,3}$	100	100
36	$M_{2,3}M_{2,3}$	20	41
	$M_{2,3}M_{4,5}$	42	69
	$M_{4,5}M_{4,5}$	100	100
49	$M_1M_{2,3}$	6.3	7.6
	$M_{2,3}M_{2,3}$	48	30
	$M_{2,3}M_{4,5}$	78	62
	$M_{2,3}M_{2,3}$	14	9.4
	$M_{4,5}M_{4,5}$	28	21
	$M_{4,5}M_{4,5}$	100	100

Source: Reprinted with permission of American Physical Society, Washington, D.C.

initial- and final-state configurations. In 1971 McGuire[58,59] computed L-shell (filling of L_2 or L_3 initial vacancies) Coster–Kronig and L_1-, L_2-, and L_3-shell Auger transition rates and compared the results with experimental term intensities for platinum, tellurium, uranium, argon, and krypton.[59] Although the agreement for the heavier elements was reasonable, discrepancies were found in the U L_2 and Ar L_1 Coster–Kronig spectra.

Shortly after McGuire's papers Walters and Bhalla[60] published their work on nonrelativistic Auger rates for the $2p$ shell. The calculations were performed using the Hartree–Fock–Slater approach with Herman, Van Dyke, and Ortenburger exchange. Table 2.9 compares a number of the theoretical relative intensities for several final-state configurations with experimental data and Fig. 2.3 illustrates the total Auger transition rates as a function of atomic number. Generally, the authors concluded that the theoretical relative intensities for different final-state configurations agree reasonably well with experimental work for intermediate Z, but configuration interaction is important below $Z = 40$. Further work on Auger and Coster–Kronig transitions to the L_1 level were carried out by Crasemann et al.[61] using non-

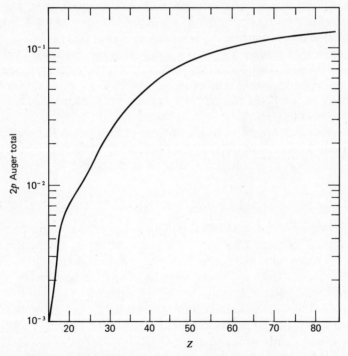

Figure 2.3. Total Auger $2p$ transition rate as a function of atomic number.[60] (Reprinted with permission of American Physical Society, New York.)

relativistic screened hydrogenic wave functions in j-j coupling, and the results compared with other researchers.

Finally, subsequent to his L-shell work McGuire[62,63] published additional work on atomic M- $(22 \leqslant Z \leqslant 90)$ and N- $(38 \leqslant Z \leqslant 103)$ shell Auger and Coster–Kronig spectra.

4. COSTER–KRONIG TRANSITIONS

As mentioned previously the transition first described by Coster and Kronig[64] involves the filling of an initial vacancy in the $L_{1,2}$ (or $M_{1,2,3}$) subshell from a higher L (or M) subshell resulting in the ejection of an electron from a higher shell than L (or M). Clearly, for the Coster–Kronig process to take place, the binding energy of the secondary electron must be less than approximately the difference in binding energy of the L (M) subshells involved in the vacancy transfer. This means that the effect can only occur in certain Z-number areas of the periodic table.

Since the Coster–Kronig process is an Auger transition, the species is left doubly positively charged after ejection of the secondary electron. Thus it is possible to create a vacancy in the $L_{2,3}$ subshell that can be filled resulting in the emission of x-radiation. However, the energy of this radiation will be different from that originating from a singly charged species, and will be present as a satellite(s) on, for example, the L x-ray. Historically, Coster and Kronig developed their theory in order to explain this type of satellite, and subsequently in the 1940s Cooper[65,66] discussed the intensity of the satellites with respect to Z value.

In order to predict the various possibilities of the Coster–Kronig process in the periodic table, ΔZ of Eq. (2.2) can be set to equal unity resulting in the following empirical equation for calculation of the energy of the secondary electron:

$$E^Z(V_p V_q Y_r) = E^z(V_p) - E^z(V_q) - E^{Z+1}(Y_r) \qquad (2.21)$$

where $V_{p,q}$ refers to the L or M shell and $p < q$. Sevier[2] has summarized the most-important allowed LLY and MMY transitions as a function of Z, and these are reproduced in Table 2.10. A recent publication has also collected theoretical L-shell Coster–Kronig energies to which the reader should consult for explicit detail.[67]

With regard to Coster–Kronig transition probabilities and rates a great deal of work was carried out before 1970 by the same workers specified in Table 2.6 using all the various coupling models. More recently, calculations regarding the Coster–Kronig process were included in the papers on L, M, and N Auger spectra mentioned in the previous section. Crasemann et al.

Table 2.10.   The Regions of Z Value where the Most-Important L and M Series Coster–Kronig Transitions Are Allowed

Transition	Z	Transition	Z
$L_1L_2M_1$	<30	$M_1M_2N_{6,7}$	$\gtrsim 93$
$L_1L_2M_2$	<33	$M_1M_3N_1$	$\gtrsim 53$
$L_1L_2M_3$	<34	$M_1M_3N_{2,3}$	<56
$L_1L_2M_4$	<42	$M_1M_3N_{4-7}$	All Z with N_{4-7} electrons
$L_1L_2M_5$	<42	$M_1M_{4,5}N_{1-7}$	All Z with N_{1-7} electrons
$L_1L_2N_1$	<70	$M_2M_3N_1$	$\gtrsim 31$
$L_1L_2N_2$	<77	$M_2M_3N_{2,3}$	$\gtrsim 35$
$L_1L_2N_3$	<82	$M_2M_3N_{4,5}$	<53
$L_1L_2N_4$	<93	$M_2M_3N_{6,7}$	All Z with $N_{6,7}$ electrons
$L_1L_2N_5$	$\gtrsim 95$	$M_2M_4N_1$	<86
$L_1L_3M_1$	<32	$M_2M_4N_{2-7}$	All Z with N_{2-7} electrons
$L_1L_3M_2$	<36	$M_2M_5N_{1-7}$	All Z with N_{1-7} electrons
$L_1L_3M_3$	<37 and >90	$M_3M_4N_1$	<56
$L_1L_3M_4$	<50 and >76	$M_3M_4N_2$	<66
$L_1L_3M_5$	<51 and >73	$M_3M_4N_3$	<72
$L_1L_3N_{1-7}$	All Z with N_{1-7} electrons	$M_3M_4N_4$	<83
$L_2L_3M_3$	>97	$M_3M_4N_5$	$\gtrsim 85$
$L_2L_3M_4$	$\gtrsim 31$ and >92	$M_3M_4N_{6,7}$	All Z with $N_{6,7}$ electrons
$L_2L_3M_5$	$\gtrsim 31$ and >90	$M_3M_5N_1$	<58
$L_2L_3N_{1-7}$	All Z with N_{1-7} electrons	$M_3M_5N_2$	<71
		$M_3M_5N_3$	<77
$M_1M_{4,5}M_{4,5}$	$\gtrsim 45$	$M_3M_5N_4$	<89
$M_{2,3}M_{4,5}M_{4,5}$	$\gtrsim 38$	$M_3M_5N_5$	<91
$M_1M_2N_1$	<48	$M_3M_5N_{6,7}$	All Z with $N_{6,7}$ electrons
$M_1M_2N_{2,3}$	<53	$M_4M_5N_{4,5}$	$\gtrsim 47$
$M_1M_2N_{4,5}$	$\gtrsim 73$	$M_4M_5N_{6,7}$	<78

Source: Reprinted with permission of Wiley-Interscience, New York.

pointed out that probability equations should include a radiative part for the Coster–Kronig process, namely, the appropriate segment of Eq. (2.16) for the L-shell process should be written as follows:

$$\omega_1 + a_1 + a_{12} + \omega_{12} + a_{13} + \omega_{13} = 1 \qquad (2.22)$$

where $f_{12} = a_{12} + \omega_{12}$ and $f_{13} = a_{13} + \omega_{13}$. However, the radiative components ω_{12} and ω_{13} are expected to be rather small. The probability of the radiationless L_1L_3Y Coster–Kronig process, a_{13}, as a function of atomic number presented by Crasemann et al.[61] is shown in Fig. 2.4.

In the last few years there have been a number of experimental measure-

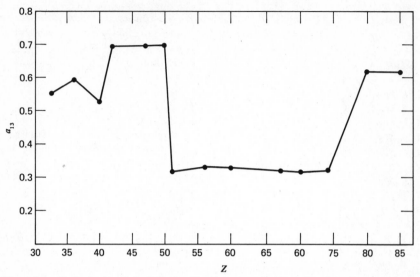

Figure 2.4.   Radiationless $L_1 L_3 Y$ Coster–Kronig process (a_{13}) as a function of atomic number (Crasemann, Chen, and Kostroun[61]). (Reprinted with permission of American Physical Society, New York.)

ments of Coster–Kronig probabilities, usually involving x-ray spectroscopy, and comparisons with theory. For example, f_{23} for $Z=80$, 81, 82, and 92 (Ref. 68) and for Ho and Tm (Ref. 69) have been measured and compared with the theoretical work of McGuire and Crasemann et al. Finally, it is interesting to note that there have been a number of suggestions that the Coster–Kronig process exhibits different behavior between the solid and the gas phase. An example of this is Matthew et al.,[70] who pointed out that the Coster–Kronig process involving the L_2 level of zinc was much more important in the solid phase leading to a significantly higher $L_3 M_{4,5} M_{4,5} / L_2 M_{4,5} M_{4,5}$ intensity ratio than for the gas phase. The effect was attributed to extra-atomic relaxation processes.

5.   FLUORESCENCE YIELD

As we have discussed previously, the Auger process is in competition with the emission of x-ray fluorescence in the relaxing of the initial vacancy. (There is, however, evidence of a radiative Auger process not discussed in this text.) In view of this relationship some general aspects of the fluorescence yield will be considered here, but detailed material of interest to the analytical chemist working with luminescent techniques is beyond the scope of this text.

With regard to the K-shell fluorescence yield, several attempts have been made over the years to describe equations that allow the prediction of ω_K; a selection of these follows:

$$\omega_K = \frac{Z^4}{Z^4 + b} \quad \text{or} \quad \left(\frac{\omega_K}{1-\omega_K}\right)^{1/4} = BZ \tag{2.23}$$

$$\left(\frac{\omega_K}{1-\omega_K}\right)^{1/4} = A + BZ + CZ^3 \tag{2.24}$$

$$\left(\frac{\omega_K}{1-\omega_K}\right)^n = A + BZ + CZ^3 \tag{2.25}$$

where A, B, and C are constants that are conventionally adjusted to fit the experimental measurements of ω_K. The wide variety of experimental methods used to measure K-shell fluorescence yields up to 1972 have been reviewed by Bambynek et al.;[3] this material is given in Table 2.11. The authors reached the conclusion that the best methods available at that time were the following:

1. Measurements with gaseous radioactive sources in proportional counters, preferably of the wall-less multiwire type.
2. Measurements of the K x-ray emission rate and the disintegration rate.
3. For high Z, simultaneous measurements of K Auger electrons and K x-rays from a weightless radioactive source, with a "windowless," cooled Si(Li) detector.

Furthermore, from a comprehensive catalog of experimentally determined K fluorescence yields the authors selected the values that can with certainty be deemed "most reliable." A detailed polynomial regressional analysis of the values indicated that Eq. (2.24) leads to a statistically better fit to the experimental data. The constants A, B, and C were calculated to be $A = 0.015 \pm 0.010$, $B = 0.037 \pm 0.0005$, and $C = -(0.64 \pm 0.07) \times 10^{-6}$. The agreement of calculated values (Refs. 52, 53, and 54) with the "most reliable" experimental values is remarkably good. In this respect, we include in this text the comparison of theoretical ω_K's with experimental values given by Bambynek et al. (Fig. 2.5). Of importance to the analytical chemist are the very low values of ω_K at low values of Z; this is the reason that x-ray fluorescence analysis is particularly "insensitive" for the lighter elements.

Finally, with regard to ω_K values, Freund[71] compiled a further list of experimental work in 1975 to which the reader of this text is directed.

The relaxing of an initial vacancy in the L shell is considerably more complex to describe than for the K shell because there are now three subshells to deal with, and, for certain Z, the Coster–Kronig vacancy transfer process

Table 2.11. Methods That Have Been Used for the Determination of K-Shell Fluorescence Yields

Number	Method	Mode of Production of Primary Vacancies	Target or Source	Detectors[a]	Atomic Numbers to which Applied	Quoted Accuracy (%)	Presently Estimated Ultimate Accuracy of the Method (%)
1	Fluorescent excitation x rays of gaseous targets	Fluorescent excitation x rays	Gaseous	ic, pc, mw	10–54	0.5–22	3
2	Fluorescent excitation x rays of solid targets	Fluorescent excitation x rays	Solid	ic, ppl, pc, NaI(Tl)	4–56	1.4–26	3
3	Auger- and conversion-electron spectroscopy	Internal conversion, electron capture	Solid	$s\pi$, sl, sd, NaI(Tl)	43–93	0.2–9.0	1
4	Auger-electron, x-, and β-ray spectroscopy	Electron capture	Solid	sd, NaI(Tl)	80	1.7	2
5	Auger-electron and K x-ray spectroscopy	Electron capture	Gaseous	pc, mw	17, 31	0.4–5.6	1
6	Auger-electron and K x-ray spectroscopy	Electron capture	Solid	pc, d, Si(Li)	12–49 / 78, 92	0.9–37 / 0.8	5 / 1
7	K x-ray and γ-ray or conversion-electron spectroscopy	Electron capture leading to metastable states	Solid	pc, NaI(Tl) anthracene, sc	27–49	5.9–8.9	5
8	Determination of K x-ray emission rate and disintegration rate	Electron capture	Solid	pc, NaI(Tl)	23–54	0.8–10.0	1

No.	Measurement of					
9	Measurement of (K x-ray)–(γ-ray) or (K x-ray)–(K conversion-electron) coincidences	Solid	pc, NaI(Tl) Ge(Li), Si(Li)	22–52	1.3–9.0	2
10	Cloud-chamber technique	x rays	Gaseous	cc	8–54	3–75	15
11	Change of ionization at K edge	x rays	Gaseous Solid	ic phc	22–53	Not quoted	20
12	Photographic emulsion technique	Electron capture	Solid	ppl	84	5.6	15
13	Charged-particle excitation	e^-, ρ, α, heavy ions	Gaseous Solid	pe, Si(Li), Ge(Li)	6–18	11–17	15

Source: Reprinted with permission of American Physical Society, Washington, D.C.

[a]The following abbreviations are used: cc, cloud chamber; ic, ionization chamber; pc, proportional counter; d, double proportional counter; mw, multiwire proportional counter; phc, photocathode; ppl, photographic film or plate; sc, semiconductor; sd, double-focusing spectrometer; sl, lens spectrometer; sπ, 180° spectrometer.

47

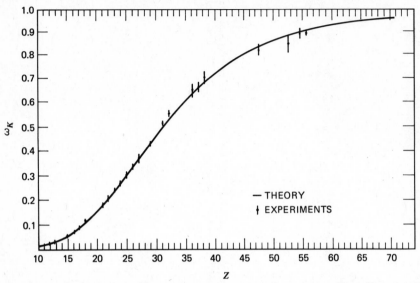

Figure 2.5.   Theoretical ω_K values according to Kostroun, Chen, and Crasemann[54] compared to experimental values collected in Reference 3. (Reprinted with permission of American Physical Society, New York.)

has to be included. A *mean* L-shell fluorescence yield, $\bar{\omega}_L$, can be defined by either the "final" vacancy distribution (V_1 in L_1, V_2 in L_2, V_3 in L_3) that results from the Coster–Kronig effect:

$$\bar{\omega}_L = \sum_{i=1}^{3} V_i \omega_i \tag{2.26}$$

or by the primary vacancy distribution (N_1 in L_1, N_2 in L_2, N_3 in L_3) altered to take account of the Coster–Kronig process by coefficients v_i:

$$\bar{\omega}_L = \sum_{i=1}^{3} N_i v_i \tag{2.27}$$

From Eqs. (2.23) and (2.24) the following relationships can be deduced:

$$\begin{aligned}
V_1 &= N_1 \\
V_2 &= N_2 + f_{12} N_1 \\
V_3 &= N_3 + f_{23} N_2 + (f_{13} + f_{12} f_{13}) N_1
\end{aligned} \tag{2.28}$$

and

$$\begin{aligned}
v_1 &= \omega_1 + f_{12} \omega_2 + (f_{13} + f_{12} f_{13}) \omega_3 \\
v_2 &= \omega_2 + f_{23} \omega_3 \\
v_3 &= \omega_3
\end{aligned}$$

Note that the coefficient v_i is the fraction of all L x-rays resulting from an initial vacancy in a particular L_1 subshell. These equations clearly show that experimental determination of ω_i for the L shell is extremely complicated. Indeed, to carry out a complete analysis of the decay of such an excited condition requires the measurement of several of the parameters that characterize the L shell. However, by using simplified procedures, for example, operating with Z value where the Coster–Kronig process is not obtained, the system can be somewhat simplified.

Much of the theoretical and experimental work discussed in References 3 and 51–62 has been summarized by Keski-Rahkonen and Krause[72] in an excellent survey of the magnitude and trends of radiative, Auger, and Coster–Kronig process rates and yields.

REFERENCES

1. E. H. S. Burhop and W. N. Asaad, *Adv. At. Mol. Phys.*, **8**, 163 (1972).
2. K. D. Sevier, *Low Energy Electron Spectrometry*, Wiley-Interscience, New York, 1972.
3. W. Bambynek, B. Crasemann, R. W. Fink, H.-U. Freund, H. Mark, C. D. Swift, R. E. Price, and P. Venugopala Rao, *Rev. Mod. Phys.*, **44**, 716 (1972).
4. R. H. A. Eade, M. A. Robb, G. Theodorakopoulos, and I. G. Csizmadia, *Chem. Phys. Lett.*, **52**, 526 (1977).
5. M. Thompson, P. A. Hewitt, and D. S. Wooliscroft, *Anal. Chem.*, **48**, 1336 (1976).
6. E. H. S. Burhop, *The Auger Effect and Other Radiationless Transitions*, Cambridge University Press, Cambridge, 1952.
7. I. Bergström and R. D. Hill, *Ark. Fys.*, **8**, 21 (1954).
8. J. J. Uebbing and H. J. Taylor, *J. Appl. Phys.*, **41**, 804 (1970).
9. T. W. Haas, J. T. Grant, and G. J. Dooley, *Phys. Rev. B*, **1**, 1449 (1970).
10. M. F. Chung and L. H. Jenkins, *Surf. Sci.*, **22**, 479 (1970).
11. C. Geffrion and G. Nadeau, *Can. J. Phys.*, **35**, 1284 (1957).
12. R. N. Yasko and R. D. Whitmoyer, *J. Vac. Sci. Technol.*, **8**, 733 (1972).
13. W. A. Coghlan and R. E. Clausing, *Surf. Sci.*, **33**, 411 (1972).
14. W. N. Asaad and E. H. S. Burhop, *Proc. Phys. Soc. Lond.*, **71**, 369 (1958).
15. M. Mladjenovič and H. Stätis, *Ark. Fys.*, **9**, 41 (1954).
16. O. Hörnfeldt, *Ark. Fys.*, **23**, 235 (1962).
17. K. Siegbahn, C. Nordling, A. Fahlman, R. Nordberg, K. Hamrin, J. Hedman, G. Johansson, T. Bergmark, S.-E. Karlsson, I. Lindgren, and B. Lindberg, *Nova Acta Reg. Soc. Sci. Upsal. Ser. IV*, **20** (1967).
18. W. N. Asaad, *Proc. Roy. Soc. Lond.*, **A249**, 555 (1959).
19. U. Gelius and L. A. Nordqvist, UUIP-495, University of Uppsala, Internal Report, August, 1966.
20. M. A. Listengarten, *Bull. Acad. Sci. U.S.S.R. Phys. Ser.*, **25**, 803 (1961).
21. M. A. Listengarten, *Bull. Acad. Sci. U.S.S.R. Phys. Ser.*, **26**, 182 (1962).

22. C. P. Bhalla and D. J. Ramsdale, *Z. Phys.*, **239**, 95 (1970).
23. F. P. Larkins, *J. Phys. B*, **9**, 47 (1976).
24. D. A. Shirley, *Phys. Rev. A*, **7**, 1520 (1973).
25. R. Hoogewijs, L. Fiermans, and J. Vennik, *Chem. Phys. Lett.*, **38**, 192 (1976).
26. R. Hoogewijs, L. Fiermans, and J. Vennik, *Chem. Phys. Lett.*, **38**, 471 (1976).
27. F. P. Larkins, *Chem. Phys. Lett.*, **55**, 335 (1978).
28. I. B. Ortenburger and P. S. Bagus, *Phys. Rev. A*, **11**, 1501 (1975).
29. R. Spohr, T. Bergmark, N. Magnusson, L. O. Werme, C. Nordling, and K. Siegbahn, *Phys. Scr.*, **2**, 31 (1970).
30. M. A. Robb, G. Theodorakopolous, and I. G. Csizmadia, *Chem. Phys. Lett.*, **57**, 423 (1978).
31. H. Ågren, S. Svensson, and U. I. Wahlgren, *Chem. Phys. Lett.*, **35**, 336 (1976).
32. I. H. Hillier and J. Kendrick, *Mol. Phys.*, **31**, 849 (1976).
33. M. T. Økland, K. Faegri, and R. Manne, *Chem. Phys. Lett.*, **40**, 185 (1976).
34. J. M. White, R. R. Rye, and J. E. Houston, *Chem. Phys. Lett.*, **46**, 146 (1977).
35. G. Wentzel, *Z. Phys.*, **43**, 524 (1927).
36. E. H. S. Burhop, *Proc. Roy. Soc.* (*London*), **A148**, 272 (1935).
37. L. Pincherle, *Nuovo Cim.*, **12**, 81, 162 (1935).
38. H. S. W. Massey and E. H. S. Burhop, *Proc. Roy. Soc.*, **A153**, 661 (1936).
39. E. G. Ramberg and F. K. Richtmeyer, *Phys. Rev.*, **51**, 913 (1937).
40. R. A. Rubenstein and J. N. Snyder, *Phys. Rev.*, **97**, 1653 (1955).
41. W. N. Asaad, *Proc. Roy. Soc.*, **A249**, 555 (1959).
42. E. J. Callan, *Phys. Rev.*, **124**, 793 (1961).
43. M. A. Listengarten, *Transl. Bull. Acad. Sci., U.S.S.R.*, **25**, 803 (1961).
44. M. A. Listengarten, *Transl. Bull. Acad. Sci., U.S.S.R.*, **26**, 182 (1962).
45. W. N. Asaad, *Nucl. Phys.*, **44**, 399 (1963).
46. W. N. Asaad, *Nucl. Phys.*, **44**, 415 (1963).
47. M. O. Krause, M. L. Vestal, W. H. Johnston, and T. A. Carlson, *Phys. Rev.*, **133**, A385 (1964).
48. W. N. Asaad, *Nucl. Phys.*, **66**, 494 (1965).
49. W. Mehlhorn and W. N. Asaad, *Z. Phys.*, **191**, 231 (1966).
50. D. Chattarji and B. Talukdar, *Phys. Rev.*, **174**, 44 (1968).
51. E. J. McGuire, *Phys. Rev.*, **185**, 1 (1969).
52. E. J. McGuire, *Phys. Rev. A*, **2**, 273 (1970).
53. D. L. Walters and C. P. Bhalla, *Phys. Rev. A*, **3**, 1919 (1971).
54. V. O. Kostroun, M. H. Chen, and B. Crasemann, *Phys. Rev. A*, **3**, 533 (1971).
55. M. H. Chen and B. Crasemann, *Phys. Rev.*, **A8**, 7 (1973).
56. W. N. Asaad and D. Petrini, *Proc. Roy. Soc. Lond.*, **350**, 381 (1976).
57. M. H. Chen, B. Crasemann, and V. O. Kostroun, *Phys. Rev. A*, **4**, 1 (1971).
58. E. J. McGuire, *Phys. Rev. A*, **3**, 587 (1971).
59. E. J. McGuire, *Phys. Rev. A*, **3**, 1801 (1971).
60. D. L. Walters and C. P. Bhalla, *Phys. Rev. A*, **4**, 2164 (1971).
61. B. Crasemann, M. H. Chen, and V. O. Kostroun, *Phys. Rev. A*, **4**, 2161 (1971).
62. E. J. McGuire, *Phys. Rev. A*, **5**, 1052 (1972).
63. E. J. McGuire, *Phys. Rev. A*, **9**, 1840 (1974).

64. D. Coster and R. De L. Kronig, *Physica*, **2**, 13 (1935).
65. J. N. Cooper, *Phys. Rev.*, **61**, 225 (1942).
66. J. N. Cooper, *Phys. Rev.*, **61**, 235 (1942).
67. NASA Technical Memo, NASA-TM-X-74159 (1976).
68. D. W. Nix and R. W. Fink, *Z. Phys.*, **A273**, 305 (1975).
69. D. G. Douglas, *Can. J. Phys.*, **54**, 1124 (1976).
70. J. A. D. Matthew, J. D. Nuttall, and T. E. Gallon, *J. Phys. C*, **9**, 883 (1976).
71. H.-U. Freund, *X-Ray Spectrom.*, **4**, 90 (1975).
72. O. Keski-Rahkonen and M. O. Krause, *At. Nucl. Data Tables*, **14**, 139 (1974).

CHAPTER

3

EXPERIMENTAL AUGER ELECTRON SPECTROSCOPY

In this chapter the instrumental aspects of modern Auger electron spectroscopy (AES) are described and, where appropriate, illustrative spectra are presented. Rather than attempt an all encompassing text, the aim is to convey to the practicising or would-be Auger spectroscopist a broad perspective of the principal instrumental considerations involved in Auger spectroscopy.

In common with all electron spectrometers, the Auger instrument is entirely contained within a high-vacuum chamber. Currently, the spectrometer as a whole is usually constructed from 304 stainless steel and μ-metal, although glass systems have been used.[1] The Auger spectrometer can logically be divided into three constituent parts: (1) A source of ionizing radiation, (2) an electron energy analyzer, and (3) a detection device plus amplification and display peripherals. Each of these parts is now described with a subsequent section giving examples of commercial Auger spectrometers and a description of the scanning Auger system.

1. THE AUGER SPECTROMETER

1.1. Ionizing Sources

1.1.1. Electron Guns

The most-common source of ionizing radiation used in AES is an electron gun producing an electron beam, which can be focused onto the sample surface. An electron gun used for surface studies is shown in Fig. 3.1. It is similar to that of a monochrome TV set, and, in fact, electron guns scavenged from old oscilloscopes make good sources for AES. Today, however, where fine focusing of the beam is an important consideration in achieving submicron spatial resolution, special guns must be purchased.

The source of electrons is a hot filament of either thoriated or pure tungsten operating at about 2500 K. For gas-phase studies pure-tungsten filaments are used since thoriated filaments can be irreversibly damaged if

53

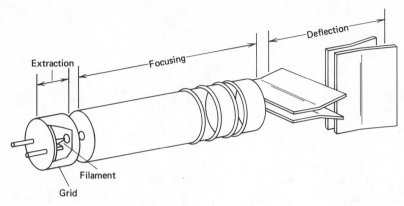

Figure 3.1.   Schematic of an electron gun.

operated at high pressures (i.e., $> 10^{-6}$ torr). The design of the electron gun for gas-phase studies is rather different from the surface-instrument gun, and an example is shown in Fig. 3.2.

Electrons leaving the filament material are accelerated along the symmetry axis of the gun by applying a large positive potential between the filament and the electron lenses of the gun. This is normally achieved in practice by "floating" the filament at a negative potential and holding the accelerating plates of the gun at or near ground potential. The electron beam thus produced is then deflected onto the sample by varying the potentials on the X and Y plates, which will deflect the beam perpendicular to the direction of propagation. If an ac voltage, or more accurately a time-base program, is applied to these plates, the electron beam will scan back and forth across the sample—this being the basis of the scanning Auger system.

The kinetic energy of Auger electrons is independent of the primary energy. This lifts any rigorous restrictions on the quality of the primary beam for nonquantitative work, which is not true for XPS and UPS. Clearly this is an advantage for AES since inexpensive power supplies can be used to run the electron gun. In fact, commercial electron-gun power supplies are very expensive indeed, but this cost can be considerably reduced by building it, since the circuitry is quite simple.[2] Care must of course be exercised in the isolation of the high-voltage power supply or more commonly the filament power supply, which will float negatively below ground by the primary energy.

The primary-beam energy must be greater than the binding energy of the core electrons, otherwise no ionization can occur. The probability for Auger ejection varies with the ratio E_p/E_B, where E_p is the primary-beam energy and E_B is the binding energy of the core electron, as shown in Fig. 3.3. For maximum yield the ratio is about three. For example, the binding energy of the $1s$

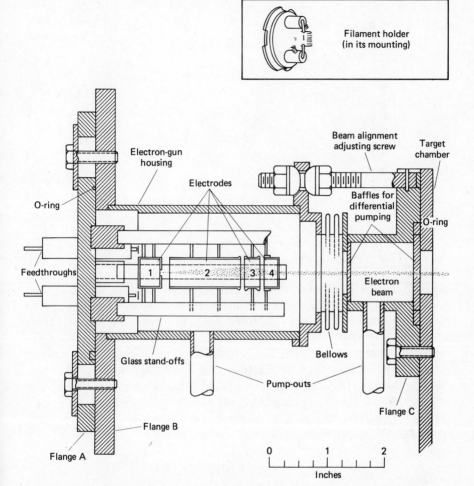

Figure 3.2. An electron gun for gas-phase Auger studies (from T.A. Carlson, *Photoelectron and Auger Spectroscopy*, Plenum Press, New York, 1975). (Reprinted with permission of Plenum Press, Plenum Publishing Company, New York.)

electrons of oxygen is about 515 eV (the exact value will depend on the chemical environment of the oxygen atoms), and so the most intense O *KLL* spectra are observed when beam energies exceed 1500 eV.

1.1.2. X-Ray Excitation

The production of a core hole can also be initiated by x-ray bombardment, which can lead to Auger emission. There are a few examples in the literature of

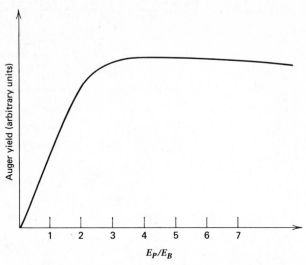

Figure 3.3.   Variation in Auger yield with the ratio E_P/E_B.

x-ray-excited AES (XAES),[3] where electron-beam damage to the sample
has been a problem. These experiments will be discussed in more detail in
Section 6, which is solely concerned with electron-beam effects. X-ray
excitation is less suitable for Auger spectroscopy than electron-beam ionization for a number of reasons. The probability of producing a core vacancy
using x-rays is lower than for electrons, the spatial resolution of x-rays is
very poor due to focusing problems, and, also, x-ray sources are somewhat
more expensive than electron guns. X-ray radiation is usually provided by a
Mg or Al source giving near monochromatic beams with energies of 1254
and 1487 eV, respectively.

1.2.   Electron-Energy Analyzers

The function of an electron-energy analyzer is to record the energy distribution of emitted electrons. There are many types of electron-energy analyzers,
which can be classified into three broad categories: (1) retarding-field or
LEED/Auger systems, (2) electrostatic analyzers, and (3) magnetic-deflection
analyzers. The latter type is seldom used, but will be discussed here. A fuller
account can be found elsewhere.[4]

1.2.1.   Retarding-Field Analyzers

Historically, the retarding-field analyzer (RFA) was one of the first types
used in electron spectroscopy. In its most simple form the analyzer is simply

a wire grid placed between the sample and the detector. The potential on the grid is varied from zero volts negatively to the primary energy of the beam. As the potential is scanned, only electrons with sufficient energy can overcome the retarding voltage and reach the detector. Thus an RFA is a high-pass energy filter, where the kinetic energy of the detected electrons is deduced from the height of the potential barrier that they surmount.

An electron of kinetic energy E ($=\frac{1}{2}mv^2$) will cross a potential energy barrier V if $E \cos^2 \theta \geq V$, where θ is the angle that the electron beam makes with the normal (i.e., at normal incidence, $\theta = 0$, $E \geq V$). Anderson et al.[5] have shown that the number of electrons crossing an energy barrier V at energies between E and $E + \delta E$, from an energy distribution $n(E)$, is given by

$$N(E)dE = \int_0^{\cos^{-1}[(V/E)^{1/2}]} n(E) \cdot dE \cdot 2\pi \sin \theta \cdot d\theta \qquad (3.1)$$

where $2\pi \sin\theta \, d\theta$ is the solid angle into which the electrons are emitted.

The height of the potential-energy barrier is not a direct measure of the total kinetic energy of the electrons, but of their momenta perpendicular to the equipotential lines running from the retarding grids. The response of an RFA in fact yields the result:

$$I(E_0) = \int_{E_0}^{\infty} N(E) \, dE \qquad (3.2)$$

where I is the current measured by the electron collector. A plot of I against the retarding potential is called the transfer characteristic. This will appear again in Section 2. If the spectrum is plotted out as a transfer characteristic (as we shall see later the second derivative d^2I/dE^2 is always used), then the data will yield electron energies at preferred values (e.g., Auger emission) from steps in the curve.

Standard designs for RFAs use parallel, cylindrical, or spherical plates. The parallel-plate RFA is the simplest to construct, but is seldom used in modern Auger spectroscopy and will not be described here. Cylindrical-grid RFAs are also unpopular since they have many associated problems: they are limited in resolution and produce asymmetric peak shapes, which are skewed to the low-energy side.

Spherical-grid-type RFAs are still in use today. In its most usual configuration the analyzer takes the form of concentric hemispheres. The resolution is given by

$$\frac{\Delta E}{E} = 2ab^{-1} \sin^2 \theta \qquad (3.3)$$

where θ is the largest angle the entering beam can make with the symmetry axis of the spheres and still be collected; a and b are the radii of the inner and outer hemispheres.

In general, RFAs suffer from common shortcomings. The spectra tend to be poor at low kinetic energies and hence difficult to interpret. Most of these problems are associated with an increasingly large secondary-electron background. In this chapter the term secondary electrons is used rather loosely, but in this case refers to electrons that have undergone multiple scattering en route to the detector and contain no information about the sample. Other problems associated with RFAs, such as variation of the work function of the electrodes, field penetration, and space-charge effects, have been well discussed by several authors.[6-8]

1.2.1.1.   LEED/Auger instrumentation. The LEED/Auger instrument became very popular in the late 1960s and early 1970s particularly with physicists studying single-crystal surfaces.[9-12] The instrument, which was primarily designed for low-energy-electron-diffraction (LEED) studies, can be converted to an RFA for Auger work quite simply with subsidiary electronics.

A schematic of a four-grid LEED/Auger system is shown in Fig. 3.4. The photocurrent from the photoemission is detected by the LEED screen, which is generally held at a positive potential (~ 50 V) to reduce production of secondary electrons. The resolution of such a device varies slightly with the manner in which the grids are interconnected. Generally, a resolving power of about 2% is achieved (i.e., the resolution of 200 eV is 4 eV).

The advantage of an RFA is that it is cheap, and it is also an attractive proposition to record LEED patterns using the same spectrometer. However, the LEED/Auger analyzer has many shortcomings when compared with electrostatic Auger spectrometers. It has poor resolution and rather low sensitivity. Since it collects over a large solid angle, it cannot be used to study angular effects unless equipped with a Faraday cup or external photomultiplier tube (see Section 3.4) and large spectral slopes make it of limited value for studies of electrons with kinetic energies below 200 eV, unless the features are strong and broad.

Some spectra recorded with a LEED/Auger instrument are shown in Fig. 3.5. They show the Auger spectra of a copper (111) crystal at various temperatures. Many of the peaks are temperature dependent, and this will be discussed in Section 4.2.[13]

1.2.2.   Dispersive Analyzers

The most widely used electron-energy analyzers in modern Auger spectrometers are the dispersive type. There are basically two types: (1) magnetic-

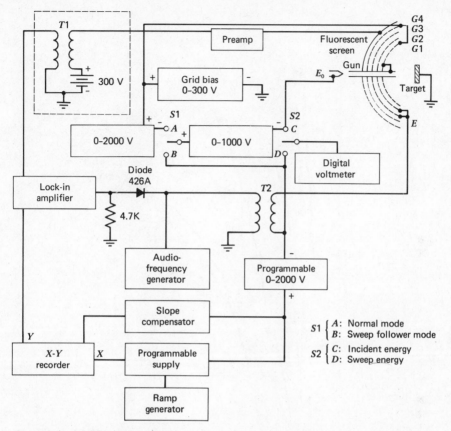

Figure 3.4.   A LEED/Auger system used for surface analysis. $G2$, $G3$ are the analyzing grids; their potential is controlled by the ramp generator and programmable supplies. The audio-frequency generator provides the ac modulation and the diode arrangement allows for tuning the lock-in amplifier to 2ω for electronic differentiation. $G1$, $G4$ are shielding grids; the secondary electrons created at these grids can be "biased out" using the grid bias supply. The electrons are collected by the fluorescent screen and detected with the lock-in amplifier. The slope compensator is useful for subtracting background slope from the data, and the sweep-follower mode produces spectra with the primary loss peaks eliminated. (From C. C. Chang, in P. F. Kane and G. R. Larrabee (Eds.), *Characterization of Solid Surfaces*, Plenum, New York, 1974, Chapter 20. Reprinted with permission.)

deflection analyzers and (2) electrostatic-deflection analyzers. The electrostatic type has been more popular for chemical analysis. Both types will now be described in some detail.

1.2.2.1. Magnetic Analyzers. Magnetic electron-energy analyzers possess one major advantage over electrostatic analyzers in that they can supply the necessary field to analyze electrons of energies greater than 5000 eV.[14]

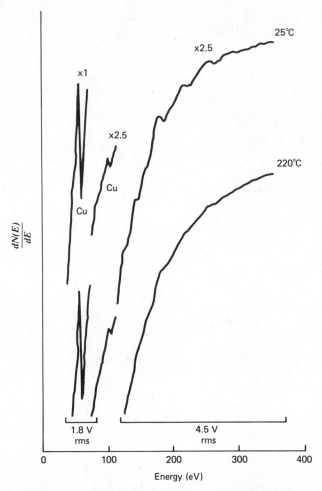

Figure 3.5. An illustration of Auger spectra from a Cu(111) surface with a LEED/Auger system.[13] (Reprinted with permission of North-Holland Publishing Company, Amsterdam.)

The great difference between electrostatic and magnetic analyzers is the cost, with the latter instrument being 5–10 times more expensive than a comparably performing electrostatic instrument.

The properties of magnetic electron-energy analyzers have been extensively treated by Siegbahn.[4] The general principle on which magnetic-type analyzers work is that when an electron passes into a uniform magnetic field that is oriented perpendicular to the direction of the particle's motion, the force exerted on the electron by the field causes it to describe a circular trajectory. In this manner, a uniform magnetic field will disperse electrons according to their momenta, focusing all electrons of the same momentum at the same point after they have traversed 180°. For an electron with kinetic energy eV moving in a uniform magnetic field B, $BeV = mv^2/r$, where r is the radius of the electron trajectory. Magnetic analyzers therefore discriminate electron energies by virtue of their momenta, and a nonlinear energy scale is obtained. Since $r = (2mV/eB)^{1/2}$, by varying the magnetic field B, electrons of different momenta will be focused at r.

In most instruments a sector magnet is used that allows for some flexibility in the positioning of the sample and the detector. As shown in Fig. 3.6, a sector magnet obeys Barber's rule, which states that the object vertex and image must lie on the same straight line.

A more-detailed account of the focusing properties of sector magnets has been given by Judd.[15] Instruments used for chemical analysis are usually iron free, and Kerst and Serber[16] have shown that in nonuniform magnetic fields electrons are focused after traversing a central angle of $\theta = \sqrt{2}\pi$.

The resolution of a magnetic analyzer is given in terms of the electron momentum, and for low energies follows the equation[17]

$$\frac{\Delta p}{p} = \frac{1}{2}\frac{\Delta E}{E} \tag{3.4}$$

where p is the momentum and E is the energy. An instrument designed by Fadley et al.[17] claimed a resolution of 0.01%. Indeed, typical magnetic analyzers are quoted to have a momentum resolution of $\Delta m/m = 0.01\%$.[18] Although the throughput of these analyzers is usually quite low, a multidetector array can be used, since the focal plane of these analyzers is well defined. A multidetector array will increase the speed of data acquisition by $n^{1/2}$, where n is the number of array elements. If parallel processing is used, an n-fold advantage occurs.

1.2.2.2. Electrostatic-Dispersion Analyzers. A large proportion of state-of-the-art electron spectrometers used in chemical analysis are equipped with electrostatic-deflection analyzers. Since these analyzers are by far the most

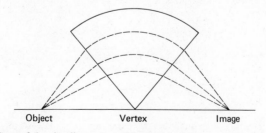

Object   Vertex   Image

Figure 3.6.   An illustration of Barber's rule for a sector magnet.

popular, they will be described in detail in this section. There are two types of dispersive electrostatic analyzers: (1) the mirror type, where electrons are reflected against a retarding potential, and (2) the deflector or prism type, where the electron trajectories follow equipotential surfaces within the analyzer.

1.2.2.3. The Parallel-Plate Dispersive Analyzer.

The parallel-plate analyzer[19-21] is the simplest example of the mirror type. It consists of two parallel metal plates across which a potential is applied (see Fig. 3.7a), with the upper plate positive. Electrons from the sample pass through a slit in the lower plate and, on entering the interspace at an angle θ, describe a parabolic path focusing onto the second slit in the lower plate. The horizontal distance χ_0 that each electron travels for a given potential across the plates depends on its initial kinetic energy. Pierce[22] and Spangenberg[23] have shown that $\chi_0 = 2eV/V' \sin 2\theta$, where $V' = -dV/dy$, the electric field strength, and

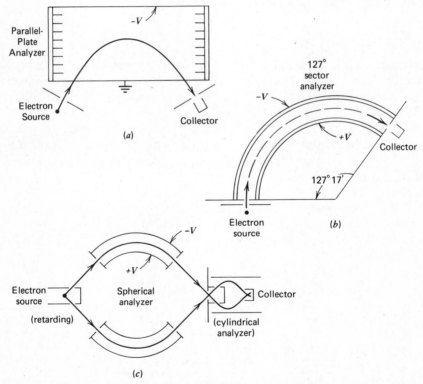

Figure 3.7. Electrostatic energy analyzers: (a) parallel plate, (b) 127° sector, and (c) spherical (center-only) analyzers. (Reprinted with permission of Pergamon Press, New York.)

y is the vertical component of the electron motion. Therefore, electrons of the same energy are focused at χ_0. The y coordinate of electron motion is given by

$$y = \frac{V'x^2}{4eV \cos^2 \theta} + x \tan \theta \tag{3.5}$$

which has a maximum at $\theta = 45°$. Thus, electrons entering the plate system at angles around $45°$ are focused upon returning to the lower plate.[19] Since there is a linear relationship between the applied voltage across the plates and the energy of an electron able to pass through the slit system, then the current reaching the detector as the applied potential is swept will represent the number of electrons reaching the detector.

The resolution achieved with this type of analyzer with slits of width ω is

$$\frac{\Delta E}{E} = \frac{\omega}{\chi_0} + (\Delta \theta)^2 \tag{3.6}$$

Improvements on this design were made by Proca and Green[24] who added another plate that produced a field-free region below the lower plate. Second-order focusing then occurs at a $30°$ entrance angle, and much higher angular resolution is obtained. The resolution is given by

$$\frac{\Delta E}{E} = \frac{1.5\omega}{\chi_0} + 4.6(\Delta \theta)^3 \tag{3.7}$$

1.2.2.4. The Cylindrical-Mirror Analyzer. The cylindrical-mirror electron-energy analyzer (CMA), which is a special case of the parallel-plate analyzer, is today the most popular for AES. It has numerous advantages including high sensitivity, speed, and good throughput. It was first described in 1957 by Blauth,[25] and Palmberg et al.[26] in 1969 showed that it was capable of recording AES of surfaces. The CMA shown in Fig. 3.8 consists of two co-axial metal cylinders, the inner one of which is grounded and connected electrically to the sample. Machined into the inner cylinder are two slots (which of course are interrupted for mechanical support) covered in fine mesh to ensure a uniform field. Fringing-field correction plates must also be placed between the ends of the inner and outer cylinders, as in parallel-plate analyzers, but are omitted from the figure for clarity.

A negative potential is applied to the outer cylinder, and, as this voltage is ramped, successive energy areas of the beam are brought into focus at the exit slit. This analyzer will focus electrons of energy E at a distance L from the entrance slit given by

$$\frac{L}{r_0} = (l_1 + l_2)\cot \theta + 2(\pi K)^{1/2} \cos \theta \exp(K \sin^2 \theta)\mathrm{erf}(K^{1/2} \sin \theta) \tag{3.8}$$

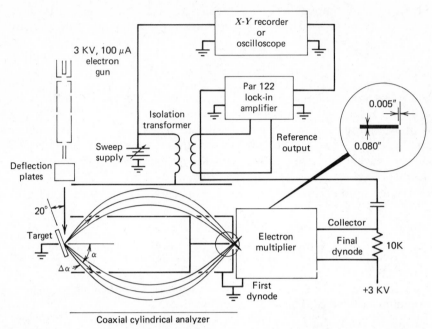

Figure 3.8.   The cylindrical-mirror analyzer (CMA). The electron gun can be mounted on the side as shown, or coaxially, inside the CMA (from C. C. Chang, in P. F. Kane and G. R. Larrabee (Eds.), *Characterization of Solid Surfaces*, Plenum, New York, 1974, Chapter 20). (Reprinted with permission of Plenum Press, Plenum Publishing Company, New York.)

where $K = (E_0/V) \ln(r_1/r_0)$
$\quad V$ = the reflecting potential
$\quad r_0, r_1$ = the radii of the two cylinders
$\quad l_1, l_2$ = the distance of the slits from the inner cylinder

The focusing qualities of a CMA are very good over a wide range of acceptance angles, and it can be usefully operated at 54.7°, the "magic" angle, where for gas-phase studies angular distributions of electrons affecting the intensity are eliminated.

The alignment of the cylinders is, as one would expect, critical; however, with a coaxially mounted electron gun producing a beam along the analyzer axis, the alignment problems are reduced. Then the only remaining variable is the position of the sample along the axis, which can be varied by means of a high-precision manipulator with x, y, z, and rotational degrees of freedom.

To summarize, a CMA is an attractively simple electron analyzer with excellent properties. It has high throughput since the entire azimuthal angle of 2π can be used and good resolution enabling the rapid acquisition of spectra. Although it is a dispersive instrument, multichannel detection is difficult, since the focal surface is approximately a cone pointing away from the source.

1.2.2.5. Prism-Type Electrostatic Analyzers. The prism-type analyzers (e.g., 127° and hemispherical) are based on rotational symmetric deflecting fields formed by cylindrical, spherical, or toroidal electrodes. The cylindrical[27] or spherical deflectors[28] are in fact special cases of the toroidal-field deflector, which has been well treated by several authors.[29,30] In a toroidal field, the central trajectory lies in the symmetry plane and has a radius of curvature r, which is also the radius of the corresponding equipotential in that plane. The trajectories of the electrons within the field can be predicted by using the Lagrange equations.[30] A cylindrical deflector gives only radial focusing after a deflection angle of $\pi/\sqrt{2}$ (127°), while the spherical deflector is space focusing at 180°.

1.2.2.6. The 127° Cylindrical Analyzer. The 127° cylindrical analyzer is one of the more-common types used today in electron spectroscopy.[31] Most of the instruments are small, with a radius of 5–10 cm. A schematic is shown in Fig. 3.7b—for comparison a spherical analyzer is depicted in Fig. 3.7c. In practice, the sample and detector are usually positioned outside the field, since the sample must have a well-defined potential, and, furthermore, any disturbance of the deflecting field has to be avoided. A 127° analyzer is well suited to line sources owing to its strong first-order focusing.

1.2.2.7. The Hemispherical Analyzer. In the hemispherical analyzer a beam entering the plate system is focused after deflection through 180° (see Fig. 3.9). For both the 127° and hemispherical analyzers the energy spectrum can be obtained by varying the plate potential, and programmable power supplies are usually built into the spectrometer package to do this. The hemispherical analyzer can be operated at constant resolution or constant resolving power. In the former, the pass energy of the analyzer is held constant and electrons are preretarded before entering the field between the concentric hemispheres. For constant resolving power the analyzer pass energy is swept and no preretardation is applied. The resolution for both analyzers is proportional to the slit width and the radius of the flight path, that is, $\Delta E/E \propto \Delta l/r$. The resolution of an ESCA 3 hemispherical analyzer (Vacuum Generators, Hastings, England) under various operating conditions is given in Table 3.1.[2]

1.2.2.8. Time-of-Flight Analyzers. For the sake of completeness there follows a brief description of the time-of-flight analyzer (TOFA). The TOFA possesses a major advantage over magnetic and electrostatic analyzers in

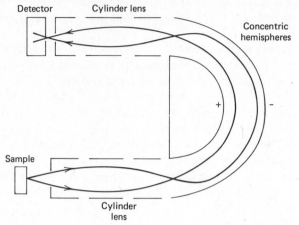

Figure 3.9.   The hemispherical analyzer with axially symmetric retarding lenses added at the entrance and exit of the analyzer.

Table 3.1.   The Resolution in eV of a Hemispherical Analyzer as a Function of Pass Energy and Slit Width

	Pass Energy (eV)				
Slit (mm)	200	100	50	20	10
0.5	0.5	0.25	0.12		
1.0	1.0	0.5	0.25	0.1	
2.0	2.0	1.0	0.5	0.2	0.1
4.0	4.0	2.0	1.0	0.4	0.2

that it is unaffected by local fields caused by contact potentials and stray magnetic fields. In contrast to the deflection methods, the determination of electron energy by timing the flight of the photoelectrons over a measured distance becomes increasingly precise at low electron energies. Also, the TOFA lends itself to multichannel detection since the continuous distribution of velocities obtained from electron spectroscopy allows for simultaneous measurements over an extended energy range.

A TOF instrument used by Tsai and co-workers[32] is shown in Fig. 3.10. A pulsed source of radiation impinges upon the sample causing photoemission. The photoelectrons are accelerated to the drift tube by a pulsed potential at grid A. The electrons are then sorted by virtue of their velocity within the drift tube. On leaving the tube, the electrons are again accelerated, collected, and detected.

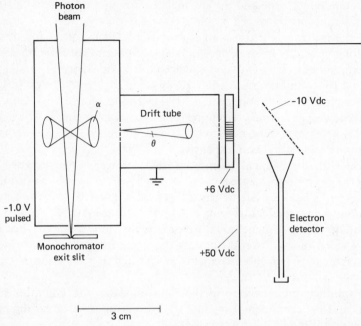

Figure 3.10.   The time-of-flight analyzer. Electrons initially injected into the solid angle α end up in the angle θ as a result of being accelerated.[32] (Reprinted with permission of American Physical Society, New York.)

1.3.   Detection

The various methods of electron detection currently in use in AES are now described. In photoelectron spectroscopy the currents arriving at the detector vary between about 10^{-18} and 10^{-7} A. RFAs, as well as using a collector plate or screen, can be equipped with a Faraday cup or the screen luminescence can be recorded with a phototube. Electron collectors have an advantage of being unaffected by adsorption of active gases and high-temperature bakeout required for ultra-high vacuum. However, the low sensitivity more than outweighs this advantage.

Most Auger electron spectrometers use some form of electron multiplier for detection. These work on the principle of ejection of secondary electrons by an incident electron. After secondary electrons collide with the surface of the electron multiplier more secondary electrons are released, until after 10–20 repetitions there are enough electrons to produce a measurable current in an amplifier.

There are two possible modes of operation, dc and ac. In the dc mode the multiplier operates as a current amplifier, while in the ac mode it provides single-electron counting. The electron multiplier is of two types, the dynode or "venetian blind" type and the channel electron multiplier[33] or "channel-tron," which is much favored for electron spectroscopy.

The channeltron is a continuous dynode consisting of a tube of semi-conducting glass, the inner surface of which has been processed to give it a high secondary-electron-emission coefficient. The resistance of these multi-pliers is $\cong 10^9 \, \Omega$ and the gain $\cong 10^8$. However, at high count rates the chan-neltron saturates resulting in peak distortion and resolution loss. Generally 150,000 counts per second is the upper limit of the channeltron. Above these levels it is necessary to use a discrete dynode multiplier and fast electronics, which are considerably more expensive.

Today, with the advent of the microprocessor, Auger systems can be highly automated with sophisticated data-manipulation software. Nevertheless, the scan and readout systems can be generally divided into three types: continuous scan, incremental scan, and multichannel systems.

The incremental-scan method increases the current through the spectrom-eter in a series of small steps, counting the signal from the detector during each increment. When the increments are plotted as a function of field, a spectrum is produced.

The multichannel-analyzer method encompasses the continuous-scan method and the incremental method, that is, scanning a large number of increments continuously between two limits of field.

1.4.   The Complete Instrument: A Summary

The Auger electron spectrometer, in its modern-day form, is a highly sophis-ticated analytical device. The three integral parts described in the previous sections are contained within a high-vacuum chamber, usually constructed from 304-type stainless steel, OFHC copper,[34] and μ metal, where magnetic shielding is needed. The spectrometer will normally be advertised as a complete package containing the pumping station and power supplies for the electron gun, detection electronics, and spectrometer scanning. Usually the spectrometer is bakeable to about 200°C where ultra-high-vacuum operation ($< 5 \times 10^{-10}$ torr) is needed (e.g., studies of clean surfaces).

Additional features that most spectrometers will be equipped with are an argon-ion source for specimen cleaning and depth profiling (see Section 3), a sample carrousel so that many samples can be processed without breaking vacuum, and a quadrupole mass spectrometer for residual-gas analysis. An invaluable aid to the Auger spectroscopist are listings of approximate Auger energies given by Chang[35] and a chart distributed by Varian.[36]*

---

*Reproduced in the fold-out chart and discussed in Appendix 2.

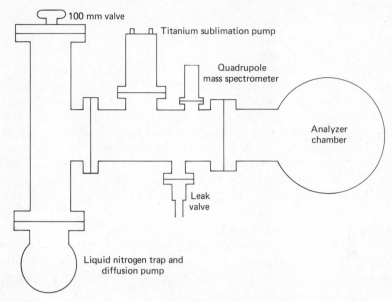

100 mm valve

Titanium sublimation pump

Quadrupole
mass spectrometer

Analyzer
chamber

Leak
valve

Liquid nitrogen trap and
diffusion pump

Figure 3.11. Schematic of an auger vacuum system.[37,38]

A schematic of the vacuum system currently in use at the University of East Anglia, Norwich, England for Auger studies of molecules adsorbed on single crystal surfaces is shown in Fig. 3.11.[37,38]

1.5.   The Scanning Auger System

As mentioned previously, the basis of the scanning Auger system is rastering or scanning the electron beam across the sample. By scanning the electron beam over the specimen surface, the strength of a particular Auger transition can be monitored and displayed. When the scan is in one direction, an Auger line scan is produced, and a two-dimensional scan produces an Auger map. By using an ion beam to etch away the sample, a three-dimensional Auger analysis or "depth profile" can be performed, and this will be discussed in Section 3. The information gained from the sample can be displayed in various ways, which will be described subsequently.

The scanning Auger system can be used as a low-resolution electron microscope and most commercial spectrometers will include this option, most commonly using emitted secondary electrons or absorbed-specimen current for imaging. This is often very useful to the spectrometer operator, allowing rapid location of points of interest on the sample.

The scanning Auger spectrometer is perhaps most commonly used in the two-dimensional mode mentioned above to form a map of the surface of interest. Normally the map is presented on a screen as a series of light and dark patches, which reflect the concentration of a particular element at any point on the surface. However, maps presented in this way are often misleading because the gain and offset for each map are set arbitrarily by the operator. This means that even though a particular region appears black on the Auger map there may still be a significant amount of that element present at the surface under study. Computer techniques have been developed to alleviate these problems, one method being to generate Auger maps from stored data and display contours of constant signal strength. Nevertheless, slight shifts or shape changes in the Auger peak due to chemical effects can cause artifacts unless the Auger current can be determined in addition to the peak height which is commonly monitored.

A useful factor in Auger spectroscopy can be the variation of electron escape depth both with electron energy and with the element that the electron is ejected from. For example, Auger transitions are generally measured within the energy range 50–2000 eV, and the escape depth usually increases with energy.[39] This means that if an element has two Auger transitions with widely differing energies, some possible information about the inhomogeneities in the depth distribution of the near surface may be available.

A short list of the more-common commercial scanning Auger systems is given in Table 3.2. The electron-beam diameters cover a wide range, and it is important to note that the time required to obtain the data will depend on

Table 3.2. Characteristics of Some of the More-Common Commercial Scanning Auger Instruments

Manufacturer	Electron Beam Diameter (μm)	Ultimate Pressure (Pa)
Vacuum Generators[a,d]	0.05	10^{-10}
Coates Welter[a]	0.05	—
Jelco	0.5	10^{-6}
Varian[b,c]	0.2	10^{-9}
Varian[b]	5	10^{-9}
Physical Electronics[b,c,d]	0.5	10^{-9}
Physical Electronics[b,d]	3	10^{-9}
Physical Electronics[b,c,d]	5	10^{-9}

[a]Field-emission electron gun.
[b]Coaxial electron gun.
[c]Computer package.
[d]Other surface analytical techniques available.

the electron beam current, as well as on other parameters. Therefore, a high-spatial-resolution scan will take much longer than a low-resolution scan, since the signal to noise improves with higher beam currents, whereas beam current must decrease as spot size decreases.

There are many examples of the use of the scanning Auger technique[40-45] on a variety of samples, as shown in the recent literature,[46-48] underlining its importance in the field of surface analysis.

In the future, improvements in the scanning Auger system will probably evolve through computer technology. Also, possible combinations with other analytical techniques could well prove important. Further advances in the theory of quantitative Auger spectroscopy (see Section 5) and the availability of better standards will make quantitative analysis more reliable, as the parameters affecting signal strength become better understood.

2. TYPICAL AUGER SPECTRA

Figure 3.12a shows the typical distribution of electrons emitted from a solid sample. The spectrum is completely dominated by a broad distribution of "slow" electrons and a rather smaller sharper peak at the primary beam energy. The broad peak is due to secondary electrons that have undergone many inelastic collisions within the sample before ejection from its surface, and the small sharp peak is caused by the elastically reflected electrons that contain the information used in low-energy-electron diffraction (LEED).[49,50]

On closer inspection of the secondary-electron distribution, small peaks are evidently superimposed upon the strongly sloping background. These peaks always occur at the same kinetic energy regardless of the primary energy and are therefore caused by Auger emission. Other small peaks on the secondary-electron background move in tandem with the elastic peak and are due to electron-energy-loss processes,[51-55], where the incident electrons lose a discrete amount of energy to an excitation process.

Since the Auger peaks are sitting on an intense and steeply sloping background, they can be extremely difficult to observe in a normal counting mode (i.e., a plot of number of electrons against kinetic energy). However, electronic differentiation of the signal makes the task easier, as shown in Fig. 3.12b. This was first demonstrated in the late 1960s.[56]

If a perturbing voltage $\Delta E = k \sin \omega t$ is superimposed on the pass energy of an electrostatic analyzer, then the total current collected can be written as a Taylor series:

$$I(E+\Delta E) = I(E) + I'(E)\Delta E - \frac{I''}{2!}(E)\Delta E^2 + \frac{I'''}{3!}(E)\Delta E^3 + \frac{I''''}{4!}(E)\Delta E^4 + \cdots \quad (3.9)$$

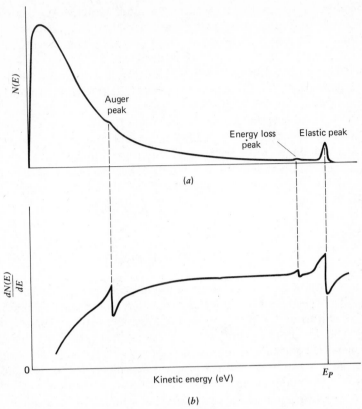

Figure 3.12. (a) Typical distribution of Auger electrons obtained from a solid surface. (b) Differentiation of signal for identification of Auger peaks.

where the prime denotes differentiation with respect to E. Hence,

$$I = I_0 + \left(I'K + \frac{I'''}{8} K^3 + \cdots \right) \sin \omega t - \left(\frac{I''}{4} K^2 + I'''' \right) \cos 2\omega t$$

$$\simeq I_0 + I'K \sin \omega t - \frac{I''}{4} K^2 \cos 2\omega t \qquad (3.10)$$

If K^3 and higher-order terms are neglected (i.e., when the modulation voltage is small), then detection at a frequency ω by use of a phase-sensitive detection system will produce a derivative spectrum (I'). The magnitude of the signal will rise as K, the modulation voltage is increased, but care must be taken not to degrade the instrumental resolution by "over modulating" the spectrometer. Typical modulation voltages will lie in the range of 0.1–2 V.

To avoid confusion, it is worthwhile to consider how differential spectra

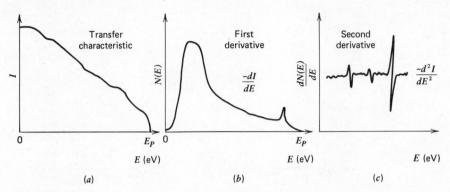

Figure 3.13. Illustration of how derivative spectra are obtained with a LEED/Auger spectrometer.

are obtained using a LEED/Auger device. When the potential on the retard grid of the RFA is swept between zero and the primary beam energy, the collector current varies as shown in Fig. 3.13a. At zero retard volts, all reflected and secondary electrons (apart from those intercepted by the grid wires) reach the collector, and at full retard volts, all reflected and secondary electrons are suppressed. The current measured at the collector as a function of electron energy is called the transfer characteristic, which is really only important as a concept and would seldom be plotted.

The first derivative of the transfer characteristic is obtained by superimposing the modulating signal on the ramped retard potential and synchronously detecting the collector current at the modulating signal frequency. The first derivative of the transfer characteristic is the familiar secondary-electron curve of Fig. 3.13b on which Auger peaks appear as small perturbations (about 10^{-12} A) on the background signal (about 10^{-7} A).

The second derivative of the transfer characteristic is obtained by synchronously detecting the collector current at the second harmonic of the modulating frequency. In this mode the "first-derivative" Auger spectrum $dN(E)/dE$ is obtained. It is important, therefore, that, unlike the case for dispersive analyzers, the derivative spectrum for LEED/Auger devices is obtained by modulating at a frequency ω and detecting at its second harmonic 2ω. This is shown schematically in Fig. 3.13c.

2.1. Direct Spectra: $N(E)$ versus Electron Energy

All gas-phase Auger spectra are reported in this form because one does not have to contend with the strongly sloping secondary-electron background produced by solids. Spectra in this form are easier to interpret than derivative

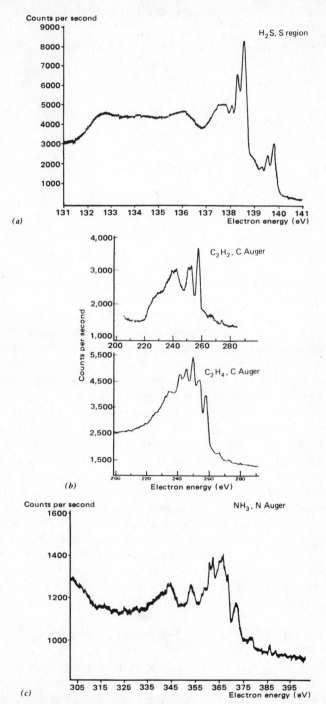

Figure 3.14. Examples of Auger spectra recorded in the counting mode.[57] (Reprinted with permission of American Chemical Society, Washington, D.C.)

spectra, since the peak area and maxima are well defined. In the next section describing derivative spectra from surfaces we will see that for ease of comparison with gas-phase spectra, the derivative spectra are often converted to direct spectra.

Some typical Auger electron spectra are shown in Fig. 3.14 recorded in a counting mode.[57]

2.2.   Auger Spectra of Surfaces

Nearly all Auger surface spectra are recorded in derivative form owing to the problems associated with the counting mode described in Section 2.1. An example of this is shown in Fig. 3.15 for a Cu(111) crystal surface before cleaning by argon-ion bombardment. In this mode, by convention, the peak position is taken as the most negative part of the derivative peak.

Unfortunately, in the derivative mode the signal contrast, enhanced by differentiation, can be at the expense of the signal-to-noise ratio.[58] To produce the original Auger spectrum the differential spectrum can be integrated. This method has been applied usccessfully by the use of minicomputers interfaced to the Auger spectrometer.[59] Alternatively, good integration can be achieved by merely scanning rapidly with a long time constant controlling the recorder pen response. Although crude, this method produces good results when checked with the elastic peak.[37,38]

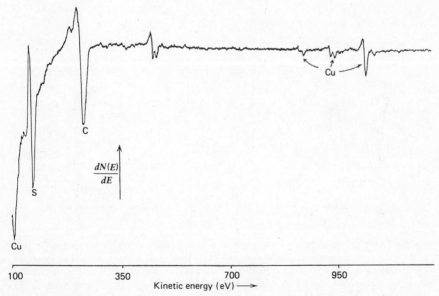

Figure 3.15.   Auger spectrum of a Cu(111) single-crystal surface before cleaning by ion bombardment.

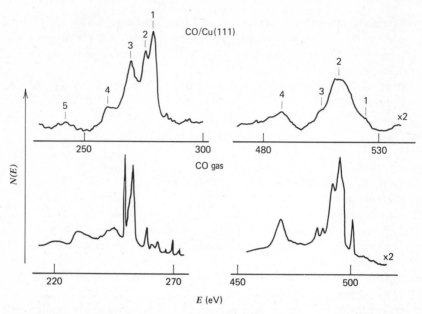

Figure 3.16. Integrated Auger spectrum of CO and Cu(111) at 1 eV resolution compared to CO gas spectrum.[37] (Reprinted with permission of North-Holland Publishing Company, Amsterdam.)

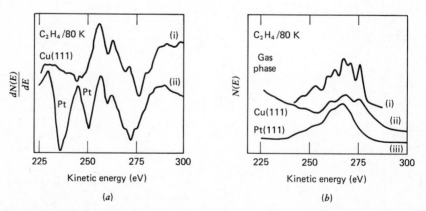

Figure 3.17. (a) Carbon Auger spectra of a monolayer of ethylene adsorbed at 80 K on (i) Cu(111) and (ii) Pt(111). Spectrometer resolution is 2 eV.[38] (b) Integrated Auger spectra. The spectrum of gas-phase ethylene shifted upward by 19 eV is included for comparison. Spectrometer resolution is 1 ev.[38] (Reprinted with permission of North-Holland Publishing Company, Amsterdam.)

76

Although Auger spectra of surfaces are quite common, those of molecular species adsorbed at surfaces are rare. It is in the latter form that the production of direct spectra is important in order to compare the condensed-phase data with the gas-phase spectra. Two examples are shown in Figs. 3.16 and 3.17 for carbon monoxide adsorbed on a Cu(111) surface and ethylene adsorbed on Cu(111) and Pt(111) surfaces along with the differential spectra.[37,38] It is evident that when the gas-phase spectra are shifted by an appropriate amount there is good agreement in the spectra, and the comparisons also show peaks that are only present in the adsorbed state. A full interpretation is given in References 37 and 38.

A further advantage of the counting mode is evident for angle-resolved spectra. This point is discussed further in Section 4.

3. CONCENTRATION DEPTH PROFILING BY AUGER SPECTROSCOPY

Concentration depth profiling (CDP) enables the determination of the variation in number, type, and concentration of species in a solid as a function of distance beneath the surface. The technique involves the removal of successive layers from the material under study (by sputtering, milling, etching, or microsectioning) and recording the Auger spectrum of the freshly exposed surface. Atomic concentrations are usually determined by comparing Auger peak-to-peak heights of the first-derivative spectrum with standard spectra of the elements under study. Usually the CMA is used for depth profiling. The data may be displayed continuously on an oscilloscope or the Auger signals from several different elements may be monitored in rapid sequence by multiplexing devices. The CMA has one disadvantage in this mode in that the Auger signal is extremely sensitive to the sample position.

There are other depth-profiling techniques that employ ion milling. These include secondary ion mass spectroscopy (SIMS),[60] ion-microprobe mass analysis (IMMA),[61] glow-discharge mass spectroscopy (GDMS),[62] and ion-scattering spectroscopy (ISS).[63] SIMS, IMMA, and GDMS detect species that have left the surface during sputtering and afford the advantage over Auger spectroscopy in that they can detect hydrogen, helium, and their isotopes. Although SIMS and IMMA are extremely sensitive, quantitative measurement with these techniques is difficult and acute problems with overlapping spectral features are quite common. Another problem associated with SIMS and IMMA is that many of the signals are artifacts due to recombination of ejected ions and are not necessarily a reflection of the surface under study. GDMS unfortunately suffers from poor spatial resolution and nonuniform sputtering, while ISS has the disadvantage in poor mass resolution of multicomponent materials, although it does possess the best monolayer sensitivity of all the techniques, since it only samples the top-most surface atoms.

AES is rather superior to any other CDP technique, and this has been reflected in the last few years with AES/sputter being the most popular method for depth profiling. Although slightly less sensitive than some of the techniques mentioned above, AES can detect 0.1 of a monolayer with ease, and is highly surface sensitive due to the low escape depth of Auger electrons[64,65] resulting in approximately the top 10 Å of the solid being sampled. It also provides excellent elemental information and in some cases surface stoichiometry. AES seldom has overlapping peaks; every element has a unique spectrum allowing for unambiguous assignments.

CDP is usually performed in a modified·XPS/AES spectrometer, where the primary electron beam and the etching beam are aimed at the same location on the sample. Argon-ion-etching beam energies between 0.5 and 100 keV are typical with an ion current of 1 μA cm^{-2}. The resolution of CDP depends on several factors. For a flat homogeneous sample the resolution is 5–10%, but in practice the resolution is much less owing to nonuniform etching by the ion beam, preferential sputtering, and surface and subsurface damage that invariably occur. The argon beam can also induce migration of species and, consequently, not only is there diminished resolution but the original surface is not sampled. In the next section the limitations of CDP are discussed in terms of the ion beam and the interactions of the beam with the surface.

3.1. The Effect of the Ion Beam

The homogeneity and energy of the argon ion beam, commonly used in CDP, affect the resolution of the technique. The composition of the beam is typically divided evenly between ions and neutrals and is energetically heterogeneous, with the center more energetic than the periphery. Thus, the center of the pit that is formed is etched more rapidly than the sides. Consequently, the electron beam is not incident on a flat base, and this combined with the difficulties in aiming the primary beam at the same location as the Ar$^+$ beam lower the spatial resolution.

Development of the surface morphology by etching with a nonuniform beam has been examined for both two- and three-dimensional cases. Two methods exist. The first assumes an initially flat surface and that the ion-beam-energy cross section has a Gaussian distribution.[66] Erosion-slowness curves are then used to generate a geometric reconstruction of the developing surface. This method can also be used to show the formation and movement of discontinuities and edges. The second method[67] suggests using a "characteristic trajectory approach." This derives from the fact that the defining equation for the time-dependent motion of the surface, which may be regarded as a wave front, is of the form of the hyperbolic or kinematic wave equation discussed in detail by Whitman.[68] These two approaches allow the decrease in spatial resolution from the ideal to be estimated and the optimum location for the electron beam to be determined.

The design of the ion source determines the homogeneity of the ion beam. Several types are commercially available, one of which is illustrated in Fig. 3.18,[69] producing an inhomogeneous beam. In this design, the flux is varied by altering the gas pressure in the gun and the beam energy is controlled by changing the grid potential. Another design that improves the homogeneity of the beam has been suggested,[70] where a saddle-field ion source produces a wedge-shaped beam with uniformity in the long direction of better than 10%.

3.2.   Physical Interactions Between the Ion Beam and the Surface

As an ion beam approaches a surface, its energy is given by the sum of its kinetic energy, its internal excitation energy (represented by the ionization energy), and an energy due to the beam–surface interaction. Hagstrum[71]

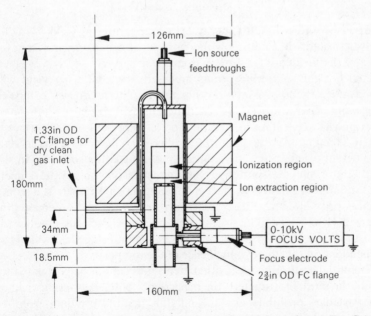

Figure 3.18.   Schematic of commercial sputter ion gun (AG2, VG Scientific Ltd., East Grinstead, U.K.). (Reprinted with permission of Vacuum Generators Company Ltd.)

has shown that ions are neutralized within 3 Å of the surface by either resonance or Auger-like processes, so that within 3 Å the internal excitation
energy must be zero. The interaction between the ion and the surface can be
approximated to the image potential,[71] $I = -e^2/4S$, where S is the ion-to-
surface distance and e is the charge on the ion. When the ion is neutralized,
$e = 0$ and $I = 0$. Therefore the energy of the atoms impinging on the surface is
entirely kinetic. This energy can be dissipated in three ways. The atoms can be
elastically scattered from the surface, they can sputter off surface atoms, or
subsurface damage can occur. Each of these possibilities depends on the
nature of the ion beam and the solid.

Of course, sputtering is the intended purpose of the ion beam, and,
ideally, the surface is peeled away at a uniform rate regardless of composition,
size or sample orientation. Sputtering results from cascades of atomic collisions in the solid when one atom hits the surface and undergoes a subsequent series of collisions. When enough energy has been acquired, the atoms
will be ejected as sputtered material while those with insufficient energy will
create subsurface damage. The sputtering yield is the ratio of atoms ejected
to the number of ions impinging on the surface. It is calculated by assuming a
random slowing down in an infinite medium.[72] This gives

$$S(E) = \frac{3}{4\pi^2} \frac{a}{\mu} \left[\frac{4m_1 m_2}{(m_1 + m_2)^2} E \right] \tag{3.11}$$

where μ is the surface binding energy, $S(E)$ is the sputtering yield, a is a dimen-
sionless constant depending on the ratio m_1/m_2, and m_1 and m_2 are the
masses of the incident and surface atoms, respectively.

When heavier sputtering beams are used, the sputtering yield will be
affected, since the approximation that moving atoms do not collide is no
longer valid. This has been discussed by Oliva-Florio et al.[73] Preferential
sputtering will occur if more than one species is present on the surface, particularly if their masses are significantly different. If the surface has crystallites
with different orientations, preferential sputtering will also occur. In single-
crystal sputtering, the sputtering yield depends on the orientation of the
crystal. If bombardment is along a channeling direction, there is a reduction
in the quantity of energy dissipated near the surface and a decrease in the
sputtering yield. Crystallites without channeling directions parallel to the
ion beam will be etched more rapidly, and the sputtering yield will of course
be dependent on the angle of incidence of the ion beam.

Preferential sputtering means that etch times cannot be correlated to the
depth of the pit. Unfortunately, the dependence of the sputtering yield on so
many variables prohibits any meaningful quantitative data from being
obtained.

3.3.   Surface Damage

Surface damage will of course be caused by the ion beam. The surface may be roughened as a result of preferential sputtering, or pyramids can be formed that have been linked with various mechanisms.[74-76] In some cases amorphization will occur due to subsurface damage (often known as "knock on" or "knock in") where the momentum of an incident ion is transferred directly to a surface atom and drives it into the bulk. The extent to which this occurs depends on the ion-beam energy.[75]

3.4.   Subsurface Damage

The depth to which damage occurs has been studied by Walls and South-worth.[77] For tungsten, they found that the damage increases with ion dose and is dependent on the bombarding ion (see Fig. 3.19). The subsurface

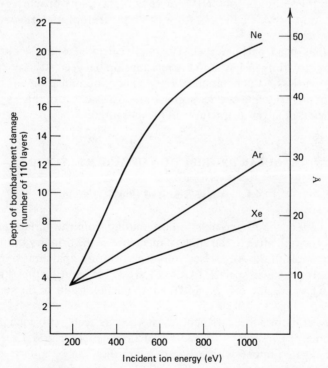

Figure 3.19.   Depth damage to tungsten metal after ion bombardment by Ne, Ar, and Xe.[77] (Reprinted with permission of Elsevier Sequoia S.A.)

damage can be repaired by high-temperature annealing, although the initial perfection is never entirely attained.[78] There are several other effects caused by ion etching, such as induced diffusion of species and, in insulators, ion impurity migration.[79]

As the ion approaches the surface of the insulator, neutralization will probably occur, as for metals. This neutralization potential is highly dependent on the ionization potential of the incoming ion, which can be shifted by charging effects. For example, the ionization potential of Ar and N_2 is so large that they still impact the surface as neutral species. Other ions, however, with lower ionization potentials impact the surface as ions and do not exchange charge with the insulator. Migration will not occur if the ion beam is composed of ions whose ionization potential is small, since the beam will impact as ions and not neutrals.

3.5. Applications

Although the technique of CDP is quite new and many problems with data interpretation prevail its application has been fruitful and a few examples are given below.

CDP has been particularly useful in the field of oxide/semiconductor interfaces providing insight into the relationship between structure and electrical properties.[80] Other applications include identification of impurities in grain boundaries and surface segregation. An example of this has been the identification of silver at fractures in nickel samples.[81]

4. ANGULAR DISTRIBUTION OF AUGER ELECTRONS

4.1. Angle-Resolved Photoemission

The intensity of the Auger emission originating from the surface of a single crystal is very sensitive to the angle of incidence of the primary beam and the collection angle of the Auger electrons. The two principal aims of recording angle-resolved Auger spectra (ARAES) are (1) to gain further knowledge on the Auger emission process and (2) to relate the results to the structure of the surface.

The experimental measurement of ARAES has been performed in various ways. Some researchers[82-84] have used three- or four-grid LEED/Auger spectrometers equipped with a Faraday cup that can be moved about the surface of the LEED screen thus giving angle-resolved spectra, while others[85] have preferred to use a movable-spot photometer outside the vacuum chamber to measure the screen electroluminescence. It is worth noting that curves obtained by the second method generally show an intensity contrast less pronounced than the first group. Nevertheless, excellent angular profiles can be obtained using sensitive photomultiplier tubes,[86] multiscanning of the screen, and signal averaging.

The typical acceptance angle for a Faraday-cup-type analyzer is a circle subtending an angle of about 5° at the specimen, which of course leads to very low signals, further impeding study since retarding-field analyzers are notorious for their low sensitivity. However, for features occurring at kinetic energies below 100 eV, the problems are minimized since these peaks are usually strong and broad. Therefore, although a Faraday cup collector may weaken such angular features, it is unlikely to smear them out completely.[83]

Another technique that has been used involved a LEED goniometer,[87] shown in Fig. 3.20, with the ability to study both the anisotropy of emission and excitation. Using this spectrometer the $N(E)$ spectra were recorded rather than the $dN(E)/dE$ derivative. The background under the peaks was assumed to be linear thus providing the advantage of better signal to noise associated with $N(E)$ spectra. It is very important to note in the derivative mode that the peak-to-peak height is only a measure of the transition probability if the lineshape does not depend on the direction of emission and if the secondary-electron background is not anisotropic. Thus, recording $N(E)$ for angular measurements has the following advantages:[84] (1) the data depend separately on the shape of the Auger line and its height; (2) the energy distribution of angular-integrated or angle-resolved lines can be studied, as well as the anisotropy of these lines; (3) the anisotropy of secondary electrons and Auger electrons can be separated.

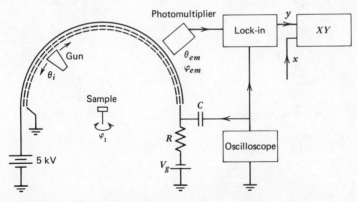

Figure 3.20.   Adaption of the LEED goniometer to study the angular distribution of secondary emission.[87] (Reprinted with permission of North-Holland Publishing Company, Amsterdam.)

Clearly, the best instrument to measure ARAES would use dispersive-type analyzers, which give much better signal-to-noise ratio than RFAs. However, studies of this type are as yet few in number.

Since this chapter is primarily designed to describe the experimental techniques used in Auger spectroscopy, a detailed description of the theory of ARAES is beyond its scope. The interested reader is thus referred to some excellent works on this subject that have appeared in the recent literature.[82–90]

4.2. Angle-Resolved Gas-Phase Auger

Angular studies of gas-phase spectra have helped in assigning spectra, although the theory is quite complex. A chamber used for gas-phase angular

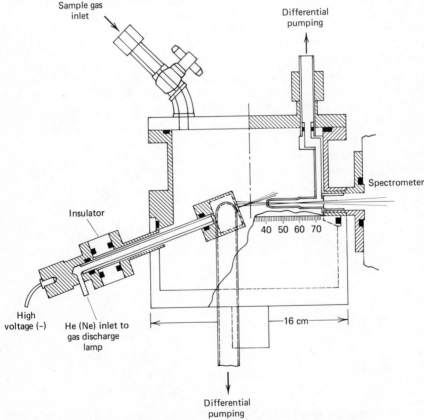

Figure 3.21. Chamber used for angular distribution measurements.[91] (Reprinted with permission of American Physical Society, New York.)

measurements is shown in Fig. 3.21.[91] Further examples of gas-phase angular spectra are given in Chapter 4.

4.3. Temperature Dependent Auger Spectra

Some peaks in the secondary emission spectra not associated with Auger transitions have been observed for various metals. The peaks are usually

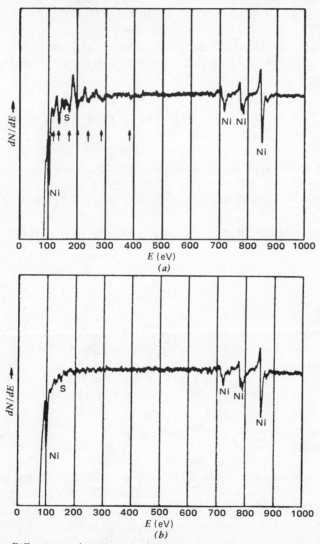

Figure 3.22.   Diffraction peaks sometimes observed in Auger spectroscopy. In (b) the peaks are suppressed where the sample temperature was 630°C as opposed to 70°C in (a).[92] (Reprinted with permission of American Physical Society, New York.)

suppressed at higher temperatures, appearing to have intensities that are unique functions of temperature. The peaks are more intense for clean well-ordered single-crystal surfaces and have been interpreted in terms of diffraction of the emitted secondary electrons.[92] An example of these peaks is given in Fig. 3.22 for a Ni(100) single crystal.[92] Since the effect is due to diffraction, the peaks will shift as the collection angle is varied. These peaks can be a nuisance in studying low-energy Auger transitions, and with an angle-resolving spectrometer it is often prudent to ensure that these peaks do not overlap with those of interest by varying the collection or excitation angle until the desired result is achieved. Alternatively, the use of a minicomputer is desirable to subtract away these annoying features from the true Auger spectrum.

5.   QUANTITATIVE AUGER ELECTRON SPECTROSCOPY

Auger electron spectroscopy, as we have seen in this chapter, is an excellent analytical tool for gaining qualitative information on solid surfaces. Today, routine quantitative analysis is still difficult owing to the many parameters affecting the Auger yield, but much headway has been made in the last few years. In the following sections, these parameters are described, along with the ways in which an Auger spectrometer can be calibrated. A few examples from the recent literature are briefly outlined.

5.1.   Theory

As a foundation for the formalism of quantitative AES a formula is required that relates the recorded Auger current with the concentration of a given element in the surface region. The Auger current from an XYZ (see Chapter 2) transition occurring in an element A is given by[93,94] (for isotropic emission)

$$I_A = (4\pi)^{-1} \int_\Omega^\infty \int_0^\infty I_p \sigma(E_p) \sec \theta \, N P_A P_T r P_E(\alpha, z) \, dz \, d\Omega \qquad (3.12)$$

where Ω is the solid angle subtended by the analyzer, I_p is the primary electron current, $\sigma(E_p)$ is the ionization cross section for an energy E_p, N is the number of atoms of the element per unit volume, P_A is the probability that an Auger process occurs after the core hole is produced, P_T is the probability that an electron passes the analyzer grids, r is the backscattering factor accounting for the effect of secondary electrons causing ionization and subsequent Auger emission, and $P_E(\alpha, z)$ is the probability that an Auger electron originating at a depth z and angle α to the surface normal will reach the analyzer without energy loss, that is, it is the escape probability. The most important parameters in the elemental sensitivity factors are the ionization cross section, the probability of an Auger process occurring, the mean escape depth, and the backscattering factor.

For an Auger electron to be ejected, a core hole must be produced by either the primary beam or the backscattered secondaries. The probability of core ionization is given by the ionization cross section (σ), and for quantitative AES this must be known for all K, L, and M shells. Ionization cross sections for K shells have been obtained by Glupe and Menhorn[95] for C, N, O, and Ne. Within experimental error, the data were identical for each gas when plotted as $\sigma(E_p)E_K^2$ against E_p/E_K. For light elements, therefore, the maximum ionization cross section decreases inversely with the square of the K-shell binding energy (E_K^2).

Cross-section data for L shells are still scarce, but it seems that the $L_{2,3}$ shell cross sections fit a power law $\sigma \propto E_p^{-\alpha}$, where $\alpha = 1.56$.

The probability that Auger emission occurs after the production of a core hole (P_A) is about 97% for the K, L, and M levels involved in Auger spectroscopy.[96] Therefore, x-ray fluorescence can be neglected. Once a core hole is produced, Auger emission can proceed via several channels. The proportion of the total Auger current in the peak used for quantitative analysis can be determined experimentally.

The escape depth of the Auger electrons is related to the inelastic mean free path of electrons in the material under study. The absolute escape depths have been measured by two experimental techniques. The first involves deposition of one element onto a surface while monitoring the Auger peaks from the two materials. The substrate peaks decay away with overlayer thickness, while the overlayer peaks asymptotically approach that of this material alone. When the overlayer is deposited at a uniform rate, the escape depths can be derived easily.[97] The second method involves absolute measurement of the Auger current from an elemental material and then comparing this with the Auger yield calculated for the constituent atoms. The number of atoms (and therefore the depth of material) contributing to the Auger current can then be determined. Both methods are subject to error. Overlayers seldom grow at a uniform rate, and backscattering effects can contribute to variations in the Auger signal with overlayer thickness. Nevertheless, the approximate escape depths for a good number of elements are now known, as shown in Fig. 3.23.

Recently, the backscattering factor has been treated using a number of theoretical models.[94,98,99] This has allowed some accurate estimates of the backscattering effect and is supported by careful experimental tests of the theory.[100,101]

Quantitative Auger spectroscopy is commonly performed by applying elemental sensitivity factors to Auger peak heights obtained by sputtering,

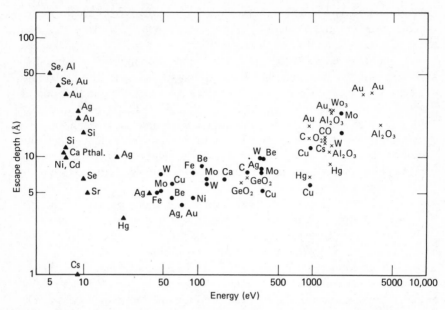

Figure 3.23.   Escape depths for a number of materials as a function of electron energy.

as outlined previously. Whether the factor is determined by experiment or by theory it is generally true that the accuracy of a quantitative Auger scan is less than 30%. Better accuracy is possible using internal standards, although matrix effects can make calibration very difficult. Generally, the relative-sensitivity-factor method is more convenient for routine analysis in the thick-film and silicon technologies.[102,103]

There are many examples in the recent literature of the application of Auger electron spectroscopy to quantitative analysis. These studies show the increased progress in calibration methods and a fuller understanding of the factors affecting the Auger yield. A few key papers are now described.

An interesting quantitative study of Ti/W films was described by Hartsough et al.[104] The Ti/W ratio of rf sputtered films was determined by AES, Rutherford backscattering (RBS), and atomic absorption (AAS). Large differences were apparent, with AES giving low Ti/W ratios because of preferential sputtering of titanium during analysis. The RBS and AAS were used as a basis for determining the sputtering yields of titanium and tungsten from the films and thence to calibrate the Auger spectra. The sputtering yield was found to vary linearly with the surface composition, usefully calibrating the Auger spectrometer. Although in this case the experimental data were used for calibration, a similar determination using theoretical fundamentals is still impossible.

The calibration of Auger data by using ion-implanted references has been described by Thomas.[105] He accurately determined the phosphorous content in glow-discharge-deposited hydrogenated amorphous silicon films. Spectra were recorded with a Physical Electronics Scanning Auger system interfaced to a Hewlett-Packard 1000 minicomputer. In order to accurately determine the fraction of phosphorous in the samples, standards were carefully prepared by ion implantation. The films were implanted at 30 kV to a dose of between 1.4×10^{15} and 3×10^{15} atoms cm^{-2}. The doses were reproducible to within 2% allowing reliable standardization since differential sputtering is small for silicon and phosphorous. The relative phosphorus sensitivity factor for the 120-eV transition was obtained from the implanted phosphorus profile (referenced to sputter-cleaned silver). Since the 351-eV peak of a sputter-cleaned silver surface is a strong and reproducible peak, it allowed the results to be compared with previously published sensitivity factors referenced in the paper. The relative sensitivity was found to be 0.60, in fair agreement with SIMS data.

The papers referenced earlier concerning the backscattering correction in quantitative Auger should allow the accuracy of the technique to improve. For example, Ichimura and Shimizu[99] have via Monte Carlo calculations determined the backscattering factors for over 25 materials including pure elements, compounds, and some alloys, which are widely used as internal or external standards for quantitative Auger. The results were obtained for a range of primary energies and angles of incidence and are invaluable to the quantitative correction of backscattering effects and improving relative sensitivity factors. The backscattering factors could be represented as a function of atomic number, although the deviation from this was quite large in some cases.

Recently, Chang[106] has commented on quantification of Auger electron spectroscopy. He points out that the area under an $N(E)$ curve can only be accurately measured to about 20%. This is due to the existence of energy-loss electrons,[51-55] which create a tail on the low-energy side of each peak in the Auger spectrum. The problem is one of an uncertainty in where to set the integration limits of the peak. The only practical approach is to choose a background function over a small energy range with the hope that beyond this range, the contribution of energy-loss electrons is negligible. It is very important to note that, without a valid way of integrating the $N(E)$ curve, useful values for the ionization cross sections cannot be determined. This underlines the fallacy in trying to model quantitative Auger with fundamental equations involving ionization cross sections, since its value can not be accurately determined.

In summary, therefore, it is clear that the most accurate quantitative analyses are achieved by the use of empirical calibrations from well-known documented standards. There is no known method that can give accurate predictions about quantitative Auger in practical applications, although empirical calculations are often extremely accurate.

6. ELECTRON-BEAM EFFECTS

The elemental analysis of a solid surface is best performed by the use of high beam currents and a finely focused electron beam. However, such excitation can have a number of perturbing effects on the surface under investigation during collection of Auger spectra. Since AES is very sensitive to the outermost layers of the solid, any beam-induced change in the concentration or chemical state of the surface will be evident in the spectrum. The electron-beam damage to a sample can sometimes result in discoloration or physical damage to the sample, large pressure bursts in the analyzer chamber, and contamination of the analyzer or adjacent samples. In this section, various types of electron-beam effects are described and examples with some suggestions to suppress them are given. Once again, the detailed theory on electron-beam effects is not presented here but is contained in the references given in this section.

6.1. Electron-Stimulated Desorption and Decomposition

The electron-stimulated desorption of a surface species[107,108] can be viewed as a process involving electronic excitation followed either by escape of the excited fragment or by deexcitation and recapture at the surface. Desorption will occur if there is a repulsive potential between the fragment and the surface. Electron-stimulated desorption is in fact a surface technique in its own right.[107,108]. For adsorbed monolayers of carbon monoxibe on single-crystal metal surfaces, Lambert and Comrie[109] discovered that desorption by the electron beam occurred about 20 times faster than dissociation.

Direct evidence for electron-beam-induced dissociation of molecular species has been given by Hooker and Grant,[110] who noted changes in the Auger spectra of CO monolayers on Ni as shown in Fig. 3.24.

In practical applications of AES there are also many examples of desorption and decomposition of surface species, which have prevented both CDP and quantitative analysis of the surface.[111,112] For example, the analysis of thermionic-oxide-emitter surfaces has been hindered by electron-beam damage.[113] Such is the problem for barium that short acquisition times resulting in poor signal-to-noise ratios have made quantitative analysis very difficult indeed and have limited the capability for Auger imaging. Coad et al.[114] have concluded that for metal oxide films AES can only provide

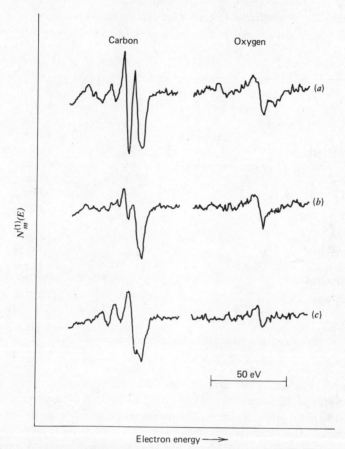

Figure 3.24. Peak-shape changes during electron-beam irradiation of a CO monolayer on Ni(110); 1.5 μA at 1.5 kV after (a) 0 min, (b) 10 min, (c) 40 min.[110] (Reprinted with permission of North-Holland Publishing Company, Amsterdam.)

useful data at electron-beam exposures of 2×10^{-3} C cm^{-2} or less. This is less than that required by most commercial Auger systems by several orders of magnitude. For example, a 5-μA beam, 0.5 mm in diameter with a 100 sec measurement time gives a 2.5×10^{-1} C cm^{-2} dose. This can be reduced somewhat by rastering the beam over a larger area.

Corrosive films on various metals have been shown to undergo electron-beam-induced decomposition,[115] such as the desorption of chlorine from an oxide film on iron. Similar electron-beam decomposition has also been observed in Auger analysis of CaF$_2$ windows, where fluorine is preferentially desorbed.[114] Finally, it is interesting to note that metallic samples are not generally affected by electron beams since electronic excitations of surface atoms are rapidly screened by the conduction electrons.

Table 3.3.   The Effect of Electron Beams on Various Materials

Material	Energy (keV)	D_0 (C/cm^2)^a	$T = 1$ mA/cm^{2b}
Si_3N_4	2	Stable	
Al_2O_3	5	10	3 h
Cu, Fe phthalocyanines	1	>1	>15 min
SiO_2	2	0.6	10 min
Li_2WO_4	≈1	0.05	8 min
NaF, LiF	0.1	0.06	60 sec
$LiNO_3$, $LiSO_4$	1	0.05	50 sec
KCl	1.5	0.03	30 sec
TeO_2	2	0.02	20 sec
$H_2O(F)$	1.5	0.01	10 sec
Native oxides	5	2×10^{-3}	2 sec
$C_6H_{12}(F)$	0.1	3×10^{-4}	0.3 sec
Na_3AlF_6	3	10^{-4}–10^{-3}	0.1 sec
$CH_3OH(F)$	1.5	2.5×10^{-4}	0.3 sec
Formvar	75	2×10^{-3}	0.2 sec

^{a}Critical dose.
^{b}Time required to yield detectable damage (beam current 10^{-3} Acm^{-2}).

Pantano and Madey[116] have shown that the threshold dose of ESD of surface monolayers is 1.6×10^{-4}–1.6×10^{-3} C cm^{-2} if all electronic excitations lead to desorption assuming a desorption cross section of 10^{-16}–10^{-17} cm^2. However, this is frequently not the case since cross sections for desorption can be very small ($\sim 10^{-18}$ cm^2) and as a result the adsorbed layer can be stable for many hours under a typical electron beam (1 μA cm^{-2} current). They also showed that the threshold for decomposition for bulk layers and thin films was 1.6×10^{-3}–1.6×10^{-2} C cm^{-2} for 10 atomic layers. Table 3.3 is taken from the paper of Pantano and Madey and shows the stability of various materials during AES.

6.2.   Beam Damage in Polymers

The study of polymers by AES is not recommended since they suffer extensive beam damage. However, this fact has been used to advantage in the field of electron-beam lithography[117] where materials that are highly sensitive and insensitive to electron beams are required. For the sensitive materials used in electron-beam lithography the typical threshold doses are quoted at 10^{-7}–10^{-5} C cm^{-2}, several orders of magnitude lower than for ESD of monolayers.

6.3. Electron-Beam Damage in Chemisorbed Species

The high cross sections for the perturbation of a chemisorbed layer by an electron beam has hindered the use of AES in this area, although the use of defocused beams and high-luminosity analyzers has produced some excellent results.[37,38,118] An alternative strategy has been adopted by Fuggle et al.,[3] who used x-rays to excite Auger emission since this excitation has little effect on chemisorbed layers. However, the backscattered secondary electrons will have some disturbing effect on the surface species, and this method although producing excellent spectra is about 100 times slower than electron excited Auger.

6.4. Charging Effects

The study of insulators by AES is complicated by a build up of charge at the surface, which can in severe cases make it virtually impossible to record an Auger spectrum. As well as deflecting the primary beam, the charging can also erode away powder compacts during AES, probably due to repulsion between charged particles in the powdered specimen. If the secondary-electron yield is greater than one, a stable condition develops whereby low-energy secondaries can return to the sample preventing excessive charging, since a positive surface potential occurs under these conditions. The secondary electron yield can be rendered greater than unity by adjusting the primary-beam incident angle, usually to between 20° and 45°, which is a very useful way of coping with insulating materials. When a stable positive surface potential is achieved, an Auger spectrum can be recorded, but the peaks will be shifted to between 2 and 15 eV lower in energy. Also, the electric field across the surface can lead to the migration of mobile ions,[116] which will obviously ruin any attempt at CDP or quantitative AES of the sample.

The use of implanted, inert gas atoms as a reference in insulating samples has been attempted.[119] For an energy shift of 20 eV due to charging, the energy of the Auger peaks was generally fixed to ± 1 eV.

6.5. Sample Heating

The power from an electron-beam source can heat the sample and, clearly, insulators will be more prone to this. In microanalysis where high beam energies and small spot sizes are used, heating can be a problem too. The examination of small or thin insulating specimens is therefore extremely difficult owing to the problems of maintaining thermal contact. Temperature rises of 250°C have been recorded for beam energies of 3 keV and current densities of 1 mA cm^{-2},[116] although the amount of heating is dependent on the sample size, sample mounting, and beam parameters.

REFERENCES

1. M. Housley and D. A. King, *Surf. Sci.*, **62**, 81 (1977).
2. M. A. Chesters, AERE R8536 (Radiation and Surface Chemistry Group, Harwell, England).
3. J. C. Fuggle, E. Umbach, and D. Menzel, *Solid State Commun.*, **20**, 89 (1976).
4. K. Siegbahn, *Alpha-, Beta-, and Gamma-Ray Spectroscopy*, North Holland, Amsterdam, 1965.
5. N. Anderson, P. P. Eggleton, and R. G. W. Keesing, *Rev. Sci. Instrum.*, **38**, 924 (1967).
6. J. A. Simpson, *Rev. Sci. Instrum.*, **32**, 1283 (1961).
7. B. Wannberg, U. Gelius, and K. Siegbahn, *J. Phys. E*, **7**, 149 (1974).
8. W. Steckelmacher, *J. Phys. E.*, **6**, 1061 (1973).
9. G. F. Amelio and E. J. Scheibner, *Surf. Sci.*, **11**, 242 (1968).
10. K. Nakayama, M. Ono, and H. Shimizu, *J. Vac. Sci. Technol.*, **9**, 749 (1972).
11. G. A. Somorjai, *Surf. Sci.*, **34**, 156 (1973).
12. D. M. Zehner, N. Barbulesco, and L. H. Jenkins, *Surf. Sci.*, **34**, 385 (1973).
13. L. McDonnell, B. D. Powell, and D. P. Woodruff, *Surf. Sci.*, **40**, 669 (1973).
14. T. A. Carlson, *Photoelectron and Auger Spectroscopy*, Plenum, New York, 1975, pp. 43–63.
15. D. L. Judd, *Rev. Sci. Instrum.*, **21**, 213 (1950).
16. D. W. Kerst and R. Serber, *Phys. Rev.*, **60**, 53 (1941).
17. C. S. Fadley, R. N. Healey, J. M. Hollander, and C. E. Miner, in D. A. Shirley (Ed.), *Electron Spectroscopy*, North Holland, London, 1972, pp. 121–140.
18. H. H. Bauer, G. D. Christian, and J. E. O'Reilly, *Instrumental Analysis*, Allyn and Bacon, Boston, 1978.
19. G. A. Harrower, *Rev. Sci. Instrum.*, **26**, 850 (1955).
20. G. D. Yarnold and H. C. Bolton, *J. Sci. Instrum.*, **26**, 38 (1949).
21. J. D. H. Eland and C. J. Danby, *J. Sci. Instrum.*, **1**, 406 (1968).
22. J. R. Pierce, *Theory and Design of Electron Beams*, Van Nostrand, New York, 1949.
23. K. R. Spangenberg, *Vacuum Tubes*, McGraw-Hill, New York, 1948.
24. G. A. Proca and T. S. Green, *Rev. Sci. Instrum.*, **41**, 1778 (1970).
25. E. Blauth, *Z. Phys.*, **147**, 228 (1957).
26. P. W. Palmberg, G. K. Bohn, and J. C. Tracey, *Appl. Phys. Lett.*, **15**, 254 (1969).
27. A. L. Hughes and V. Rojansky, *Phys. Rev.*, **34**, 284 (1929).
28. E. M. Purcell, *Phys. Rev.*, **54**, 818 (1938).
29. H. Ewald and H. Liebl, *Z. Naturforsch.*, **10a**, 872 (1955).
30. H. Wollnik, in A. Septier (Ed.), *Focussing of Charged Particles*, Academic Press, London, 1967.
31. A. D. Baker and D. Betteridge, *Photoelectron Spectroscopy—Chemical and Analytical Aspects*, Pergamon, Toronto, 1972.
32. B. Tsai, T. Baer, and M. L. Horovitz, *Rev. Sci. Instrum.*, **45**, 494 (1974).
33. A. Barrie, *Instrumentation for Electron Spectroscopy*, Kratos Ltd., Scientific Instrument Division.

34. W. Espe, *Materials of High Vacuum Technology*, Pergamon Press, London, 1966, Vol. 1.
35. C. C. Chang, *Surf. Sci.*, **25**, 53 (1971).
36. Y. E. Strausser and J. J. Uebbing, *Varian Chart of Auger Electron Energies*, Varian USA, 1971.
37. M. D. Baker, N. D. S. Canning, and M. A. Chester, *Surf. Sci.*, **111**, 452 (1981).
38. N. D. S. Canning, M. D. Baker, and M. A. Chesters, *Surf. Sci.*, **111**, 441 (1981).
39. C. R. Brundle, *Surf. Sci.*, **48**, 99 (1975).
40. S. Ichimura, M. Shikata, and R. Shimizu, *Surf. Sci.*, **108**, L393–398 (1981).
41. T. E. Brady and C. T. Hovland, *J. Vac. Sci. Technol.*, **18**, 339 (1981).
42. R. H. Stulen, *Appl. Surf. Sci.*, **5**, 212 (1980).
43. G. A. Haas and R. E. Thomas, *Surf. Sci.*, **4**, 64 (1968).
44. G. Eng and H. K. A. Kan, *Appl. Surf. Sci.*, **8**, 81 (1981).
45. T. W. Haas and J. T. Grant, *Appl. Surf. Sci.*, **2**, 322 (1979).
46. R. J. Blattner, *Microstruct. Sci.*, **8**, 63 (1980).
47. L. A. Harris, *J. Vac. Sci. Technol.*, **11**, 23 (1974).
48. R. H. Stulen, *Appl. Surf. Sci.*, **5**, 212 (1980).
49. M. Prutton, *Sci. Prog. Oxf.*, **65**, 209 (1978).
50. R. W. Joyner and G. A. Somorjai, *Surface and Defect Properties of Solids*, American Chemical Society, Washington, D.C., 1973, Vol. 2.
51. F. C. Tompkins, *Chemisorption of Gases on Metals*, Academic Press, London, 1978.
52. H. Froitzheim, "Topics in Current Physics," in H. Ibach (Ed.), *Electron Spectroscopy for Surface Analysis*, Springer, Berlin, 1977, Vol. 4.
53. G. W. Rubloff, *Solid State Commun.* **26**, 523 (1978).
54. H. Ibach, H. Hopster, and B. Sexton, *Appl. Surf. Sci.*, **1**, 1 (1977).
55. V. D. Meyer, A. Skerbele, and E. N. Lassettre, *J. Chem. Phys.*, **43**, 805 (1965).
56. L. A. Harris, *J. Appl. Phys.*, **39**, 1419 (1968).
57. M. Thompson, P. A. Hewitt, and D. S. Wooliscroft, *Anal. Chem.*, **48**, 1336 (1976).
58. J. E. Houston, *Appl. Phys. Lett.*, **24**, 42 (1974).
59. M. A. Chesters, B. J. Hopkins, P. A. Taylor, and R. I. Winton, *Surf. Sci.*, **83**, 181 (1979).
60. K. Wittmaack, *Surf. Sci.*, **89**, 668 (1979).
61. J. A. Cookson, J. W. McMillan, and T. B. Pierce, *J. Radioanal. Chem.*, **48**, 337 (1979).
62. T. C. Tisone, B. F. T. Bolker, and S. T. Latos, *J. Vac. Sci. Technol.*, **17**, 415 (1980).
63. H. J. Kang, R. Shimizu, and T. Okutani, *Surf. Sci.*, **116**, L173 (1982).
64. M. L. Tarng and G. K. Wehner, *J. Appl. Phys.*, **44**, 1534 (1973).
65. J. W. T. Ridgway and D. Haneman, *Surf. Sci.*, **24**, 451 (1971).
66. A. Carter, D. G. Armar, S. E. Donnelly, D. C. Ingram, and R. P. Webb, *Rad. Effects Lett.*, **53**, 143 (1980).
67. R. Smith and J. M. Walls, *Surf. Sci.*, **80**, 557 (1979).
68. G. B. Whitman, *Linear and Non-Linear Waves*, Wiley, New York, 1974, p. 65.
69. J. M. Walls, E. Braun, and H. N. Southworth, Proceedings of the 6th International Vacuum Congress, *Jpn. J. Appl. Phys. Suppl.*, **2**, Pt 1, 355 (1974).

70. L. Bradley, Y. M. Bosworth, D. Briggs, V. A. Gibson, R. J. Oldman, A. C. Evans, and J. Franks, *Appl. Spec.*, **32**, 175 (1978).

71. H. D. Hagstrum, *Phys. Rev.*, **96**, 336 (1954).

72. P. Sigmund, *Phys. Rev.*, **184**, 383 (1969).

73. A. R. Oliva-Florio, E. V. Alonso, R. A. Baragiola, J. Ferron and M. M. Jakas, *Rad. Eff. Lett.*, **50**, 3 (1979).

74. J. L. Whitton, L. Tanović and J. S. Williams, *Appl. Surf. Sci.*, **1**, 408 (1978).

75. O. Auciello, R. Kelly and R. Iricibar, *Rad. Eff.*, **46**, 105 (1980).

76. G. Carter, M. J. Nobes, G. W. Lewis, and J. L. Whitton, *Rad. Eff. Lett.*, **50**, 97 (1980).

77. J. M. Walls and H. N. Southworth, *Surf. Technol.*, **4**, 255 (1976).

78. J. M. Walls, H. N. Southworth, and E. Braun, *Vacuum*, **24**, 471 (1974).

79. D. V. McCaughan, R. A. Kushner, and V. T. Murphy, *Phys. Rev. Lett.*, **30**, 614 (1973).

80. C. T. Hovland, A. C. Johnson, and W. M. Riggs, *Ind. Res. Dev.*, **20**, No. 9, 124 (1978).

81. F. J. Szalkowski, *J. Colloid. Interf. Sci.*, **58**, 199 (1977).

82. S. P. Weeks and A. Liebsch, *Surf. Sci.*, **62**, 197 (1977).

83. S. J. White, D. P. Woodruff, and L. McDonnell, *Surf. Sci.*, **72**, 77 (1978).

84. D. Aberdam, R. Baudoing, E. Blanc, and C. Gaubert, *Surf. Sci.*, **71**, 279 (1978).

85. T. Matsudaira, N. Nishijima, and M. Onchi, *Surf. Sci.*, **61**, 651 (1976).

86. J. D. Place and M. Prutton, *Surf. Sci.*, **82**, 315 (1979).

87. G. Allié, E. Blanc, and D. Dufayard, *Surf. Sci.*, **57**, 293 (1976).

88. D. P. Woodruff, *Surf. Sci.*, **53**, 538 (1975).

89. V. U. Kitov and E. S. Parilis, *Surf. Sci.*, **107**, 363 (1981).

90. J. M. Plociennik, A. Barbet, and L. Mathey, *Surf. Sci.*, **102**, 282 (1981).

91. T. A. Carlson and A. E. Jonas, *J. Chem. Phys.*, **55**, 4913 (1971).

92. G. E. Becker and H. D. Hagstrum, *J. Vac. Sci. Technol.*, **11**, 284 (1974).

93. P. W. Palmberg, *J. Vac. Sci. Technol.*, **13**, 214, 1976.

94. A. Jablonski, *Surf. Sci.*, **87**, 539 (1979).

95. G. Glupe and W. Melhorn, *Phys. Lett.*, **25A**, 274 (1967).

96. H. E. Bishop and J. C. Rivière, *J. Appl. Phys.*, **40**, 1740 (1969).

97. H. Tokutaka, K. Nishimori, and T. Takashima, *Surf. Sci.*, **86**, 54 (1979).

98. F. Pons, J. Le Héricy, and J. P. Langeron, *Surf. Sci.*, **69**, 565 (1977).

99. S. Ichimura and R. Shimizu, *Surf. Sci.*, **112**, 386 (1981).

100. F. Pons, J. Le Héricy, and J. P. Langeron, *Surf. Sci.*, **69**, 547 (1977).

101. S. Ichimura, R. Shimizu, and T. Ikuta, *Surf. Sci.*, **115**, 259 (1982).

102. P. M. Hall, J. M. Morabito, and D. K. Conley, *Surf. Sci.*, **62**, 1 (1977).

103. P. M. Hall and J. M. Morabito, *Surf. Sci.*, **67**, 373 (1977).

104. L. D. Hartsough, A. Koch, J. Moulder, and T. Sigmon, *J. Vac. Sci. Technol.*, **17**, 392 (1980).

105. J. H. Thomas III, *J. Vac. Sci. Technol.*, **17**, 1306 (1980).

106. C. C. Chang, *J. Vac. Sci. Technol.*, **18**, 276 (1981).

107. D. Menzel, *Surf. Sci.*, **47**, 370 (1975).

108. D. Menzel, *Angew. Chem. Intern. Ed.*, **9**, 255 (1970).

109. R. M. Lambert and C. M. Comrie, *Surf. Sci.*, **38**, 197 (1973).

110. M. P. Hooker and J. T. Grant, *Surf. Sci.*, **55**, 741 (1976).

111. S. Ichimura and R. Shimizu, *J. Appl. Phys.*, **50**, 6020 (1979).
112. J. Ahn, C. R. Perleberg, D. L. Wilcox, J. W. Coburn, and H. F. Winters, *J. Appl. Phys.*, **46**, 4581 (1975).
113. C. G. Pantano, D. B. Dove, and G. Y. Onoda, Jr., *J. Vac. Sci. Technol.*, **13**, 414 (1976).
114. J. P. Coad, M. Gettings, and J. C. Rivière, *Faraday. Disc. Chem. Soc.*, **60**, 269 (1976).
115. G. T. Burstein, *Mat. Sci. Eng.*, **42**, 207 (1980).
116. C. G. Pantano and T. E. Madey, *Appl. Surf. Sci.*, **7**, 115 (1981).
117. L. F. Thompson, L. E. Stillwagon, and E. M. Doerries, *J. Vac. Sci. Technol.*, **15**, 938 (1978).
118. F. P. Netzer and J. A. D. Matthew, *J. Elec. Spectrosc. Rel. Phenom.*, **16**, 359 (1979).
119. P. J. K. Paterson, P. H. Holloway, and Y. E. Strausser, *Appl. Surf. Sci.*, **4**, 37 (1980)

CHAPTER

4

GAS-PHASE AUGER ELECTRON SPECTROSCOPY

Compared with the wide use that is made of the technique of Auger electron spectroscopy for the examination of solid surfaces and the research and development effort that is being put into this aspect of the technique, the study of gas-phase Auger electron spectra is a minor aspect of the subject. So far, there have been about 100 papers published concerning gas-phase spectra (compared with about 4500 for the technique as a whole); and these deal with the theoretical aspects of about four vapor-phase species (although this number has recently been considerably increased by the publication of a study of the spectra of more than 20 organic silicon compounds by Siegbahn's group at Uppsala[1]). Of the numerous reviews of Auger spectroscopy in the literature, few have been concerned with gas-phase work. The amount of space in the review—no doubt reflecting the particular reviewer's interests—varies from complete omission in a recent review by Hofmann[2] to about one-third of the total as in Carlson.[3] This latter review covers the literature up to 1977 and concentrates mainly (at least as far as gas-phase work is concerned) on the work on small molecules carried out at Carlson's own laboratory, Oak Ridge. As there are no well-established analytical applications of gas-phase Auger electron spectrometry, this aspect of the technique tends to be omitted from reviews or surveys covering analytical applications of Auger spectroscopy (as, for example, are published in *Analytical Chemistry*) except when the catalytic properties of surfaces in gas-phase reactions is being considered, although, strictly speaking, this is not true gas-phase work.

Interest in gas-phase Auger spectra has mainly been from the theoretical point of view, since a number of simplifications (over the solid-state case) are possible. There is no need to utilize the differentiated spectrum, because gas-phase spectra, generally, have much superior signal-to-background characteristics; this, in turn, simplifies identification and deconvolution of peaks as well as enabling low-intensity transitions to be detected. Effects due to charging of the sample specimen (particularly severe for instruments using an electron-gun source when a nonconducting sample is under examination), secondary collisions, and solid-state broadening are also avoided. High-resolution gas-phase spectra thus provide the experimental data against which to test the various theoretical models of the Auger effect and associated satellite processes. They also provide information about shifts of Auger lines in different chemical environments, as well as a measure of electron binding energies.

In this chapter we will consider the work on gas-phase Auger electron spectroscopy under the broad headings of noble gases, molecular species, and free metal atoms, and, finally, we will examine the analytical potential of this aspect of the technique.

There are a number of processes other than the basic Auger process (Chapter 2, Section 1) that have been studied in the gas phase. These other processes are known in general as satellite processes and give rise to satellite lines in the Auger spectrum. Since there may be many more satellite lines than "normal" lines, for example, the K-Auger spectrum of neon[4] was found to contain some 85 lines in addition to the "normal" lines, some consideration will be given to the way in which these processes have been classified. As with most newly developing fields of scientific study each research group has devised its own notation, which makes the comparison of one study with another a bit awkward initially. Confusion can also arise when the interactions under consideration are not specified as being between holes or between electrons. No attempt will be made here to present a comprehensive survey of all possible transitions of the Auger type or to suggest a unified approach to the classification of lines, transitions, and states.

1.   CLASSIFICATION OF TRANSITIONS

1.1.   Normal Auger Process

It is generally accepted that this process is one in which a single inner-shell vacancy is created while all the other electrons remain in their original orbitals. The vacancy is then filled by the interaction of two electrons in outer shells, one filling the vacancy and the other being ejected. As was pointed out earlier, the process can be considered as a transition involving holes. It should also be remembered that the "initial" state in any Auger transition is the ionized state with the core vacancy and the "final" state involves the single, double, or triply ionized state of the atom together with the ejected Auger electron in the positive-energy continuum.[5] The notation used to describe this type of transition and the designation of the initial and final states have already been fully discussed in Chapter 2, Section 1. These transitions were designated "A" lines by Carlson and co-workers in an examination of the K-Auger of neon[4] and KWS, KWW, and KSS for the designation of normal K-Auger transitions for simple molecular species[6], where W and S stand for weakly and strongly bound valence electrons. In the designation of lines in the molecular spectra, the letters B, C, and D are used to describe the regions of the spectrum where these normal lines may be found, the A designation being used for the high-energy satellites. The same designation K, W, and S is reproduced in Carlson's book,[3] where for the case of a free atom W is equivalent to $L_{2,3}$ and S to L_1. The final charge on the ion is, of course, $+2$.

1.2. Core-Electron Excitation and Autoionization

In this process the core electron is not removed to give the ionized initial state but is promoted to a vacant bound orbital (above the first ionization level). The initial state, although containing a core vacancy, is thus neutral. There are two modes of decay following this excitation; either the electron takes part in the subsequent process or it does not, in which case it is referred to as a spectator electron. In either case the overall charge on the final state is $+1$ and the process is known as autoionization. In either case a high-energy satellite line results. It is more probable that the excited electron will remain a spectator, since this orbital will overlap less with the initial vacancy than a more closely bound orbital. In Reference 4 these transitions are designated $B\beta$ and $B\alpha$, respectively, and in References 3 and 6 as $Ke - W$ and $Ke - WWe$, respectively, where the symbol e represents an electron in an excited bound orbital. It should perhaps be pointed out at this stage that the summary diagram given in Reference 6 contains a number of errors (which do not occur in the text), and these are propagated in the reproduction of this diagram in Reference 3. The excitation of the core electron into the bound excited orbital may be produced by absorption of photons of exactly the right energy (resonance absorption) or by electron impact. Since the photon absorption process can only occur if the resonance condition obtains, the use of monochromatic x-rays, as the ionizing radiation, with energy above the core-level binding energy will result in an Auger spectrum in which the autoionization satellites are absent, whereas they will be present if electron impact ionization is used. Thus by examining the spectra of the same Auger region of the same atom or molecule obtained with the two different types of source, the lines due to autoionization can be identified.

1.3. Monopole Excitation and Ionization

There is quite a high probability (at least for K-shell ionization) of the order of 20% that the first ionization to produce the initial state is simultaneously accompanied by the excitation or ionization of a second electron, usually in a weakly bound orbital. This results from a sudden change in the central potential due to the rapid change in electron shielding. In the first case, which is sometimes known as electron shake up, the subsequent decay can either involve the excited electron or not; in the latter case the electron in the excited orbital is again referred to as a spectator electron. In Reference 4 these transitions are designed $C\beta$ and $C\alpha$, respectively, in Reference 6 as $KWe\text{-}WW$ and $KWe\text{-}WWWe$, respectively, and in Reference 3 as $KLe\text{-}LL$ and $KLe\text{-}LLLe$. In both cases the charge on the final-state ion is $+2$. If the excited electron participates in the transition ($KLe\text{-}LL$), then the satellite is usually high energy, whereas if the excited electron behaves as a spectator, the satellite is usually low energy. As with autoionization, it is more probable that the excited electron will behave as a spectator, since the excited orbital will overlap less with the core vacancy orbital than will a more tightly bound orbital.

If the core vacancy is created with simultaneous ionization of a second electron, the process is known as shake off and is the subsequent rearrangement designated as giving rise to D satellite lines in Reference 4, $KK\text{-}WWW$ in Reference 6, and $KL\text{-}LLL$ in Reference 7. The final charge on the ion is $+3$, and the satellite appears on the low-energy side of the parent line. Information concerning the high-energy satellites arising from monopole excitation (shake up) can be obtained from photoelectron spectra. It has been demonstrated that monopole excitation is the same whether it is produced by electron impact or photoionization.[8] Thus if electron shake up occurs, extra photoelectron lines will appear at lower energies than the normal photoelectron lines, the energy difference being the amount of excitation. Thus if the process $KLe\text{-}LL$ occurs, the Auger line will have an energy equal to that for the $K\text{-}LL$ plus that due to monopole excitation.

1.4. Double Processes

In this category of process are satellite lines from excitation processes involving more than two electrons, for example, double shake off, and those from decay process involving more than two electrons, for example, the double Auger process. The two electrons may be ejected or one may be ejected and the other excited. If both electrons are ejected, the final charge on the ion is $+3$, and in either case the energy of the Auger electrons is less than that of the normal process. This process is described as $K\text{-}WWW$ in Reference 6 and as F lines in Reference 4.

1.5. Coster–Kronig Transitions

Although these processes have been described already (Chapter 2, Section 4), they are included here for the sake of completeness. As was pointed out previously, these processes are normal Auger processes with the property that the initial vacancy is filled by an electron from the same shell with the ejection of an electron from a higher shell (e.g., a LLM or an MMN process). For the process to occur at all the binding energy of the ejected electron must be less than the difference between energies of the two subshells involved. The occurrence of Coster–Kronig transitions can give rise to spectator vacancies in subsequent Auger transitions. For example, the filling of an L_1 initial vacancy by an $L_1L_3M_{4,5}$ Coster–Kronig transition would give rise to an L_3 vacancy and an $M_{4,5}$ vacancy. The L_3 vacancy could then be filled by an L_3MM Auger process sometimes designated as $L_3MM(M_{4,5})$ and sometimes as $L_3M_{4,5}$-$MMM_{4,5}$. In the latter case the notation would not distinguish between this type of Auger vacancy transition and electron shake off in the M shell nor, of course, would the transitions be distinguishable energetically.

1.6. Summary Diagram

The processes described in the preceding sections are summarized in Fig. 4.1. Argon ($Z = 18$ and electronic structure $1s^2 2s^2 2p^6 3s^2 3p^6$) is used as an example and for each type of process, normal or satellite, only a limited number of transitions are shown. The complexity of a "complete" Auger spectrum can be appreciated when it is realized that there are several permutations of the involvement of the various shells and subshells possible, to give additional transitions that are not shown in the diagram. The atomic number of argon, 18, places the element in a region where L-S coupling predominates so the L_1, L_2, L_3, etc., notation is not strictly appropriate (since this refers to the j-j coupling scheme). Under pure L-S coupling the L and M shells give rise to six possible states each, and although not all possible transitions are observed (some being forbidden in the pure L-S scheme), the reasons for the complexity of the spectrum can be readily seen.

2. NOBLE GASES

Interestingly enough, Auger's original work in 1923 concerned the gases neon, argon, krypton, and xenon, and the English text of an address regarding this work is reported in the literature.[9] For the reasons which were outlined earlier, developments in the applications of the Auger effect centered on the examination of the electrons ejected from surfaces, and it was not until 1960 that a study of a noble gas was next reported[10] by Mehlhorn at the University of Munster in West Germany. Between 1965 and 1977, both at Munster and later at Freiburg, Mehlhorn and his co-workers published about 10 papers

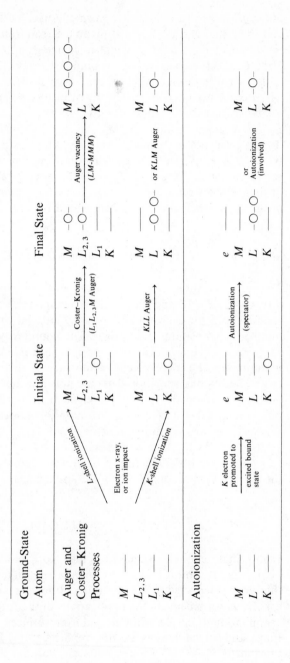

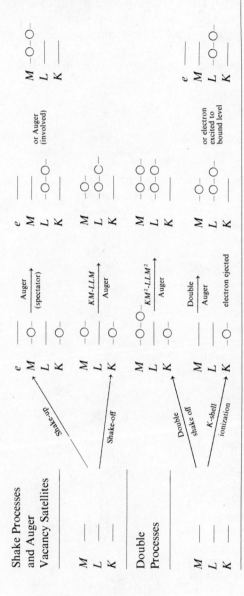

Figure 4.1. Summary of Auger and Related Processes. (The transitions shown in the diagram do not represent all possible transitions for an atom with full KLM shells, only representative examples are given.)

concerning the AES of the noble gases. Two other research groups have been active in this field, namely, that of The University of Uppsala in Sweden and that at the Oak Ridge National Laboratory, Tennessee. Other contributions have come from the University of Tokyo, and the Universities of Bergen and Oslo in Norway.

The literature on the AES of the noble gases is summarized in Table 4.1.

Table 4.1.   Auger Spectra of Rare Gases[a]

Element	Research Group	Subject of Study	Year	Reference
He	Uppsala	Autoionization	1971	11
Ne	Munster	*KLL* spectrum	1966	12
	Munster	Transition probabilities of *KLL* spectrum	1968	13
	Uppsala	*KLL* spectrum	1967	191
	Uppsala	*KLL* spectrum	1971	11
	Oak Ridge	Shake off as a function of electron-impact energy	1970/1971	15, 16
	Oak Ridge	Complete *KLL* spectrum and satellites	1971	4
	Austin	High resolution: H^+ and He^+ bombardment	1973	17
	Uppsala	*KLL* spectrum: relaxation and continuum interaction	1976	18
	Virginia	Calculated *K* Auger rates	1975	19
	Bergen/Oslo	Correlation effects	1979	20
Ar	Karlsruhe	*LMM* spectrum	1960	10
	Munster	Coster–Kronig and Auger spectra of *L* shell	1968	21, 22, 23
	Munster	*LMM* spectrum	1968	13
	Oak Ridge	Shake off	1970	15
	Oak Ridge	*KLL* spectrum	1975	24
	Uppsala	*LMM* spectrum	1971/1973	11, 25
	Uppsala	*KLL* and *KLM* spectra	1977	7
	Tokyo	*LMM* spectrum	1973	26
	US Academy of Sciences	Excitation cross sections	1973	27
	Berlin	*LMM* spectrum with ion impact (H^+, D^+, H_2^+, He^+)	1974	28
	Albuquerque	Configuration-interaction calculations	1975	29
	Orsay/Grenoble	*KLL* relativistic Dirac–Fock calculations	1976	30
Kr	Uppsala	*MNN* spectrum	1967/1971	11, 191

Element	Research Group	Subject of Study	Year	Reference
	Uppsala	*LMM* and *MNN* spectra	1972	31
	Munster	*LMM* and Coster–Kronig spectra	1965	32
	Freiburg	Correlation effects in *MNN* spectrum	1972	33
	Oak Ridge	*LMM* spectrum x-ray excitation	1965	34
	Albuquerque	Calculation of *MNN* and satellite spectra	1975	35
Xe	Uppsala	*MNN* and *NOO* spectra	1971 1972	11, 31
	York	*MNN* spectrum of solid	1974 1975	36 37
	Freiburg	*MNN* spectrum, correlation effects	1974	38
	Oregon	*MNX* spectra, calculation with mixed coupling	1979	39

*a*This table does not give an exhaustive list of references but contains a mixture of theoretical and experimental studies. The experimental work is mainly for electron-impact excitation, but some work on ion and x-ray excitation is included.

2.1. Helium ($Z = 2$, $1s^2$)

Obviously with only two electrons and both in the K shell it is not possible for helium to exhibit a normal Auger spectrum. However, it is possible to observe helium autoionization under the action of electron impact, though the initial state involves excitation of both electrons, the $2s^2$ state being the lowest excitation configuration that can autoionize. This may be regarded as a double shake up. The existance of autoionizing levels, that is, excited levels between the first and second ionization limit, was discovered only a short time after the Auger effect itself was discovered.[40,41] The observation of resonant states in helium due to double electron excitation was first reported in 1934.[42] Although a number of workers have developed a general theoretical treatment of the nature of autoionizing levels,[43-48] the major theoretical application has been to the interpretation and understanding of the autoionization of helium, which might at first sight have been thought to be the simplest of all possible cases of autoionization. Part of the autoionization spectrum of helium excited by electron impact with 4-keV electrons at a pressure of 1 torr is shown in Fig. 4.2. The lines are asymmetric

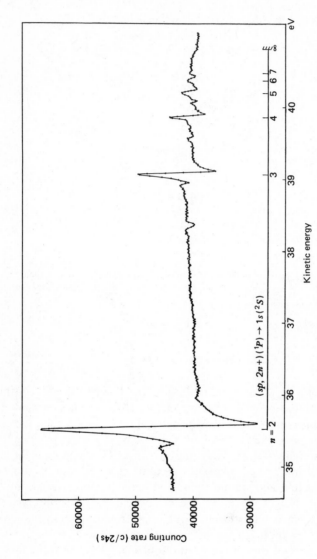

Figure 4.2. Autoionization electron spectrum from helium in the energy range 35–41 eV, excited by electron impact. The position of the first members and the limit of the series corresponding to the transitions $(sp, 2n+)(^1P) \rightarrow 1s(^2S)$ are indicated.[11] (Reprinted with permission of North-Holland Publishing Company, Amsterdam.)

because of configuration mixing in forming the stationary state of double excitation. The designation of the levels as $sp2n+$ is due to a classification by Cooper et al.[44] of the levels observed in the photoabsorption spectrum by Madden and Codling.[49] The $+$ quantum number corresponds to radial motions of the two electrons in step with each other (a $-$ quantum number would designate radial motions out of step).

Additional discussion of autoionization is beyond the scope of the present text. Further explanation of the theory may be found in Sevier's book,[50] autoionization spectra of other gases have been reported by Siegbahn et al.;[11] and a classification has been given by Madden and Codling.[51] The intensity of the lines in autoionizing spectra is rather interesting; for example, the series of lines corresponding to the $3s3p^6np \rightarrow 3s^23p^5$ ($^2p_{1/2}$ or $^2p_{3/2}$) transition in argon gives rise to less-intense peaks than the background (i.e., negative or absorption peaks), whereas the peaks for the corresponding transitions from the $3s3p^6nd$ levels have a positive intensity. An explanation, in strictly theoretical terms, for the peak shapes will be found in Reference 50.

2.2.   Neon $(Z = 10\ 1s^22s^22p^6)$

2.2.1.   KLL Spectrum

The first reports of the KLL spectrum (the only normal spectrum that can be obtained) of neon were by Körber and Mehlhorn in the mid-1960s. [12,52] In these experiments both x-ray and electron-impact ionization were used, but in the later report,[12] the resolution of the spectrometer used is 0.12% as opposed to 1.2% for the earlier work with x-ray excitation. The five transitions allowed in the Russell–Saunders coupling scheme were observed, and the intensities were in good agreement with values calculated by Asaad[53] taking configuration interaction into account. Ten lines of the KL-LLL spectrum were also identified and measured. The Auger Spectrum produced by ion impact was also reported by Edwards and Rudd,[54] though the main object of this work was to study the autoionization spectra (in the range 11–39 eV), since the use of ion impact allows the excitation of transitions that are optically forbidden. Sixty lines in the spectrum were observed and 43 were classified into 13 Rydberg series, 10 of which had not previously been reported. The five lines of the neon KLL Auger spectrum were reported, and 14 lines of the KL-LLL (shake-off) spectrum were identified and their intensities were measured, though some of these lines were not well resolved. In 1970, Krause et al.[55] reported on the use of x-ray and electron-impact excitation and showed that the Auger electron spectra (normal lines and satellites) were essentially the same regardless of mode of excitation. Intensity values were quoted for six diagram lines, the transition to the ^3P state being observed with 0.6 of the intensity of the most intense line (transition to $^1D_2$ final state).

In 1971 an analysis of the complete spectrum of neon containing some 90 lines was reported by Krause et al.[4] The authors used two double-focusing electrostatic energy analyzers to record the spectrum using both x-ray and electron excitation. The spectrometers were both operated with a resolution of about 0.06%, and excellent agreement was found between results obtained from the two instruments. The incident electron energy was between 4 and 5 keV, the x-ray energy was 1.5 keV (Al $K\alpha$), and the source pressure was between 10^{-3} and 10^{-1} torr. The spectrum obtained between 720 and 810 eV is shown in Fig. 4.3. The energies and intensities of the normal lines (designated A lines) are shown in Table 4.2. The values are referred to the most intense line ($^1D_2$ final state), which has an energy of 804.3 eV, normalized to an intensity of 100. The forbidden transition is corrected for the interference of a $B\alpha$ line (see Section 4.1.2 for details of nomenclature) by obtaining the spectrum produced by x-ray photons. The interference from the $B\alpha2$ line probably accounts for the greater violation of the L-S coupling scheme that was inferred from a previously reported spectrum.[55]

In 1973, Stolterfoht et al.[56] showed that the spectrum produced by collisions with 4.2-MeV protons was essentially similar to that produced by electron or x-ray excitation. Transitions to the forbidden ^3P state of the $2s^2 2p^4$ configuration are not reported. The spectrum produced was compared with the spectra from electron and x-ray impact reported by Krause et al.[55] The authors compared the relative satellite intensity produced by different ionizing mechanisms and showed that the use of 0.3 MeV (Edwards and Rudd[54]) produced the highest ratio of satellite intensity to total intensity, indicative of the strong multiple-ionization effects of slow protons.

In the same year, Matthews et al.[17] reported the high-resolution KLL spectrum obtained with either 250-keV protons or 1.0-MeV He$^+$ ions. The 36-cm double-focusing electrostatic analyzer used had a resolution of 0.02%. Forty distinct lines were observed, and 30 of them were assigned. It was also observed that double $2p$ vacancies in the initial state were produced by He$^+$, whereas H$^+$ only produced one. Furthermore, He$^+$ ionization favored the the production of single-vacancy satellites by a factor of 4 over H$^+$ bombardment, possibly indicative of a Z^2 dependence for the satellite production probability.

2.2.2. Theory of the KLL Spectrum

Although a detailed discussion of the theoretical calculations of the energies and intensities of the Auger spectrum are considered to be beyond the scope of this book, it is thought that some indication as to the present state of the theoretical calculations should be given.

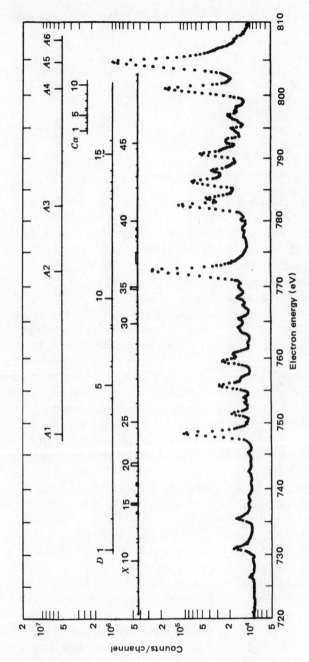

Figure 4.3. Neon K Auger spectrum excited by 4.5-keV electrons. The A lines are diagram Auger lines, $C\alpha$ are shake-up processes, and the D lines are from electron shake off.[4] (Reprinted with permission of Les Éditions de Physique, Les Ulis, France.)

111

Table 4.2.   Normal K-Auger Lines of Neon

Final-State Configuration	L-S state	$E - E_0$ (eV)	I/I_0
$2s^0 2p^6$	$^1S_0$	-56.2	9.95
$2s2p^5$	$^1P_1$	-32.7	28.2
$2s2p^5$	$^3P_{0,1,2}$	-22.2	10.2
$2s^2 2p^4$	$^1S_0$	-3.7	15.6
$2s^2 2p^4$	$^1D_2$	0	100
$2s^2 p^4$	$^3P_{0,2}$	$+3.1$	0.07

As far as transition energies are concerned, one of the most recent contributions is by Briancon and Desclaux[30] who reported the results of multiconfiguration relativistic Dirac–Fock calculations for Ne, Ar, U, and Am. They showed that by this method it was possible to calculate entirely *ab initio* Auger energies to within a few electron volts in intermediate coupling. The results of their calculations are shown in Table 4.3 in which is also shown the results of two semiempirical calculations and the experimental values as measured by Körber and Mehlhorn in 1966. The authors also give a brief summary of earlier calculations of this type, which provides a useful guide to the literature on the theoretical calculations of Auger energies, in particular, the review articles of Briancon,[57] Sevier,[50] Burhop and Asaad,[59] and Asaad.[58] Interestingly enough the authors give only the briefest mention of the earlier calculations by Kelly,[19] who calculated the energies of the lines in the L-S coupling scheme (although the main purpose of the paper was the calculation of intensities) as the difference between the Hartree–Fock calculations of the final and initial energies and taking account of relativistic and correlation effects. The results of this calculation are also shown in Table 4.3.

As far as intensities of the transitions are concerned Kelly's paper[19] is one of the more recent and also gives a summary of previous contributions in this area. The importance of configuration mixing or interaction, first pointed out by Asaad,[53] is underlined, though it is pointed out that the first efforts to include the configuration interaction of the $2p^6$ ^1S and $2s^2 2p^4$ ^1S final states (by Bhalla[61]) still produced a result in rather poor agreement with experiment. A ratio of 0.99 for the intensity of the $2s^2 2p^4$ line to that of the $2p^6$ line was predicted, whereas the experimentally determined value is 1.5 (see, for example, Reference 55). The results of Kelly's calculations based first on the Hartree–Fock approximation and second by including correlation effects by using many-body perturbation theory[62] are shown in Table 4.4. L-S coupling was used and spin-orbit effects were neglected.

Table 4.3. Calculated *KLL* Auger Energies for Neon

Final Configuration	L-S State	Semiempirical Ref. 60	Ref. 58	Relativistic Dirac–Fock, Ref. 30	Hartree–Fock, Ref. 19	Experimental, Ref. 12
$2s^0 2p^6$	$^1S_0$	751	754.1	747.5	248.15	748.0
$2s^1 2p^5$	$^1P_1$	774	776.1	771.6	771.71	771.4
	$^3P_0$	785	787.9	783.7	782.45	782.0
	$^3P_1$			783.1		
	$^3P_2$					
$2s^2 2p^4$	$^1S_0$	802	805.4	801.8	801.27	800.4
	$^1D_2$	806	807.5	806.9	804.51	804.1
	$^3P_0$	809	810.3	810.3		
	$^3P_2$	809	810.3	810.4		

Table 4.4. Ratio of Auger Transition Rates to $1s - 2s2s(^1S)$ Rate for Ne

Method	Reference	Final State, $2s2p^5(^1P)$	$2s2p^5(^3P)$	$2s^22p^4(^1S)$	$2s^22p^4(^1D)$
Hartree–Fock	19	2.139	0.830	0.480	5.979
Hartree–Fock plus correlation effects	19	2.789	1.004	1.572	10.067
Bhalla	61	2.98	0.99	0.99	8.3
Experiment 1	12	3.06	0.98	1.67	13.1
Experiment 2	63	2.73	1.00	1.55	9.31
Experiment 3[a]	55	2.87	1.06	1.5	10.00
Experiment 4[b]	55	2.92	1.06	1.5	10.13

[a]Electron impact, 3–10 keV.
[b]X-rays, 1.5 keV.

As can be seen from the table the method predicts relative intensities in close agreement with experimental intensity ratios. However, the agreement with absolute transition rates was poor; the calculated values being about 1.4–1.5 times higher than the experimental values.

2.2.3. KLL Spectrum Satellites

2.2.3.1. K electron Excitation (the B satellites). Krause et al.[4] found that transitions $1s \rightarrow np$ were dominant at 3.5-keV excitation energy and that electron excitation only occurred in about 2% of cases (ionization occurring in the other 98%). Of the cases where electron excitation occurred, the subsequent decay mode in which the excited electron remained a spectator ($B\alpha$ lines) was much more prominent than the process in which the excited electron participated ($B\beta$ lines). The ratio of the total intensity of the latter to the total intensity of the former being 0.15, that is, the $B\beta$ lines represented about 13% of the total B satellite processes. The high-energy side of the $^1D_2$ Auger line (not shown) delineates the difference between electron and x-ray excitation in that the $B\alpha$ autoionizing lines corresponding to $1s$-ns, np, nd transitions are only seen with electron excitation. The positions of the $B\beta$ lines (excited electron participates in transition) overlap considerably with the $C\beta$ lines.

2.2.3.2. Shake-up (the C satellites). Either the excited electron participates in the transition ($C\beta$) or it does not ($C\alpha$). The $C\alpha$ lines will be grouped together on the immediate low energy of the parent lines. The position of the $C\alpha$ lines associated with the main $A4$ and $A5$ peaks are shown in Fig. 4.3. The $C\beta$ lines appear about 40 eV above the parent line (not shown). As for the B lines the intensity ratio of $\Sigma C\beta/\Sigma C\alpha$ is small, about 0.13. The shake-up probability of a $2p$ electron going to the $3p$ level was deduced to be about 6% by comparing ΣC with the sum of the $A4$ and $A5$ line intensities. As was pointed out in Section 1.3, the high-energy $C\beta$ satellites will have corresponding low-energy satellites in the photoelectron spectrum.[63,64]

2.2.3.3. The Shake-off Satellites (D lines). The positions of the shake-off satellites are shown in Fig. 4.3. The probability for KL ionization as opposed to K ionization was calculated as 13.8% by computing the intensity ratio $\Sigma D/(\Sigma D + \Sigma A)$. Shake-off satellites have also been studied by the Munster group[12,65] and by Krause et al. as a function of energy of electron impact,[15] when it was found that the ratio of satellite to main lines was independent of the energy of the exciting electrons except when close to the ionization threshold (about 880 eV).

2.2.3.4. Other Satellites. Krause et al. list the energies and intensities of lines. Of these, 6 are tentatively assigned as due to double Auger processes which one electron is ejected and another excited. 2 to double shake-off processes, 3 to K-shell excitation with the electron remaining a spectator ($B\alpha$), and 19 to shake-up with subsequent spectator electron ($C\alpha$).

2.3.   Argon ($Z = 18$, $1s^2 2s^2 2p^6 3s^2 3p^6$)

The literature concerning the experimental measurement of the argon Auger electron spectrum is outlined in Table 4.1. Apart from Auger's original work in 1926 the first contribution was by Mehlhorn at Karlsruhe and concerned the LMM spectrum.[10] Siegbahn's group has made one of the most recent contributions,[7] a study of the KLL and KLM spectra including M- and L-shell shake off.

2.3.1.   The KLL Spectrum

The KLL spectrum has only been published twice.[7,24] In both cases the excitation was by electron impact using 6-keV electrons. Siegbahn and co-workers used a magnetic ESCA instrument described in Reference 14 with a gas pressure of about 20 Pa. Krause used an electrostatic analyzer with a resolution of $\Delta E/E$ of 0.1 or 0.05% with a gas pressure of about 0.5 Pa. The resolution of the two instruments would appear to be comparable, since both studies reported line widths of 1–2 eV. The spectrum obtained by Siegbahn et al. is shown in Fig. 4.4. The main lines corresponding to the normal KLL process are also indicated on the diagram. The energies of these lines were

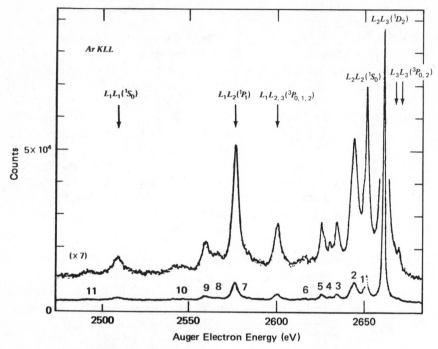

Figure 4.4.   The *KLL* Auger electron spectrum of argon. (Reprinted with permission of The Royal Swedish Academy of Sciences.)

calculated by a relativistic *ab initio* method in the intermediate-coupling scheme.[30,66] Reference to Fig. 2.1 indicates that argon is nearer pure *L-S* than intermediate in terms of coupling, and this is borne out both by the calculations and the results, although transitions to the L_3L_3 final state (forbidden in the pure *L-S* scheme) are observed (though at very low intensity). Thus it would be more appropriate in the case of argon to use the *L-S* notation for the final state rather than the *j-j* notation. In the calculation, wave functions based on *j-j* states *are* used that describe intermediate coupling if a linear combination of the configurations are taken as the total wave function. The variational principle was then applied; the Hamiltonian being the sum of one-electron Dirac operators and the classical Coulomb repulsion. The coefficients and the numerical radial function are optimized for the desired eigenstate, and when self-consistency is achieved, the full Breit operator is applied as a first-order perturbation.[67] The results of the calculations together with the experimental results for the normal transitions are shown in Table 4.5. The table also shows Krause's values for the intensity

Table 4.5. Argon KLL Spectrum: Initial State: $1s2s^2 2p^6$ configuration, ^2S state, parity +

Final State Configuration	L-S state	j-j Notation	Parity	Intensity[a]	Energy[b]	Calculated Energy[b]	Intensity[a]
$2s^0 2p^6$	$^1S_0$	$L_1 L_1$	+	5.1	2508.9	2506.4	5.6
$2s^1 2p^5$	$^1P_1$	$L_1 L_2$	−	184	2575.7	2573.6	20.6
	$^3P_0$	$L_1 L_2$	−	⎱		2597.4	
	$^3P_1$	$L_1 L_3$	−	7.7	2599.7	2598.3	5.5
	$^3P_2$	$L_1 L_3$	−	⎰		2599.7	
$2s^2 2p^4$	$^1S_0$	$L_2 L_2$	+	7.5	2680.91	2651.0	7.1
	$^1D_2$	$L_2 L_3$	+	59.9	2660.51	2659.9	60.0
	$^3P_0$	$L_3 L_3$	+	0.4	2666.9	2666.3	0.1
	$^3P_2$	$L_3 L_3$	+	1.0	2669.1	2668.8	1.1

[a]Values taken from Reference 24.
[b]Values taken from Reference 7.

117

of the transitions (area values) and Chen and Craseman's theoretical values based on the intermediate-coupling configuration interaction model.[68] The parity values are as given in Reference 24 and would seem to contradict the selection rule normally quoted, namely, that transitions from an initial K vacancy state to a final $2s^2 2p^4$ configuration are forbidden by the nonconservation of parity (see, for example, Reference 50, p. 55).

2.3.1.1. KLL Spectrum Satellites. Siegbahn et al. also report on the satellites due to shake off in the L and M shells, that is, $KL\text{-}LLL$ and $KM\text{-}LLM$ transitions.

2.3.1.2. $KL\text{-}LLL$ Transitions. The two possible terms of the initial state, that is, $1s2s$ or $1s2p$ hole configurations are[1,3] S and [1,3]P, and the splitting between ^1P and ^3P has been calculated as 12.0 eV and between ^1S and ^3S as 17.0 eV. The allowed transitions in the L-S scheme (due to parity reasons) are shown in Table 4.6 and fall into three groups.

Table 4.6. $KL\text{-}LLL$ Transitions

Initial Configuration	Final configuration	L-S Notation	j-j notation	Assignment Peak in Fig. 4.4
$1s^2 p^5$	— $2s^0 2p^5$	$^3P - {^2P}$ $^1P - {^2P}$	$KL_{2,3} - L_{2,3}L_1 L_1$	11
$1s2s$	— $2s^0 2p^5$	$^3S - {^2P}$ $^1S - {^2P}$	$KL_1 - L_1 L_1 L_{2,3}$	
$1s2p^5$	— $2s2p^4$	$^3P - {^2P}$ $^3P - {^2S}$ $^1P - {^2P}$ $^3P - {^2D}$ $^1P - {^2S}$ $^1P - {^2D}$ $^3P - {^4P}$	$KL_{2,3} - L_{2,3}L_1 L_{2,3}$	7–10
$1s2s$	— $2s2p^4$	$^3S - {^2S}$ $^3S - {^2D}$ $^1S - {^2S}$ $^1S - {^2D}$	$KL_1 - L_1 L_{2,3} L_{2,3}$	3–6
$1s2p^5$	— $2s^0 2p^3$	$^3P - {^2P}$ $^3P - {^2D}$ $^1P - {^2P}$ $^1P - {^2D}$	$KL_{2,3} - L_{2,3}L_{2,3}L_{2,3}$	

First, an initial shake off in $2p$ leads to two possible transitions to a final state with two vacancies in the $2s$ orbital (peak 11). The second group corresponds to $KL_1L_{2,3}$ transitions with an L_1 vacancy (peaks 7–11). By analogy with the neon spectrum it was suggested that only five of the transitions would be of any interest because the initial states ^3S and ^1S would be not strongly populated. Of the transitions with $^{1,3}P$ initial state the first is probably peak 10 assigned as $^3P \to {}^2P$. The other two are $^1P \to {}^2P$ and $^3P \to {}^2D$, the former probably overlaps the M shake-off peak 9 and the latter is assigned as peak 8. The third group of satellite peaks are those for $KL_{2,3}L_{2,3}$ transitions with a vacancy in either the L_1 or $L_{2,3}$ subshells (peaks 3–6). Again by analogy with neon, the transition $^3P \to {}^2D$ should be the most intense and is probably peak 5 since peak 3 corresponds to the M satellite of the $^1S_0$ peak. The $^3P \to {}^2P$ transition is assigned to peak 6, $^1P \to {}^2P$ to peak 4, and $^1P \to {}^2D$ is thought to be merged into peak 3.

2.3.1.3. KL-LLM Transitions.

Shake off in the M shell can be considered as a case of the atom having a different chemical surrounding. The normal KLL spectrum would be expected (with respect to intensity ratios), but shifted to lower kinetic energy as the binding energies of the K and L electrons will have increased owing to the removal of the M-shell electron. This shift can be estimated from the ionization potentials of K^+ ($Z=19$) and Ca^{2+} ($Z=20$), which are, of course, isoelectronic with Ar, as follows. The energy of a normal Auger electron is given by $E = E_K - E_{LL}$ and of an Auger electron in the case of M-shell shake off will be $E^M = E_{KM} - E_{LLM}$ the difference is thus given by $E - E^M = E_K - E_{KM} - (E_{LL} - E_{LLM})$. The difference between E_K and E_{KM} is the binding energy of the M electron in an ion with a K vacancy, and this is approximately equal to the ionization potential of K^+. The difference between E_{LL} and E_{LLM} is the binding energy of an M electron in an ion with a double vacancy in the L shell and can be approximated as the ionization potential of Ca^{2+}. The model for calculating the magnitude of the shift in the KLL spectrum in the case of M-shell shake off is known as the equivalent core model.[69] The ionization potentials for K^+ and Ca^{2+} are 31.81 and 51.21 eV, respectively,[70] giving a value of the shift of -19.4 eV. The experimentally determined value was -17.0 eV. The results of the *ab initio* calculations for the satellites for the $^1D_2$ and $^1S_0$ peaks show the same shift from the $^1S_0$ line as from the $^1D_2$ line. The theoretical energies were normalized to be the most intense peak ($^1D_2$) by correcting the values by 0.6 eV, the difference between theory and experiment for this transition. This satellite structure is expected to appear for each of the main lines, and thus peak 11 in Fig. 4.4 is the satellite for the $^1S_0$ line but may also have contributions for the L-shell shake off as well. Peak 9 is assigned to the ^1P satellites and is also merged with L-shell shake-off satellites, which is also the case for the ^3P satellites in peak 7. The satellite corresponding to the ^3P peak is very weak and cannot be resolved. As the intensity ratio of the satellite for the $^1D_2$ peak defined as $I_{sat}/(I_{sat} + I_{main})$ is about 16%, the probability for the M-shell shake off is deduced to be 16%.

2.3.2. LMM Spectrum

There have been considerably more studies of the argon LMM spectrum reported (see Table 4.1) than of the KLL spectrum probably because the LMM spectrum covers the energy region 150–220 eV (approximately) and the KLL spectrum occurs at much higher energies, approximately 2450–2680 eV, posing problems in the instrumentation for measuring electron energies of this magnitude with a gas-phase sample.

Some of the earliest reports of the LMM spectrum were by Mehlhorn et al. at Munster[21,22] in 1968. The L_1MM, L_2MM, and L_3MM spectra were reported and a comparison with the theoretical intensities calculated by Rubenstein[71] made. Fairly good agreement for the $L_2M_{2,3}M_{2,3}$, $L_3M_{2,3}M_{2,3}$, and L_1MM groups of transitions was reported, but a systematic difference for the $L_2M_1M_1$, $L_3M_1M_1$, $L_2M_1M_2$, and $L_3M_1M_{2,3}$ transitions was reported. A value for the binding energy of L_1 was determined to be 326.5 ± 6 eV, showing Rubenstein's value of 287 eV to be incorrect, and this was suggested as a possible reason for the discrepancies observed for some of the Coster–Kronig (LLM-type) transitions. Siegbahn et al. reported on the $L_{2,3}MM$ spectrum in 1973,[25] and this appears to be the most-recent and the most-detailed study of this part of the spectrum, particularly regarding the assignment of satellite peaks. Stolterfoht et al.[28] reported the LMM spectrum produced by ion impact (this method of ionization has been used by a number of other workers[27,72,73]) in 1974 and the most-recent contributions, concerning postcollision interaction, is from Hanashiro et al.[74] in 1979.

2.3.2.1. L_1MM Spectrum.

As Coster–Kronig transitions convert most of the L_1 vacancies to $L_{2,3}M$ configurations, the intensity of the L_1MM spectrum is low. That recorded by Stolterfoht et al.[28] using ion-impact ionization (500-keV protons) is shown in Fig. 4.5. The assignment of the lines is that due to Mehlhorn.[21] The transitions being analogous to the LKK series with the same restrictions on which transitions are allowed in the pure Russell–Saunders coupling scheme (the transition to the ^3P states of the s^2p^4 configuration being forbidden). The intensities were used to calculate the Auger yield and Coster–Kronig yield for the L_1 shell. The Coster–Kronig yield was calculated to be about 96%, the value varied with energy of the exciting protons with a decrease observed as the proton energy was

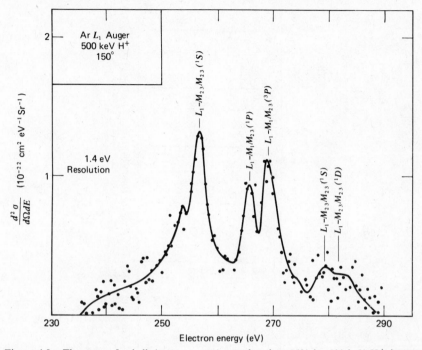

Figure 4.5. The argon L_1-shell Auger spectrum produced at 150° by 500-keV H$^+$ impact. Energy resolution is 1.4 eV.[28] (Reprinted with permission of American Physical Society, New York.)

increased. Since the value calculated is averged over different charge states of the outer shell, a decrease is to be expected because increasing the proton energy will increase the production of outer-shell vacancies. However, when more than two vacancies are produced, the Coster–Kronig yield is zero because the transitions are no longer energetically possible. The calculated value was in good agreement with the theoretical value of McGuire,[75] although a discrepancy with the value calculated from Mehlhorn's results was noted, accounted for by the fact that in Mehlhorn's study (using electron-impact ionization) results were obtained for transitions with no M-shell vacancy.

2.3.2.2. $L_{2,3}MM$ Spectrum. The $L_{2,3}MM$ Spectrum was first reported by Mehlhorn and Stalherm[22] who used a 4-keV electrons and measured the spectrum using an electrostatic analyzer with a resolution of 0.17% at a gas pressure of 2×10^{-3} torr. The energies and intensities of the lines are given in Table 4.7. Siegbahn et al. have also reported the $L_{2,3}MM$ spectrum

Table 4.7. Argon $L_{2,3}MM$ Spectrum

Transition and L-S Designation		Energy (eV) L_2MM Munster	Energy (eV) L_2MM Uppsala	Energy (eV) L_3MM Munster	Energy (eV) L_3MM Uppsala	Intensity[a] L_2MM Munster	Intensity[a] L_2MM Uppsala[c]	Intensity[a] L_3MM Munster	Intensity[a] L_3MM Uppsala[c]	Theory Ref. 71[b]	Theory Ref. 36[c]	Theory Ref. 76[b]	Theory Ref. 22[d]
M_1M_1	$^1S_0$	179.93	180.06	177.79	177.91	7.1	4.8	7.2	5.3	1.9	1.9	1.6	3.2
$M_1M_{2,3}$	$^1P_1$	184.30	189.30	187.16	187.33	8.4	8.4	8.0	8.2	16.8	17.0	14.1	17.0
	$^3P_0$	192.84	193.02			} 21.7	} 18.2	} 19.1	} 21.7	} 33.6	} 33.8	} 26.4	} 34.0
	$^3P_1$	192.90	193.13	190.76		13.3	9.8	11.1	13.5	16.8	16.8	12.3	17.0
	$^3P_2$		190.91	191.13									
$M_{2,3}M_{2,3}$	$^1S_0$	203.01	203.23	200.87	201.09	12.4	14.7	12.2	15.6	9.3	9.5	11.3	8.1
	$^1D_2$	205.40	205.62	203.26	203.47	48.0	48.5	46.0	42.0	49.3	49.2	47.7	49.9
	$^3P_0$	206.95		204.81		} 100	} 100	} 100	} 100	} 100	} 100	} 100	} 100
	$^3P_1$	207.23	207.27	204.87	205.21	39.6	36.8	41.8	42.4	41.4	41.3	41.0	42.0
	$^3P_2$	207.14		205.00									

[a]In both spectra (L_2 and L_3) the intensities are normalized to the $M_{2,3}M_{2,3}$ transitions as 100. This gives the somewhat misleading impression that the two spectra are the same intensity, which is not true.
[b]Taken from Reference 25.
[c]Values calculated by present authors.
[d]Including configuration interaction.

excited by electron impact.[11,25] (Reference 25 is a full discussion of the results presented in their book, Reference 11), and the results of their study obtained with electron excitation between 3 and 5 keV and a spectrometer resolution of 0.06% are also shown in Table 4.7. It is not clear from Reference 25 what pressure the collision chamber was maintained at, the source chamber being maintained at 4×10^{-3}–10^{-1} torr and the analyzer at 10^{-5} torr. Kondow et al.[26] have also reported the $L_{2,3}MM$ spectrum obtained by electron impact (860 eV), and, although a resolution equivalent to 0.08% was used, the spectra reported do not appear as well resolved as that of Siegbahn et al. Relative intensities were not reported, but the authors observed four extra peaks when the spectrum was recorded at 6° rather than 15°. Siegbahn et al. recorded their spectra at 90° to the incident beam and Mehlhorn et al. at 54.5°. The spectrum has also been produced by ion impact excitation.[27,28,72,73] The most-recent contribution in this area would appear to be by Stolterfoht et al.[28] who used 50–600-keV H^+, D^+, H_2^+ and He^+. The electrons ejected at 150° to the incident beam were measured with a parallel-plate analyzer with a maximum resolution of about 0.17%. The energies and intensities of the peaks were not reported specifically even though the complete spectrum between 120 and 245 eV is reported. The main object of the work was to report on branching ratios and to study the satellite spectrum to determine probabilities for the excitation of multiple vacancy states and to evaluate the L_1 Coster–Kronig yield. The complete spectrum as recorded by the Uppsala group is shown in Fig. 4.6, together with an assignment of the main lines as given in Table 4.7. The table also shows the intensities of the lines as measured by the two groups, Uppsala and Munster, and gives the theoretical values calculated by a number of authors. It should be noted that the intensity values have been quoted relative to the $M_{2,3}M_{2,3}$ transitions normalized to a value of 100. This gives the misleading impression that the L_2 and L_3 spectra are of comparable intensity. As can be seen from Fig. 4.6 the intensity ratio of the L_3 to L_2 spectrum is is about 2:1, the value varies from 1.47:1 to 1.96:1 depending on which peaks are chosen. The ratio $\sum L_3/\sum L_2$ is 1.76, based on peak heights. The failure of the intermediate-coupling theory[25] to account for the M_1M_1 line intensity could be that the lines gain intensity from the $M_{2,3}M_{2,3}$ lines through configuration interaction. However, the results of including configuration interaction in the calculation are also shown in Table 4.7,[22] and it can be seen that this explanation does not account for the discrepancies entirely. The low measured intensity of the $M_1M_{2,3}$ transitions may be due to the occurrence of shake up.

The problem of the discrepancy between theoretical and actual intensities has been discussed by McGuire,[29] who showed that a better agreement with theory could be obtained by reassigning the $L_{2,3}M_1M_{2,3}(^1P)$ and

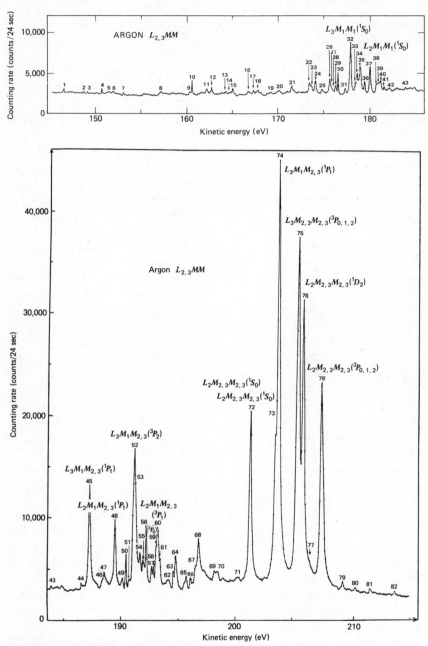

Figure 4.6. The $L_{2,3}MM$ Auger spectrum of argon.[25] (Reprinted with permission of The Royal Swedish Academy of Sciences.)

124

$L_{2,3}M_1M_1(^1S)$ doublets. Peaks in the spectrum will be doublets separated by the L_2L_3 splitting of 2.16 eV. McGuire identified nine such doublets in the region where the $L_{2,3}M_1M_1$ doublet would be expected. By comparing calculated and observed[70] positions of the $3p^4$ and $3s3p^5$ levels, McGuire deduces strong configuration interaction for the ^1P position and assigns the $L_2M_1M_1$ transition to line 32 and $L_3M_1M_1$ to line 26 (Siegbahn's assignment was $L_2M_1M_1 - 37$, $L_3M_1M_1 - 32$) McGuire reassigns line 37 to one of the satellites of the $L_2M_1M_{2,3}$ transition. However, McGuire attributes a relative intensity of 5 to line 32, whereas it really is 34, and thus the agreement between experimental and calculated values is just as poor in respect of the ^1S lines as was reported by Siegbahn. The line also appears in McGuire's table of assignments as part of the $L_3M_1M_{2,3}(^1P)$ transitions, adding further confusion to the issue.

2.3.2.3. The $L_{2,3}MM$ Spectrum Satellites. Transitions from doubly ionized states appear to dominate the satellite spectrum. The energy regions for the various transitions are given in Table 4.8. The energies for the various double-hole initial states have been calculated by Siegbahn et al. using a relativistic Hartree–Fock–Slater program,[14,77] and the results compared with Mehlhorn's values obtained from $L_1L_{2,3}M_{2,3}$ and $L_1L_{2,3}M_1$ Coster–Kronig spectra.[21] The values are shown in Table 4.9. However, the values obtained are spread over too wide a range to provide an unambiguous assignment, and it it has only been possible to assign transitions to the $M_{2,3}M_{2,3}M_{2,3}$ final states. This is aided by the fact that the separation of the ^2P and ^2D states is 1.71 eV,[70] and several "pairs" of lines in the spectrum are separated by this value, that is, 49, 55; 50, 56; 54, 61; 58, 63. From a consideration of the ionization cross section of L_1 and $L_{2,3}$ shells (expected to be about the same) and Lotz's empirical formula[78] the ratio of $L_{2,3}$ to L_1 vacancies is expected to be about 4:1. The L_1 vacancies decay predominantly through Coster–Kronig transitions, which, although having a fairly high probability of occurring with shake off (about 19%),[79] will produce $L_{2,3}M$ vacancies totaling about 20% of the $L_{2,3}$ vacancies. Since

Table 4.8. Satellites in $L_{2,3}MM$ Auger Spectrum of Argon

Transition	Energy Region (eV)
$L_{2,3}M_{2,3} - M_1M_1M_{2,3}$	155
$L_{2,3}M_{2,3} - M_1M_{2,3}M_{2,3}$	175–185
$L_{2,3}M_{2,3} - M_{2,3}M_{2,3}M_{2,3}$	190–200

Table 4.9.   Energies of $L_{2,3}M_{2,3}$ Double-Hole States in Argon

| Term | Calculated | | Measured |
	Siegbahn	Mehlhorn	Siegbahn
$^1P_1$	0.00	0.00	0.00
$^3D_3$	0.29	0.30	0.30
$^3D_2$	1.33	1.15	1.47
$^3S_1$	1.76	1.65	
$^3P_2$	2.89	2.46	
$^3D_1$	3.56	2.56	2.63
$^3P_0$	4.00	3.28	
$^3P_1$	5.03	3.79	
$^1D_2$	5.93	4.56	4.56
$^1S_0$	8.17	6.42	6.53

the probability of shake off in the M shell during primary $L_{2,3}$ ionization (the other mechanism for producing $L_{2,3}M$ double vacancies) is about 10%, it is deduced that the dominating parent process producing the $L_{2,3}M$ double vacancies is the $L_1L_{2,3}M$. Coster–Kronig transitions and the relative intensities of the Coster–Kronig transitions would be reflected in the relative intensities of the $L_{2,3}M$-MMM satellite process. It is expected that transitions to ^3D final states (of the Coster–Kronig process) will have the greatest intensities,[31,76] and so the satellite lines initiating from ^3D states will have the greatest intensity. Thus lines 50, 54, and 58 correspond to transitions $L_{2,3}M_{2,3}(^3D)$ to the $M_{2,3}M_{2,3}M_{2,3}(^2P)$ and the lines 56, 61, and 63 to transitions $L_{2,3}M_{2,3}(^3D)$ to $M_{2,3}M_{2,3}M_{2,3}(^2D)$. Transitions to the $4S$ state could not be identified. Lines 49 and 55 are separated by the upper state ^2D-^2P splitting of 1.71 eV, and they are assigned to transitions from the $^1P_1$ state. Also, pair 64, 67 is assigned to $^1D_2$-^2P and $^1D_2$-^2D transitions.

Of the remaining satellites in the energy region 185–200 eV most are assigned to either $L_{2,3}M_{2,3}(^{1,3}P) \rightarrow M_{2,3}M_{2,3}M_{2,3}$ or to transitions from $L_{2,3}M_1$ double vacancies.

The satellites in the 170–180-eV region are assigned to shake up in the Auger transition, because many are doublets with the L_2-L_3 energy separation of 2.15 eV, indicating the same initial states as the Auger lines. Four shake-up lines associated with the $L_{2,3}$-$M_{2,3}M_{2,3}$ transitions are tentatively assigned. McGuire[29] has also discussed the $L_{2,3}M$-MMM spectrum of argon.

2.3.3. The KLM Spectrum (KMM Spectrum) and Satellites

The *KLM* spectrum recorded by Siegbahn et al.[7] is shown in Fig. 4.7. By comparing the intensities of the *KLL* and *KLM* spectra's strongest lines, it was deduced that the occurrence of a *KLL* event is seven times more probable than a *KLM*. (The *KMM* spectrum constitutes about 1% of the total *K* Auger

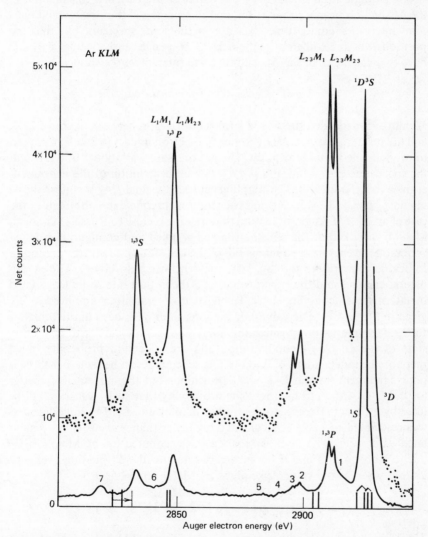

Figure 4.7. The *KLM* Auger electron spectrum of argon. Calculated transition energies are indicated by bars.[7] (Reprinted with permission of The Royal Swedish Academy of Sciences.)

transition rate.) The possible kinetic energies of the Auger electrons derived from the calculations are also shown in Fig. 4.7. Transitions to a final state $L_{2,3}M_{2,3}$ were omitted because these (to a $^{1,3}P$ state) are forbidden in the L-S scheme owing to parity reasons. As can be seen from Fig. 4.7 good agreement was obtained from the $M_{2,3}$ final-state vacancy, but the agreement for the M_1 peaks was poor. The reason for this is that the M_1 vacancy undergoes strong configuration interaction. The satellites are assigned to shake off in the M-shell on primary ionization, and it is suggested that they would have total intensities comparable to those in the KLL spectrum but that the positions would be different to the KM-LLM case because the final state will have two vacancies in the M shell that will interact very strongly.

2.3.4. Coster–Kronig Transitions

Mehlhorn[21] recorded the LMM spectrum of argon between 38 and 54 eV, and the results are discussed by Sevier.[50] Electron impact (4.1 keV) was used to produce the initial ionization. The gas pressure was about 10^{-3} torr and the spectrometer resolution was 1.2%. A full description of the instrument is given in Reference 32. The coupling between the final $L_{2,3}M$ states following an $L_1L_{2,3}M$ Coster–Kronig transition is intermediate, although in the case of an L_1MM Auger transition the coupling is close to Russel–Saunders. Mehlhorn looked at the effects of mixed coupled and configuration interaction on the energies and intensities of the transitions. Maxima were found to be due primarily to the $^1S_0$, $^1D_2$, and $^3D_3$ lines, respectively. The experimental energies and the experimental and theoretical intensities (calculated in intermediate coupling using Rubenstein's[71] transition amplitudes) are given in Table 4.10. The values for the intensities have been quoted relative to the $^3D_3$ line as unity, since this final state cannot mix with any other state of the $L_{2,3}M_{2,3}$ configuration and the transition probability must therefore be independent of the coupling scheme. It can be seen that there is poor agreement for the $L_1L_{2,3}M_1$ transitions but the agreement is better for the $L_1L_{2,3}M_{2,3}$ transitions. The probability that an L_1 vacancy will be filled by a Coster–Kronig process rather than by and L_1MM Auger process was calculated to be 94.8%, considerably lower than Rubenstein's[71] calculated value of 97.0%. The discrepancy was accounted for by Mehlhorn as due to Rubenstein's use of the incorrect value for the L_1 binding energy of 287 eV, whereas the value determined by Mehlhorn was 326.5.

2.4. Krypton $(Z = 36\ 1s^2 2s^2 2p^6 3d^{10} 4s^2 4p^6)$

The K-shell electron binding energy is approximately 14.3 keV and the L_1, L_2, and L_3 subshell electron binding energies are approximately 1.92, 1.73, and

Table 4.10. Argon Coster–Kronig Spectrum

Transition	L-S designation	Energy	Intensity	
			Experiment	Theory
$L_1L_{2,3}M_1$	$^1P_1$	28.7	0.62	1.84
	$^3P_0$			0.007
	$^3P_1$	30.5	0.47	2.05
	$^3P_2$			0.035
$L_1L_{2,3}M_{2,3}$	$^1S_0$	41.1	0.82	0.475
	$^3S_1$	45.91	0.34	0.297
	$^1P_1$	47.6	0.067	0.103
	$^3P_0$	44.3	0.13	0.212
	$^3P_1$	43.8	0.077	0.117
	$^3P_2$	45.1	0.12	0.242
	$^1D_2$	43.0	0.57	0.521
	$^3D_1$	45.0	0.37	0.353
	$^3D_2$	46.41	0.65	0.538
	$^3D_3$	47.26	1.0	1.0

1.65 keV so the KLL Auger electron spectrum would have energies of the order of 12 keV (semiempirical calculations give values ranging from 10.4 keV for the KL_1L_1 transition to 10.9 keV for the KL_3L_3 transition[14]). The KLL spectrum has not been recorded experimentally because of the difficulties associated with measuring electron energies of this magnitude emitted from a gaseous target material. Both the LMM and MNN spectra have been reported at high resolution[11,31,33] and also the MMN Coster–Kronig spectrum.[32] According to Sevier,[50] all the L Coster–Kronig spectra are allowed apart from $L_1L_2M_{1,2,3}$, $L_1L_3M_{1,2}$, $L_2L_3M_{4,5}$, but none have been measured experimentally. The same is true for the LNN transitions and the allowed MMM transitions ($M_{1,2,3}M_{4,5}M_{4,5}$).

2.4.1. The LMM Spectrum and Satellites

The first report of the krypton LMM spectrum was by Krause[34] in 1965. An electrostatic analyzer was used with a resolution of 2%. Initial ionization was by x-rays from a molybdenum tube. The target chamber pressure was 10^{-2} torr, while the analyzer and detector pressures were approximately 4×10^{-5} torr. The high-resolution spectrum has been reported by Werme et al.[31] obtained with a double-focusing electrostatic spectrometer with a resolution of 0.06% using electron-impact (3–5 keV) ionization. The collision chamber pressure was 10^{-1} torr and the analyzer was held at 10^{-5} torr.

Although Krause suggested assigning contributions to two of the lines in his spectrum from L_1MM transitions, there has been no further mention in the literature of the L_1 spectrum. It would be expected that the L_1MM spectrum would be very weak because of filling of L_1 vacancies by Coster–Kronig transitions.

The $L_{2,3}M_{2,3}M_{4,5}$ spectrum obtained by Werme et al. is shown in Fig. 4.8. It can be seen that the spectrum consists of six groups of lines, three of these correspond to transitions from an L_2 subshell vacancy and three to an L_3 subshell vacancy. The separation between the lines with the same final state gives a value of the spin-orbit splitting for the $2p$ level of 52.7 ± 0.2 eV, in good agreement with values obtained by other methods[11,80] The identification was made by comparing the results with calculated average energies by a self-consistent-field, restricted Hartree–Fock–Slater method[14,77] The assignment of the 28 lines observed between 1182 and 1518 eV is given in Table 4.11. As can be seen from the table, the satellite lines are assigned either to shake up accompanying the Auger transition or to transitions from initial double vacancies, produced either by shake off in the initial ionization or to Coster–Kronig transitions filling initial L_1 vacancies.

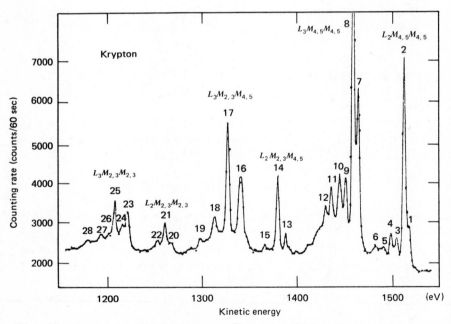

Figure 4.8. Krypton $L_{2,3}MM$ Auger spectrum.[31] (Reprinted with permission of The Royal Swedish Academy of Sciences.)

Table 4.11. Krypton *LMM* Spectrum and Satellites

Line No.	Energy	Normal Lines	Satellites
1	1518.2		
2	1513.2	$L_2M_{4,5}M_{4,5}$	
3	1505.5		
4	1499.4		$L_2N_1 - M_{4,5}^2N_1$; $L_2N_{2,3} - M_{4,5}N_{2,3}$
5	1491.4		$L_2M_{2,7} - M_{4,5}^2M_{2,3}$; $L_2M_{4,5} - M_{4,5}^3$
6	1482.7		$L_2M_{4,5}M_{4,5}$ + shake up of N_1, $N_{2,3}$
7	1466.0		
8	1460.4	$L_3M_{4,5}M_{4,5}$	
(most intense)			
9	1452.6		
10	1446.1		$L_3N_1 - M_{4,5}^2N_1$; $L_3N_{2,3} - M_{4,5}^2N_{2,3}$
11	1437.2		$L_3M_{2,3} - M_{4,5}^2M_{2,3}$; $L_3M_{4,5} - M_{4,5}$
12	1431.0		$L_3M_{4,5}M_{4,5}$ + shake up of N_1, $N_{2,3}$
13	1386.6	$L_2M_{2,3}M_{4,5}$	
14	1380.0		
15	1366.5		$L_2M_{2,3} - M_{2,3}^2M_{4,5}$ $L_2M_{4,5} - M_{2,3}M_{4,5}^2$
16	1341.7	$L_3M_{2,3}M_{4,5}$	
17	1327.4		
18	1313.4		$L_3M_{2,3} - M_{2,3}^2M_{4,5}$; $L_3M_{4,5} - M_{2,3}M_{4,5}$
19	1299.6		$L_3M_{2,3}M_{4,5}$ + shake up of N_1, $N_{2,3}$
20	1269.1		
21	1262.2	$L_2M_{2,3}M_{2,3}$	
22	1254.8		
23	1222.9		
24	1216.9	$L_3M_{2,3}M_{2,3}$	
25	1209.8		
26	1202.9		
27	1196.5		$L_3N_{1,2,3} - M_{2,3}^2N_{1,2,3}$; $L_3M_{4,5} - M_{2,3}M_{4,5}$
28	1182.1		$L_3M_{2,3} - M_{2,3}$; $L_3M_{2,3}$ + shake up of $N_{2,3}$

As far as the intensity of the lines is concerned, Werme et al. discuss the agreement between experiment and calculations based on *j-j* and *L-S* coupling and conclude that neither gives particularly good agreement and that an intermediate-coupling scheme is to be applied, although attempts with Asaad's and Burhop's method[5] did not result in any improvement.

2.4.2. The MNN Spectrum and Satellites

Of the M spectra the $M_{4,5}NN$ has received the most attention. High-resolution studies have been published by Werme et al.[31] and by Mehlhorn et al.[33] and a detailed theoretical interpretation given by McGuire.[35] The $M_{2,3}NN$ spectrum was measured by Mehlhorn[32] (between 118 and 150 eV, but only just, since the competing process the $M_{2,3}M_{4,5}N_{2,3}$ Coster–Kronig transitions were found to be about considerably more intense. The Coster–Kronig transitions with an N-shell vacancy (presumably due to shake off in the initial ionization) also had a greater intensity than the $M_{2,3}NN$ Auger process. Mehlhorn also identified the energy region of the M_1NN spectrum (228.3–253.5 eV) and, although the spectrum obtained from the appropriate region is reported, there is not really any sign of the Auger transitions.

The high-resolution $M_{4,5}NN$ spectrum obtained by Werme et al. is shown in Fig. 4.9. Mehlhorn's spectrum is very similar. Both these authors assign the peaks numbered 35 and 40 in Fig. 4.9 to the $M_{4,5}N_1N_1$ transitions, while pointing out that an unambiguous designation of the peaks is very difficult. McGuire[35] argues that this assignment is incorrect and that the peaks numbered 51 and 52 should be assigned to the $M_{4,5}N_1N_1$ transitions and that 35 and 40 use components of the $M_{4,5}N_1N_{2,3}(^1P_1)$ transitions split off from the main components (lines 27 and 30, respectively) by configuration interaction with the $(4p)^3\ ^2D(4d)\ ^1P_1$ level. McGuire's assignments are given in Table 4.12 together with the intensity data from References 31 and 33. McGuire indicates that Mehlhorn's data is slightly in error and the corrected values (from a private communication) are given in the table. The intensities are normalized so that the sum of all three are equal. The calculated values are in good agreement for the $N_{4,5}N_{2,3}N_{2,3}$ transitions but differ by about 10–15% for the $M_{4,5}N_1N_{2,3}$ and are 40% too low for the $M_{4,5}N_1N_1$ transitions.

Werme et al. assign some of the satellite lines to shake-up processes, and McGuire treats the satellite spectrum in some detail, assigning some lines to shake off in the initial ionization and including a discussion of the mechanism of production of initial double-vacancy states.

2.4.3. Coster–Kronig Spectra

Part of the spectrum $(M_{2,3})$ was measured by Mehlhorn.[32] The assignments of the lines also includes Auger transitions in the presence of an N vacancy. The initial double vacancy could be produced either by shake-off or Coster–Kronig transitions from an initial M_1 vacancy.

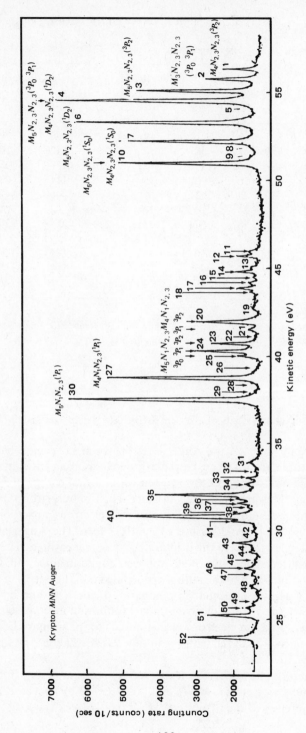

Figure 4.9. Krypton $M_{4,5}NN$.[31] (Reprinted with permission of The Royal Swedish Academy of Sciences.)

133

Table 4.12.   Krypton $M_{4,5}$ Auger Spectrum

Final State		Line Number		Intensity (M_4)			Intensity (M_5)		
		M_4	M_5	a	b	c	a	b	c
$N_{2,3}N_{2,3}$	$^3P_2$	1	3	21	19	25	90	77	55
	$^3P_{0,1}$	2	4	38	37	35	32	37	30
	$^1D_2$	4	6	118	115	117	138	138	165
	$^1S_0$	7	10	100	95	75	111	102	107
$N_1N_{2,3}$	$^3P_2$	20	23	52	61	39	46	41	24
	$^3P_1$	21	24	10	10	7.8	62	61	29
	$^1P_1$	27	30	157⎫	165⎫	⎫	211⎫	234⎫	⎫
				⎬287	⎬300	⎬364	⎬392	⎬420	⎬515
	$^1P_1$	35	40	130⎭	135⎭	⎭	181⎭	186⎭	⎭
N_1N_1	$^1S_0$	51	52	79	74	48	114	115	68
$N_1N_{2,3}$	$^3P_0$		25			0.5	21	23	12

[a]Values from Reference 31.
[b]Values recalculated by McGuire[35] from revised data from Reference 33.
[c]McGuire's[35] calculated values.

2.5.   Xenon ($Z=54\ 1s^22s^22p^63s^23p^63d^{10}4s^24p^64d^{10}5s^25p^6$)

As indicated in Table 4.1, the Auger electron spectra of xenon have been obtained by three groups. The Uppsala group[31] used an instrument with resolution with electron-impact ionization with a beam energy of 3–5 keV. The gas pressure was approximately 0.1 torr. The Freiburg group[38] used an instrument with a resolution of 0.13%, also with electron-impact ionization (at 4 keV) but with a gas pressure of 3×10^{-4} torr. The Auger electrons ejected at an angle of 150° to the primary beam were measured. The spectra were also measured at an angle of 90° and were found to agree with the 150° spectrum to within 5%, and the authors concluded that the effects of angular distribution would be neglected in comparing theoretical line intensities. The Oregon workers[39] used electron-impact ionization at both 900 eV and 3 keV and measured the electrons ejected at 60° to the primary beam with a cylindrical-mirror analyzer with resolution 0.09%. The gas pressure was approximately 10^{-3} torr.

The Freiburg group reported on the MNN spectrum (510–540 eV), the Uppsala group reported the MNN and NOO (5–40 eV) spectra, and the Oregon group reported the MNN and MNO (560–610 eV) spectra.

2.5.1. $M_{4,5}N_{4,5}N_{4,5}$ Spectrum

The spectrum obtained by Werme et al.[31] is shown in Fig. 4.10. The two most intense line groups are the M_4 and M_5, which just overlap. The peaks in the region 370–395 eV were interpreted as the $M_{4,5}N_1N_{4,5}$ transitions, the peaks in the region 430–460 eV as the $M_{4,5}N_{2,3}N_{4,5}$ transitions and the peaks between 560 and 600 eV as the $M_{4,5}N_{4,5}O$ transitions. The spectrum obtained by Hagmann et al.[38] is shown in Fig. 4.11. This spectrum has a continuous background subtracted to account for the contributions from the $M_{4,5}O_i$-$N_{4,5}N_{4,5}O_i$ and $M_{4,5}N_{4,5}N_i$ satellite spectra, each of which contains a large number of lines and were represented by a trapezoidal-shaped continuous distribution. The details of the estimation of the background are given in Hagmann's thesis,[81] but do not appear to have been published. The initial double vacancies were considered to have occurred by either shake off in creating an $M_{4,5}$ vacancy in the primary process or by a primary $M_{2,3}$ vacancy decaying via $M_{4,5}N$ and $M_{4,5}O$ Coster–Kronig processes (the contribution from Coster–Kronig decay of initial M_1 vacancies was neglected due to the relatively small M_1 vacancy production), the rates of these processes having been calculated by McGuire.[82] The relative populations of the initial double vacancies produced by shake off were obtained from Nestor et al.,[83] and the relative $M_{4,5}$ and $M_{2,3}$ subshell ionization cross sections were obtained according to the Grysinski formula.[84] If it is assumed that the spectator vacancies do not affect the decay of a single $M_{4,5}$ vacancy, then the relative populations of initial double vacancies also represents the relative intensities of the different satellite groups. The total satellite spectrum was obtained by adding together the different satellite groups weighted with their relative intensities. The Oregon group adopted a different approach to correcting for the background. First, by obtaining spectra of 900 eV, the possibility of ionization of the 3p electrons (the $M_{2,3}$ subshell) is eliminated because the binding energy of these electrons is greater than 900 eV. This considerably reduces the satellite spectra owing to transitions from initial double-hole states created by Coster–Kronig processes. The only processes that can lead to satellite lines are shake up and shake off. A linear background was then substracted, having different slopes under the M_4 and M_5 groups and being adjusted for height on the high- and low-energy sides of the groups and at the minimum between the groups. The drawback with excitation at energies near the ionization threshold is that the ionization cross section decreases rapidly and the background due to inelastically scattered electrons increases, necessitating the use of a high-transmission spectrometer. The spectrum obtained was very similar to that in Fig. 4.11.

The results of the group's experiments are shown in Table 4.13 for energies

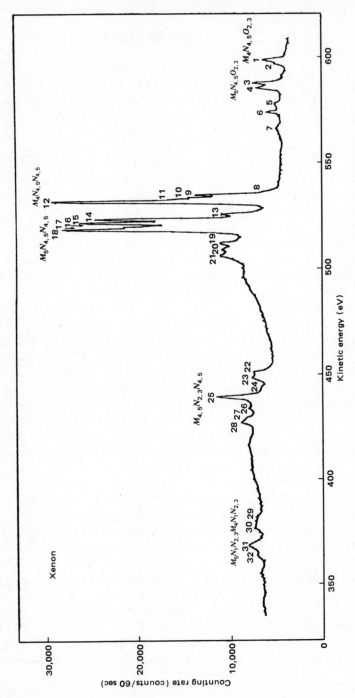

Figure 4.10. Xenon $M_{4,5}NN$ Auger spectrum.[31] (Reprinted with permission of The Royal Swedish Academy of Sciences.)

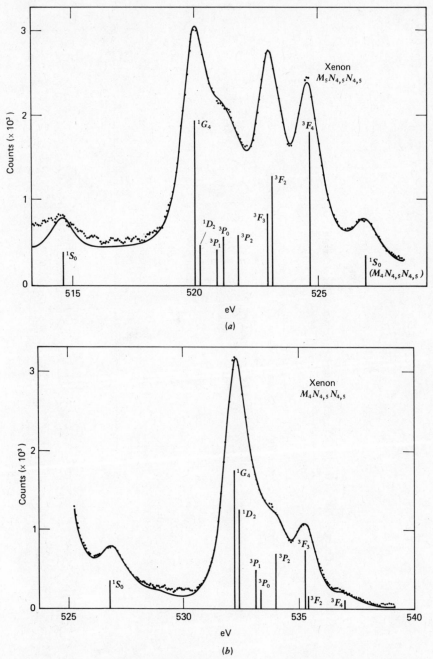

Figure 4.11.   Xenon Auger spectra.[38] (a) $M_5N_{4,5}N_{4,5}$; (b) $M_4N_{4,5}N_{4,5}$. (Reprinted with permission of Springer-Verlag, New York.)

Table 4.13. Energies (eV) of the $M_{4,5}N_{4,5}N_{4,5}$ Auger Spectrum Lines of Xenon

Transition	$M_{4,5}$			M_4		M_5		M_{45}
	Uppsala	Freiburg	Oregon	Uppsala	Freiburg	Uppsala	Freiburg	Oregon
$^3F_4$	5.22	5.33	5.18	4.75	4.89	4.76	4.75	4.76
$^3F_2$	3.43	3.73	3.42	} 3.02	3.26	} 3.18	3.23	} 3.15
$^3F_3$	3.60	3.62	3.48		3.13		3.02	
$^3P_2$	1.53	2.08	1.84	1.73	1.90	1.71	1.81	1.72
$^3P_0$	0.67	1.26	0.89	} 0.97	1.22	} 0.95	1.18	} 0.92
$^3P_1$	0.91	0.98	0.70		0.95		0.92	
$^1D_2$	-0.01	0.28	-0.06	} 0.00	0.20	} 0.00	0.23	} 0.00
$^1G_4$	0.00	0.00	0.00		0.00		0.00	
$^1S_0$	-6.10	-5.13	-6.13	-5.47	-5.36	-5.53	-5.47	-5.71

138

of the lines, together with the calculated values. The Uppsala group used a relativistic Hartree–Fock–Slater calculation with intermediate coupling, the Freiburg group used an intermediate coupling plus configuration-interaction calculation, and the Oregon group used a mixed-coupling scheme with j-j coupling applied to the initial state and intermediate coupling to the final state (using L-S wave functions as the basis set—the true wave functions being linear combinations of this basis set). The Coulomb matrix elements were taken from Slater[85] the spin-orbit elements from Condon and Shortly,[86] and for the numerical values of the Slater integrals the values tabulated by Mann were used[87] and the values of the spin-orbit parameter were from the calculations of Huang et al.[88] As can be seen from the table only minor discrepancies exist between the three sets of theoretical values and the experimental values. This, however, is not the case for the relative intensities of the lines as can be seen from Table 4.14, which includes the values obtained by McGuire[82] (Sandia Labs, Albuquerque) using L-S coupling in the final states. Although the agreement between the experimental values of Uppsala group and the other two groups is poor (possibly due to the corrections for the satellite and scattered background contribution made by the Freiburg and Oregon groups), the agreement between theory and experiment is quite good for the intermediate-coupling and intermediate-coupling-plus-configuration-interaction cases showing the L-S coupling in the final state to be an incorrect assumption in this case. The small effect of configuration interaction on the theoretical values and the agreement between the intermediate-coupling theory and experiment shows that correlation effects play only a minor role in the $M_{4,5}N_{4,5}N_{4,5}$ spectrum.

It is interesting to note that although there is good agreement between the Oregon group's calculations and those of the other groups for the inter-mediate-coupling case, the agreement between the Oregon group's calculations and McGuire's for the L-S case is poor; the Oregon calculations for the L-S case have been omitted from the table. Closer inspection of the tabulated values reveals that the Oregon calculations have the values for the $^3F_2$ and $^3P_2$ final states the wrong way around; or possibly McGuire's values are the wrong way around. The Oregon group (the later publication) make no comment on this discrepancy even though they cite McGuire's paper.

Hagmann et al. also calculated the relative intensities of the M_4 to M_5 line groups from their experimental results and found a value of 0.69 ± 0.05 (the Oregon group reported values of 0.67 and 0.60 at 3 keV and 0.9 keV, respectively), which they equated to the relative probabilities of M_4 to M_5 ionization. This agrees well with the theoretical value of 4/6, the ratio of $2J+1$ values for state produced by removal of $4d$ electron. They also pointed out that lines due to double Auger transitions, as are observed in the $M_{4,5}NN$

Table 4.14. Relative Intensities of the $M_{4,5}N_{4,5}N_{4,5}$ Auger Spectrum Lines of Xenon

Transition	Theory				Experiment			
	Albuquerque	Freiburg	Uppsala	Oregon	Freiburg	Uppsala	Oregon[a]	Oregon[b]
M_4								
$^3F_4$	4.6	2.0	2.0	1.9	2.1	1.2	1.4	1.6
$^3F_2$	17.5 ⎱ 33.0	2.7 ⎱ 18.4	3.8 ⎱ 19.3	3.5 ⎱ 19.0	3.1 ⎱ 15.7	13.8	15.3	14.2
$^3F_3$	15.5 ⎰	15.7 ⎰	15.5 ⎰	15.5 ⎰	12.6 ⎰	15.5	13.1	13.2
$^3P_2$	8.8	12.5	11.7	11.9	12.0			
$^3P_0$	3.5 ⎱ 12.3	1.4 ⎱ 10.2	1.6 ⎱ 10.4	1.6 ⎱ 10.4	4.3 ⎱ 12.8	16.7	10.4	13.4
$^3P_1$	8.8 ⎰	8.8 ⎰	8.8 ⎰	8.8 ⎰	8.5 ⎰			
$^1D_2$	12.4	23.2	23.2	23.3	21.3	42.5	51.7	49.7
$^1G_4$	25.3 ⎱ 37.7	28.2 ⎱ 51.4	27.9 ⎱ 51.1	27.9 ⎱ 51.2	29.9 ⎱ 51.2			
$^1S_0$	3.5	5.5	5.4	5.5	6.3	10.3	8.2	8.0
M_5								
$^3F_4$	23.8	26.2	25.9	26.0	21.6	21.1	23.1	23.2
$^3F_2$	3.3 ⎱ 13.8	16.1 ⎱ 26.8	15.2 ⎱ 25.7	15.5 ⎱ 26.0	15.4 ⎱ 25.6	22.2	25.3	24.5
$^3F_3$	10.5 ⎰	10.7 ⎰	10.5 ⎰	10.5 ⎰	10.2 ⎰	14.0	10.4	11.9
$^3P_2$	13.7	9.3	10.1	9.9	7.1			
$^3P_0$	1.6 ⎱ 7.5	3.2 ⎱ 9.1	3.0 ⎱ 8.9	3.0 ⎱ 8.9	6.9 ⎱ 12.0	15.4	10.4	11.0
$^3P_1$	5.9 ⎰	5.9 ⎰	5.9 ⎰	5.9 ⎰	5.1 ⎰			
$^1D_2$	12.4	3.5	4.1	4.0	5.7	23.8	28.2	25.7
$^1G_4$	25.3 ⎱ 37.7	23.5 ⎱ 27.0	23.2 ⎱ 27.3	23.1 ⎱ 27.1	23.2 ⎱ 28.9			
$^1S_0$	3.5	1.8	2.1	2.1	4.8	3.5	2.6	3.6

[a]Spectrum excited by 3-keV electrons.
[b]Spectrum excited by 0.9-keV electrons.

spectrum of krypton, were not observed. Both groups discuss line widths, briefly noting that the observed values are less than the calculated values.

Thomas et al. (Oregon) also report the $M_{4,5}N_{2,3}N_{4,5}$ spectrum. The theoretical and experimental energies are given in Table 4.15. The energies being calculated as outlined previously for the $M_{4,5}N_{4,5}N_{4,5}$ spectra. The theoretical intensities are also compared with the experimental in Table 4.15. In this case, the intensities are calculated by applying a mixed-coupling scheme: j-j coupling to the initial state and intermediate coupling to the final state. Transition amplitudes in L-S coupling are calculated from Shore and Menzel,[89] and with the aid of transformation coefficients from L-S coupling in final and initial states to L-S coupling in the final and j-j coupling in the initial state, the transition amplitudes in mixed coupling are then calculated. Finally, the mixing coefficients obtained from the eigenvectors of the energy matrices are used to calculate the transition rates of different Auger components. Although the theory does predict the major features of the spectrum, the agreement between the theoretical and experimental values is poorer than for the $M_{4,5}N_{4,5}N_{4,5}$ case. The authors consider that interaction between the $4p^5 4d^9$ and $4p^6 4d^7 4f$ configurations could be responsible for this poor agreement, and since the shifts and line broadenings of the photo-electron spectrum are absent, the super Coster–Kronig processes $(N_2 N_{4,5} N_{4,5})$, which are energetically possible for elements around $Z = 54$, do not occur.

2.5.2. MNO Spectra

These spectra are reported by the Oregon group and the experimental and theoretical intensities and energies are given in Table 4.16. The $N_{4,5}$ splitting is 12.6 eV and the binding energies for the $5s(O_1)$ electrons are about 11 eV higher than the $5p(O_{2,3})$ electrons, so the $M_4 N_{4,5} O_1$ and $M_5 N_{4,5} O_{2,3}$ line groups are expected to overlap making identification and assignment difficult. Again the agreement between theory and experiment is poorer than for the $M_{4,5}N_{4,5}N_{4,5}$ spectra, possibly because of the difficulties of resolving the overlapping spectra and possibly because of the greater importance of configuration interaction for the $M_{4,5}N_{4,5}O_{1,2,3}$ spectra (involving outer electrons).

2.5.3. NOO Spectrum

The spectrum has only been considered by one group of workers, that at Uppsala.[31] The spectrum is shown in Fig. 4.12 covering the energy range 8–36 eV approximately. Comparison work shows many similarities between

Table 4.15. Xenon $M_{4,5}N_{2,3}N_{4,5}$ Auger Transitions

Transition	Energy (eV)[a]			Relative Intensities					
				M_4		M_5		Experiment	
	Theory[b]	Theory[c]	Experiment	Theory[d]	Theory[e]	Theory[d]	Theory[e]	M_4	M_5
$^1F_3$	0.00	0.00	0.00	27.0 ⎱ 36.5	34.2 ⎱ 48.1	27.0 ⎱ 36.5	16.5 ⎱ 20.8	46.2	30.6
$^1P_1$	1.36	0.14	1.95	9.5 ⎰	13.9 ⎰	9.5 ⎰	4.3 ⎰		
$^3P_2$	10.21	5.67	4.55	2.0	14.8	11.9	2.5	2.6	5.4
$^3D_1$	5.99	6.39	7.15	11.3	1.9	0.8	8.2	0.0	5.9
$^3D_3$	11.07	13.63	9.36	1.9	1.0	18.2	22.0	8.2	6.3
$^3D_2$	12.71	15.18	10.66	11.9	0.7	6.0	11.9		
$^3P_0$	13.16	15.63	11.83	4.0 ⎱ 26.3	4.0 ⎱ 15.8	0.0 ⎱ 6.7	0.0 ⎱ 13.5	32.8	16.5
$^3F_2$	9.36	16.34		10.4 ⎰	11.1	0.7 ⎰	1.6 ⎰		
$^3F_3$	14.30	19.15	13.00	8.4 ⎱ 16.7	2.2 ⎱ 15.5	5.1 ⎱ 7.5	11.8 ⎱ 12.0	10.2	14.3
$^3P_1$	11.69	19.92	16.02	8.3 ⎰	13.3 ⎰	2.4 ⎰	0.2 ⎰		
$^1D_2$	14.78	21.49	19.89	4.9	2.5	5.9	7.5	0.0	6.4
$^3F_4$	20.85	23.33	20.93	0.4	0.4	13.5	13.5	0.0	14.6

[a] Relative to $^1F_3$ transition.
[b] L-S coupling plus diagonal contributions from the spin-orbit interactions.
[c] Intermediate coupling.
[d] L-S coupling for final state.
[e] Intermediate coupling for final state.

Table 4.16. Xenon $M_{4,5}N_{4,5}O_{1,2,3}$ Auger Transitions

Transition		Energy^a (eV)			Relative Intensities					
		Theory^b	Theory^c	Experiment	M_4 Theory^d	M_4 Theory^e	M_4 Experiment	M_5 Theory^d	M_5 Theory^e	M_5 Experiment
O_1	$^1D_2$	0.00	0.00	0.00	40.3	66.2	68.4	40.3	7.7	8.0
	$^3D_1$	−0.60	0.28		29.0	29.0		0.6	0.6	
	$^3D_2$	−0.19	1.44	2.80	29.4	3.5	31.6	13.6	46.2	92.0
	$^3D_3$	1.38	2.25		1.3	1.3		45.5	45.5	
$O_{2,3}$	$^3D_1$	−0.07	−1.07		11.4	12.4		0.8	0.6	
	$^1P_1$	−0.39	−1.05	−1.56	8.8	3.4		8.8	9.9	
	$^3F_2$	−0.88	−0.81	−0.91	11.8	12.4	26.3	0.8	0.8	18.5
	$^3P_1$	−0.19	−0.18		8.5	12.9		2.4	1.6	
	$^3P_0$	−0.55	−0.15		4.0	4.0		0.0	0.0	
	$^1F_3$	0.00	0.00	0.00	23.8	11.2	64.8	23.8	26.3	25.9
	$^3P_2$	0.18	0.28		2.0	2.5		12.1	10.7	
	$^3F_3$	0.35	0.91	1.43	9.6	23.9	8.9	5.8	0.4	9.8
	$^3D_2$	0.74	0.96		12.0	9.3		6.1	7.5	
	$^3F_4$	1.99	2.39		0.5	0.5		15.4	15.4	
	$^3D_3$	1.94	2.57	2.08	2.0	0.4	0.0	18.4	21.3	45.9
	$^1D_2$	1.36	2.57		5.5	7.1		5.5	5.6	

Braced (grouped) theory totals:

- O_1 — M_4 Theory^d: {$^1D_2$,$^3D_1$}=69.3, {$^3D_2$,$^3D_3$}=30.7; M_4 Theory^e: 95.2, 4.8; M_5 Theory^d: 40.9, 59.1; M_5 Theory^e: 8.3, 91.7.
- $O_{2,3}$ — M_4 Theory^d: {rows $^3D_1$–$^3F_2$}=32.0, {$^3P_1$–$^3P_2$}=38.3, {$^3F_3$,$^3D_2$}=21.6, {$^3F_4$–$^1D_2$}=8.0; M_4 Theory^e: 28.2, 30.6, 33.2, 8.0; M_5 Theory^d: 10.4, 38.3, 11.9, 39.3; M_5 Theory^e: 11.3, 38.6, 7.9, 42.3.

^{a} Relative to $^1D_2$ line for $M_{4,5}N_{4,5}O_1$ transitions and to $^1F_3$ line for the $M_{4,5}N_{4,5}O_{2,3}$ transitions.
^{b} L-S coupling plus diagonal contributions from the spin-orbit interaction.
^{c} Intermediate coupling.
^{d} L-S coupling for final state.
^{e} Intermediate coupling for final state.

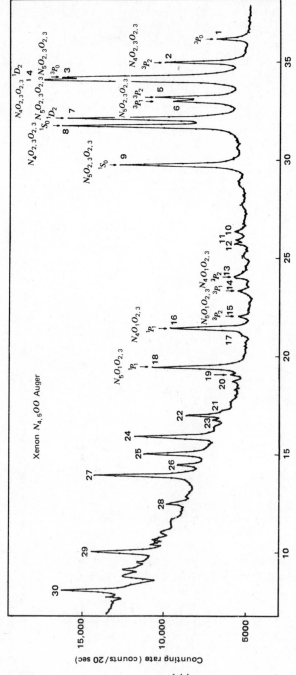

Figure 4.12. Xenon $N_{4,5}OO$ Auger spectrum.[31] (Reprinted with permission of The Royal Swedish Academy of Sciences.)

the Xe $N_{4,5}OO$ and the Kr $M_{4,5}NN$ spectra, which is to be expected since they both originate with d-subshell vacancies and have final states in the outermost orbitals of atoms having the same group state configuration. As for Kr, the difficulty is assigning the lines corresponding to the final, state configuration s^0p^6. The transitions with final states s^2p^4 and sp^5 are identified from spectroscopic data. Many satellite lines are observed in the spectrum separated by the spin-orbit coupling splitting of the $4d$ shell (approximately 1.94 eV), and the picture is further confused by the occurrence of peaks (and dips) due to autoionization. The assignment of the spectrum is given in Table 4.17. The assignment of the lines 24 and 27 to the $5s^05p^6$ $^1S_0$ lines is tentative. Lines 22 and 25 are interpreted as shake-up lines to final states $5s^25p^3(^2P)7s:^1P_1$. It is interesting to note that in the assignment the features of the N_4 and N_5 spectra are not the same, in particular, line 6 is assigned to the $^3P_1$ final state of the N_5 transitions. This transition is forbidden under any coupling scheme.

2.5.4. $M_{4,5}N_{4,5}N_{4,5}$ Spectrum of Solid Xe

In an attempt to understand solid-state broadening processes the Auger electron spectrum of solid Xe was recorded by Nuttall and Gallon.[36] The

Table 4.17. The Xenon $N_{4,5}OO$ Auger Electron Spectrum

Line Number	Relative Energy		Relative Intensity		Assignment			
	N_4	N_5	N_4	N_5	N_4		N_5	
1	0.00		15		$O_{2,3}O_{2,3}$	$^3P_0$		
2	1.21		43		$O_{2,3}O_{2,3}$	$^3P_2$		
3		0.0		86			$O_{2,3}O_{2,3}$	$^3P_0$
4	2.13		104		$O_{2,3}O_{2,3}$	$^1D_2$		
5		1.00		48			$O_{2,3}O_{2,3}$	$^3P_2$
6		1.22		39			$O_{2,3}O_{2,3}$	$^3P_1$
7		2.12		97			$O_{2,3}O_{2,3}$	$^1D_2$
8	4.49		100		$O_{2,3}O_{2,3}$	$^1S_0$		
9		4.48		73			$O_{2,3}O_{2,3}$	$^1S_0$
13	12.17		11		$O_1O_{2,3}$	$^3P_2$		
14	12.86		10		$O_1O_{2,3}$	$^3P_1$		
15		12.17		7			$O_1O_{2,3}$	$^3P_2$
16	14.78		62		$O_1O_{2,3}$	$^1P_1$		
18		14.76		80			$O_1O_{2,3}$	$^1P_1$
24	20.28		90		O_1O_1	$^1S_0$		
27		20.26		134			O_1O_1	$^1S_0$

instrument used 2-keV electron-impact ionization and was operated with a background pressure of 10^{-8} torr with the xenon condensed onto the base of the cryostat at 10 K. The resolution was 0.083%. The spectrum obtained is shown in Fig. 4.13, and a comparison with gas-phase spectra readily shows that considerable broadening has taken place. The authors interpret peak a as the $M_4N_{4,5}N_{4,5}$ peak consisting mainly of unresolved ^1D and ^1G peaks. Because the other structure is less intense and mainly on the high-energy side, the authors deduce a value for the halfwidth of the peak, by extrapolating the low-energy side of the peak, as 4 eV. The corresponding gas-phase value is just over 1 eV. Peak b is interpreted as the $M_5N_{4,5}N_{4,5}$ transitions and peak c as the broadened $M_5N_{4,5}N_{4,5}$ ^1S peak. The separation between peaks a and b is approximately 12 eV, which agrees with the value for the spin-orbit splitting of the $3d$ subshell. The authors conclude that phonon broadening produced by fluctuations in atomic potential caused by vibration of the lattice[90] is the most probable process.

The authors also reported the spectrum of solid Ar $(L_{2,3}MM)$ and Kr $(M_{4,5}NN)$,[37] and by simulation of the spectra based on the gas-phase data with considerably increased line widths were able to assign the peaks observed in the solid spectrum and to conclude that the transitions were quasiatomic. Line broadening in going from the gaseous state to the solid state will be discussed further in the next section.

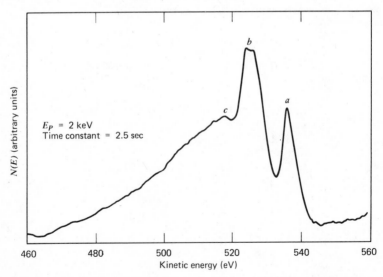

Figure 4.13. $M_{4,5}N_{4,5}N_{4,5}$ Auger spectrum of solid Xe at 10 K.[36] (Reprinted with permission of Pergamon Press, New York.)

3. FREE-METAL ATOMS

In the preceding section, studies of the Auger electron spectra of the rare
were considered in some detail, since it is the spectra of these elements that
are used as the test of the various theories of the Auger process. However,
correct prediction of Auger transition energies and intensities for rare-gas
atoms is only part of the picture, and the theory must be able to deal with the
solid state as well, in which chemical effects (particularly extra-atomic
relaxation) may be important.

The problem that started the study of the spectra of free-metal atoms was
the fact that the intensities of the KL_1L_1, $KL_1L_{2,3}$, and $KL_{2,3}L_{2,3}$ groups
taken relative to either the KL_1L_1 group or the total KLL intensity showed a
rapid change on going from neon to sodium or magnesium. The magnitude
the experimental values and their dependence on Z were both in disagreement
with the theory. As the experimental evidence available at that time (1974)
had been obtained from gaseous neon and solid Na_2O and MgO targets, it
was suggested that solid-state and chemical effects might be responsible
for the sudden change in group intensities (particularly as the intensities of
the F KLL transitions had been shown to have been influenced by chemical
effects in LiF, NaF, and MgF_2[91]) and accordingly the K Auger spectrum
of the free sodium atom was obtained.[92] This was followed rapidly by
studies of magnesium,[93] zinc,[94,95] and cadmium.[96]

It was found that not only were the relative line group intensities different
for the metal in the vapor phase, but also there was also a considerable shift
(of the order of 10 eV) in the energies of the lines and, furthermore, that
some species in the gas phase give rise to extremely sharp lines and considerable fine structure in their Auger spectrum which is absent in the spectrum
of the solid even at high resolution. In recent years, a number of studies of
free-metal species have been reported with a view to obtaining information
and gaining understanding of extra-atomic relaxation and line-broadening
processes. In other words, the Auger spectra of the free species have been
used as the reference spectra for the studies of solid-state effects that are
manifested in the Auger spectrum. Spectra that involve valence-band
electrons are also of interest because it is not entirely clear under what
conditions Auger spectra of solids should be interpreted in terms of quasiatomic transitions and under what conditions they should be interpreted
as bandlike transitions.

Studies of free-metal spectra have also been used for the determination
of core-level binding energies. Prior to the availability of the results of these
studies, most tabulations of binding energies (for example, Reference 97)
referred to solids, measured relative to the Fermi level. Although there had
been attempts to evaluate core-level binding energies of free atoms by com-

bining the optical data for valence orbitals with x-ray data for core levels in metals,[98] these values have been partly in error because it is assumed that x-ray energies do not change in going from the solid to the gaseous state. ESCA has also been used to study binding energies, and both ESCA and Auger have been used to study correlation effects.[99] The chemical shifts that have been observed in both Auger and ESCA spectra have been considered in some detail and a parameter has been proposed on the basis of the difference in energy between an Auger line and a photoelectron line, which would be unique for each individual compound.[100]

A summary of the experimental work published concerning the Auger spectra of free-metal atoms is shown in Table 4.18. As can be seen from the table the two most active research groups in this area are that at Freiburg in West Germany and that at Oulu in Finland, with considerable collaboration between the two groups. Other contributions have come from Ghent in Belgium, Aarhus in Denmark, and Munster in Germany (presumably prior to Mehlhorn's and Schmitz's move to Freiburg).

It is not intended to discuss the spectra of each metal species in as much detail as was done for the rare gases; a few examples of low, medium, and high Z-numbers will be discussed.

In general, the literature consists of reports of studies of just one metal, with few reports of several metals from the point of view of discussing solid-state effects for a number of metals. There are also few reports in which the spectrum of the solid and the vapor phases have been obtained with the same instrument, although there is one report in which they have been obtained simulatneously.[106]

3.1.   Sodium ($Z = 11$, $1s^2 2s^2 2p^6 3s^1$)

The KLL spectrum of sodium was one of the first free-metal-atom species to be studied.[92] The same group of workers reported on the L-Auger and autoionizing spectra three years later in 1977.[102] In the study of the KLL spectrum the sodium vapor was produced from an oven maintained at 547°C, at which temperature the sodium vapor pressure is 8.5 torr and the fraction of Na_2 molecules is about 5%. The target-atom beam was produced from a stainless-steel nozzle, at the end of the oven, kept at 627°C. By measuring the intensity of the elastically scattered primary electrons against an argon target of known pressure, the pressure in the sodium beam was found to be about 10^{-3} torr. The primary electron beam of 3.5 keV and 500 μA crossed the sodium beam about 3 mm from the nozzle. The secondary electrons were analyzed in a 127°C electrostatic cylindrical analyzer with an energy resolution of 0.16%.

Table 4.18. Auger Electron Spectra of Free-Metal Atoms

Metal	Group	Subject Studied	Year	Reference
Be, B, C	Aarhus	Measurement of Free-atom K-shell binding energies	1978	101
Na	Munster	K Auger spectrum	1974	92
	Freiburg	L Auger and autoionizing spectra	1977	102
Mg	Freiburg	K Auger spectrum	1974	93
	Freiburg	L Auger, Coster–Kronig, and satellites	1976	103
	Ghent	KLL relaxation processes (no experiments on free metal)	1977	104
	Oulu	KLL solid-state effects	1977	105
Zn	Freiburg	$L_{2,3}MM$ spectrum	1974	94
	Oulu	$L_{2,3}M_{4,5}M_{4,5}$ spectrum	1974	95
	Oulu	$L_{2,3}M_{4,5}M_{4,5}$ spectrum	1977	105
	Oulu	Direct measurement of vapor–metal shifts (also XPS spectrum)	1979	106
Cd	Freiburg/ Oulu	$M_{4,5}N_{4,5}N_{4,5}$ and $M_{4,5}N_{4,5}O_1$ spectra	1974	96
	York	$M_{4,5}N_{4,5}$ spectrum at liquid-nitrogen temperature, comparison with spectrum simulated from results of Reference 96	1975	37
	Oulu	Solid-state effects in the $M_{4,5}N_{4,5}N_{4,5}$ spectrum	1977	105, 107
	Oulu	Vapor–metal shifts and XPS spectrum	1979	106
Hg	Oregon	$N_{6,7}O_{4,5}O_{4,5}$ spectrum	1977	108
Various	Freiburg	Core-level energies (Li, Na, Mg, K, Ca, Zn, Sr, Ba); also XPS studies (Mg, Zn, Sr, Cd)	1977	109
	Orsay	Study of ionization efficiency curve of multicharged ions (Sr, Ba, Mn, Ag, In, Ag)	1971	110
	Houston	Vapor–solid shifts for AES and XPS, proposal of Auger parameter (no experimental)	1975	100
Molecular Species	Tampere/ Oulu	Solid-state effects (molecular spectra of $InCl_3$, $SnCl_2$, Sb_4, and Te_2 taken as "free-"metal spectra)	1979	111
	Oulu	Comparison with theory for Sb_4 and Te_2	1979	112
	Oulu/ Oregon	I_2 and Xe comparison with theory; completes run from $Z=$ 48 to 54	1979	39

The spectrum obtained from averaging 216 runs is shown in Fig. 4.14 after smoothing and subtraction of the background (assumed to vary linearly with energy). The energy scale was calibrated against the ^1D transition of neon. The spectrum shows the normal Auger lines (numbers without primes), the M_1 shake-off satellite lines (single-primed numbers), and the M_1 shake-up ($3s \rightarrow 4s$) satellites (double-primed numbers). Line A is assigned to the double Auger transition $1s2s^2 2p^6 3s \rightarrow 1s^2 2s^2 2p^4 4s$ (one electron is ejected and the other excited). The assignment of the lines is given in Table 4.19. In designating the initial and final states the authors appear to ignore the ejected Auger electron, as the initial state $1s2s^2 2p^6 3s$ is designated as $^{1,3}S$ and the final states of the normal lines are designated $(^1S)^2S$; $(^1P)^2P$, $(^3P)^{2,4}P$; $(^1S)^2S$, $(^1D)^2D$, $(^3P)^{2,4}P$. Yet the selection rules $\Delta S = \Delta L = \Delta J = 0$, $\pi_i = \pi_f$ are still quoted, apparently making most of these transitions forbidden.

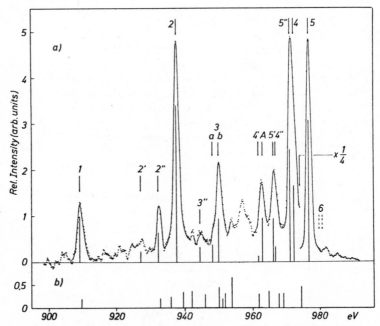

Figure 4.14. K Auger spectrum of the free sodium excited by electrons ($E_p = 3.5$ keV, $I_p = 500$ μA, $p_{Na} = 9.10^{-4}$ torr, energy resolution of spectrometer 0.16%). The spectrum has been smoothed by five-point averaging; the background has been substracted. Unprimed numbers mark KLL lines, single primed numbers mark $KM_1 - LLM_1$ lines, and double primed numbers characterize $KM_1^* - LLM_1^*$ transitions, where M_1^* stands for a $3s \rightarrow 4s$ excitation. Line A is due to the double Auger transition $1s2s^2 2p^6 3s^{1,3}S \rightarrow 1s^2 2s^2 2p^4 (^1D)4s^2 D$. The relative intensities of these lines as well as of some $KL - LLL$ satellite lines are given as bar diagram. The solid curve shows the sum of intensities of the lines used in resolving the spectrum graphically.[92] (Reprinted with permission of Springer-Verlag, New York.)

Table 4.19. KLL Auger Spectrum and Satellites of Sodium Vapor

Line No.	Initial Configuration	Final Configuration		Energy (eV)		Intensity[b]		
				Vapor	Solid[a]	Vapor	Solid[a]	Theory[c]
1	$Ne^{-1}3s$	$1s^2 2s^0 2p^6 3s$	^2S	909.1	922.8	1.0	1.0	1.0
2	$Ne^{-1}3s$	$1s^2 2s^1 2p^5 3s$	^2P	937.6	950.8	3.4	4.3	3.25
3a	$Ne^{-1}3s$	$1s^2 2s^1 2p^5 3s$	^2P	948.1		0.39		0.98
3b		$1s^2 2s^1 2p^5 3s$	^4P	949.9	963.4	0.93	1.5	
4	$Ne^{-1}3s$	$1s^2 2s^2 2p^4 3s$	^2S	972.2	985.6	1.65	1.7	1.08
5	$Ne^{-1}3s$	$1s^2 2s^2 2p^4 3s$	^2D	976.7	989.8	12.3	16.6	9.4
2'	Ne^{-1}	$1s^2 2s^1 2p^5$	^1P	927.0		0.26		
3'	Ne^{-1}	$1s^2 2s^1 2p^5$	^3P					
4'	Ne^{-1}	$1s^2 2s^2 2p^4$	^1S	961.6		0.13		
5'	Ne^{-1}	$1s^2 2s^2 2p^4$	^1D	966.1		0.94		
2"	$Ne^{-1}4s$	$1s^2 2s^2 2p^5 4s$	^2P	932.4		0.64		
3"	$Ne^{-1}4s$	$1s^2 2s^2 2p^5 4s$	^4P	944.0		0.32		
4"	$Ne^{-1}4s$	$1s^2 2s^2 2p^4 4s$	^2S	966.7		0.33		
5"	$Ne^{-1}4s$	$1s^2 2s^2 2p^4 4s$	^2D	971.1		2.4		
A	$Ne^{-1}3s$	$1s^2 2s^2 2p^4 4s$	^2D	962.8		0.96		

[a] Reference 113, Na_2O.
[b] Relative to line 1.
[c] Reference 68.

151

To avoid the use of the *KLM* notation, which is not applicable for $Z=11$, the electron configurations of the initial and final state are given in the table (with the abbreviation Ne^{-1} representing $1s2s^22p^6$) together with the final-state designations for L-S coupling including the single outer $3s$ or $4s$ electron as appropriate. As can be seen from the table, the relative intensities found for the free atom are in better agreement with the theory (calculated with numerical Hartree–Fock–Slater wave functions for the bound and continuum states and corrected to allow for configuration interaction between the ^2S final states, that is, lines 1 and 4), than those found for sodium in Na_2O.[113] Hillig et al. suggest that the remaining discrepancy (10–20%) may be attributed partly to experimental errors and partly due to incomplete consideration of the correlation effects in the theoretical treatment. However, the intensity of the transitions to the $^{2,4}P$ final states, which should not undergo configuration interaction, relative to the total *KLL* intensity is still in error by about 12%. There is no doubt that a solid state and/or chemical effect acts on the relative intensities.

The absolute energies of the lines of the free atom are 13.1 eV smaller than the corresponding lines for the Na_2O referred to the Fermi level, indicative of a large chemical and solid-state effect. The shake-off satellite lines are 10.6 eV lower than the parent. From the known energy of the $1s^22s^22p^43s$ ^2D configuration of sodium (from optical data[70]) of 101.9 eV and the experimental value to the Auger transition of this configuration of 976.7 eV, the K binding energy of the free sodium atom is calculated to be 1078.6 ± 3 eV. This may be compared with 1072.1 ± 0.5 eV measured on Na_2O relative to the Fermi level by x-ray photoelectron spectroscopy[114] and with 1079 eV calculated with relativistic Hartree–Fock–Slater wave functions.[14]

The total shake probability (shake up + shake off) for a $3s$ electron as a fraction of the $1s$ ionization probability was found experimentally to be 0.188 ± 0.02 in reasonable agreement with the theoretical value of 0.197. No discussion of the change in line widths between the solid and vapor phase was given, although it was found that the experimental line widths were considerably less than the theoretical values. Although this work was the first study of a free-metal species, it is difficult to use the results as a basis for examining vapor–solid effects as there are no results available for solid sodium metal, and the results have been compared with sodium oxide.

In the later paper concerning the L-Auger and autoionizing spectrum the authors used a similar experimental arrangement though a modified oven was also used in which the primary electron beam was actually passed through the oven. Some of the spectra were obtained with a 150° spherical-plate spectrometer with resolution 0.12% as well as with the 127° cylindrical analyzer used for the *KLL* spectrum studies. A variety of primary-electron-beam energies were used (0.75–16 keV), and the spectrum measured over the energy range 18–75 eV. The assignment of the lines was done by using the results of the photoabsorption spectrum, and the ejected-electron spectra for excitation by methods other than electron impact and by comparison with theoretical values. The following processes were discussed:

1. $2s$ ionization followed by Coster–Kronig transitions.
2. $2s \rightarrow np$ excitation followed by autoionization.
3. $2p$ excitation leading to autoionizing states $2s^2 2p^5 3sns$ and $2s^2 2p^5 3snd$.
4. $2s$ ionization plus excitation.
5. Coster–Kronig transitions $2s2p^6 3s \rightarrow 2s^2 2p^5$.
6. Angular distribution.

In particular the Coster–Kronig transitions (5) above yield information on the energies, widths, relative excitation, and decay probabilities of the ^1S, ^3S initial states.

3.2. Magnesium ($Z = 12$, $1s^2 2s^2 2p^6 3s^2$)

The KLL spectrum and L-shell spectra have been studied by Breuckmann et al.,[93,103] in almost exactly the same way as their study of sodium. The magnesium atom has only closed shells and small spin-orbit coupling, thus the treatment of normal and satellite lines may be made within the L-S coupling scheme.

The Mg vapor was produced in a furnace kept at 310°C (corresponding to a target gas pressure of 5×10^{-5} torr) and the primary electron beam (3.8 keV, 1 mA) passed through the furnace. The secondary electrons were analyzed by the same analyzer as for sodium. The KLL spectrum is shown in Fig. 4.15, and it can be easily seen from a comparison with Fig. 4.14 that there is a considerable similarity between the two spectra. The energy difference between the two spectra reflecting the greater increase in $1s$ binding energy compared with the L-shell electrons on going from sodium to magnesium. The assignment of the lines is similar to that given in Table 4.19 for sodium. The unprimed numbers referring to the normal lines, the double primes to the $3s$ shake off satellites, and the single primes to the $3s \rightarrow 4s$ shake up satellites. Again the differences between the vapor-phase intensities and energies as compared with Mg in MgO show considerable chemical and/or solid-state effects with the experimental intensities of the vapor phase showing better agreement with the theoretical values. As for sodium, the satellite spectrum shows that shake up is more probable than shake off.

The solid-state effects in going from the vapor to the solid are discussed in detail by Hoogewijs et al.[104] They criticize Breuckmann and co-worker's

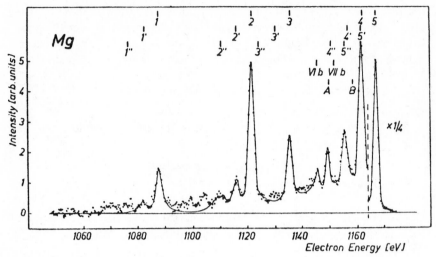

Figure 4.15. K Auger spectrum of the free Mg atom.[93] (Reprinted with permission of Springer-Verlag, New York.)

theoretical approach on the grounds that the use of the nonrelativistic Hartree–Fock method to calculate the spectrum is not very accurate because of neglect of quantum-electrodyamical effects, and point out that the close agreement between theory and experiment obtained by Breuckmann et al. is due to an accidental cancellation of errors.[115] In Hoogewij's approach a term $\Delta E_R^a(L'\bar{L})$ is introduced into the Auger energy equation to represent the variation of the "dynamic relaxation energy" accompanying the emission of the L' electron due to the presence of a spectator hole $\bar{L}$. The overall equation is based on that due by Asaad and Burhop[5] and Shirley[116] considering the Auger emission as a two-step process, namely,

$$E_{kin}^a = E_B^a(K) - E_B^a(L) - E_B^a(L') - F(LL') + R^a(LL') + \Delta E_R^a(L'\bar{L})$$

where E_B^a terms represent atomic binding energies, F describes the coupling energy of the two holes in the final state, and $R^a(LL')$ is the so-called "static relaxation energy" describing the lowering of the L' electron binding energy with respect to the frozen orbital value, owing to the relaxation which takes place during the creation of the $\bar{L}$ hole.

For the case of the solid metal, the Friedel model[117] is used to account for the *extra*-atomic relaxation processes. (The model was originally introduced to calculate the electron redistribution in the vicinity of a positively charged impurity in a monovalent metal.) It is assumed that the extra-atomic screening of a core hole is done through a semilocalized state consisting of an electron–hole pair, created by the capture of a conduction-band electron in an empty level which, owing to the presence of the core hole, has dropped below the Fermi level. In metals, high fields are screened over short distances, so the screening orbital is the first unoccupied orbital, in this case the $3p$. In the case of an Auger transition the screening of the final double-hole state is accomplished by two conduction-band electrons occupying the first two unoccupied atomic orbitals. The application of this model is discussed further by Hoogewijs et al. in a later publication.[118]

Application of these two models gave energies corresponding to the experimental values of the KLL energies to within 1 eV for the vapor-phase free metal and to within 3 eV for the solid metal. In addition a value of the static extra-atomic relaxation energy of 10 eV was calculated (the theoretical value is 11.1 eV), and it was noted that the calculated total double-hole extra-atomic relaxation energy is about four times the corresponding single-hole value, in agreement with Mott and Gurney's model[119] which predicts that the extra-atomic relaxation energy should vary as the square of the excess charge considered.

Väyrynen et al.[105] present a similar argument in deducing that the solid–vapor shift for the ^1D line in the KLL spectrum of magnesium metal is 15.5 eV. They found experimentally a value of 16.0 eV. The spectra were recorded on the same spectrometer with a resolution of approximately 0.05%. They also discuss broadening of the line components (the other important solid-state effect), pointing out that the spectrometer broadening function should be known accurately and also different backgrounds for the vapor and the solid cause further complications; for example, the solid spectrum may have peaks broadened by overlap with plasmon loss peaks. A value of 0.20 eV is found for the increase in peak width on going from the vapor to the solid.

To a first approximation the widths of the Auger lines would be expected to include the width of the initial state and a convolution of the final states multiplied by the transition probability, and thus if the transition probability remains constant throughout the final-state band, the broadening of the Auger lines should be about twice the width of the final-state band. The broadening observed for Mg is less than that for Zn (LMM) and Cd (MNN) as the $2s$ and $2p$ levels of Mg are energetically deeper in the core than the $3d$ or $4d$ levels of Zn and Cd, and thus the Mg spectrum is more atomlike than the Zn or Cd spectra. Despite the apparent higher resolution of their spectrometer, Vayrynen and co-worker's spectrum of Mg vapor does not show nearly as much satellite structure as shown in Fig. 4.14 (due to Breuckmann and Schmidt).

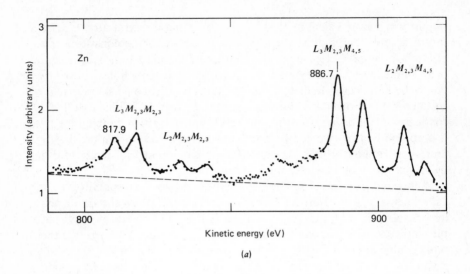

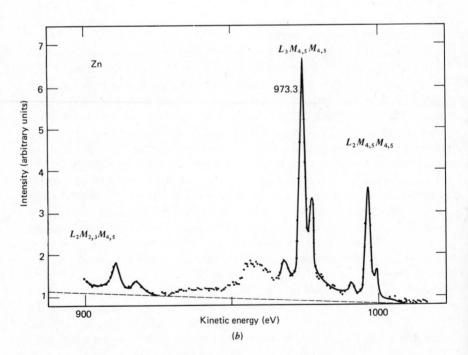

Figure 4.16. The $L_{2,3}MM$ Auger electron spectrum from zinc vapor excited by 3 keV electrons.[94] (Reprinted with permission of North-Holland Publishing Company, Amsterdam.)

3.3. Zinc ($Z = 30$, $1s^2 2s^2 2p^6 3s^2 3p^6 3d^{10} 4s^2$)

Aksela and co-workers have measured the $L_{2,3}MM$ spectra of the free zinc atom[94,95] and the spectra of vapor phase and solid zinc with the same spectrometer.[105,106] The LMM spectrum of the free atom from 800 to 1,000 eV is shown in Fig. 4.16. Three groups of lines can be seen, namely, $L_{2,3}M_{2,3}M_{2,3}$, $L_{2,3}M_{2,3}M_{4,5}$, and $L_{2,3}M_{4,5}M_{4,5}$, and each group is clearly split by the L_2, L_3 subshell splitting. The spectrometer used was a cylindrical $127°$ electrostatic analyzer with a resolution of 0.14%. The pressure of Zn in the target region was 5×10^{-4} torr, and the primary electron beam was 300 μA at 3 keV. The solid–vapor shift was found to be about 18 eV by comparison with previous spectra of the zinc metal, considerably in excess of the value calculated for extra-atomic relaxation in the solid. The $L_{2,3}M_{4,5}M_{4,5}$ spectrum was examined in more detail in the later publication,[95] where the spectrometer resolution was 0.05%. Good agreement between experimental and calculated results were found for calculations based on a mixed-coupling scheme (j-j coupling in the initial states and intermediate coupling with configuration interaction in the final states). The line-width values were found to be about the same for both the L_2 and L_3 groups, namely, $0.5 \pm .15$ eV (somewhat smaller than McGuire's theoretical value of 0.76 eV[76]); this was not the case for photoelectron measurements on the solid,[120] which may mean that the $L_2 L_3 M_{4,5}$ Coster–Kronig transition is possible in the metal because of the extra-atomic relaxation, whereas it is not possible in the free atoms.

When the $L_{2,3}M_{4,5}M_{4,5}$ spectra were obtained with the same instrument,[105] the vapor–solid shift was found to be 13.7 eV, showing the previously reported value of 18 eV to be considerably in error and highlighting the difficulty in comparing results obtained with different instruments in which different reference levels have been used. Also, of course, vapor spectra will have been calibrated against the vacuum level and solid spectra against the Fermi level. An increase in line width of 0.45 eV was observed in going from the vapor to the solid, and it is also noted that the intensity ratio of the L_3 group to the L_2 group increases from 2.3:1 to 2.8:1 on going from the vapor to the solid as a result of the $L_2 L_3 M_{4,5}$ Coster–Kronig process, which is energetically possible for the solid but not for the vapor. This process increases the number of L_3 vacancies in the solid state over those created by electron impact.

The spectra of the vapor and solid have also been recorded simultaneously,[106] giving the definitive value of 13.10 eV as the vapor–solid shift. The experiment involved observing (a) the spectrum from the surface of a thin, partially cooled rod located inside the oven on to which the vapor was condensing and (b) the spectrum from the surrounding vapor. The experimental setup was also used for the recording of ESCA (x-ray photo-electron) spectra. The photoelectron and Auger energy shifts may have several sources: extra-atomic relaxation of the conduction electrons associated with the formation of core holes, changes in electron configuration, initial-state chemical shifts of neutral species due to re-arrangement of valence electrons, and possible electrostatic potential differences between the solid and vapor phases. For zinc there should be no configuration changes and chemical shifts will be small, so that if the difference between the Auger and photoelectron shifts are considered and the effects of different static potentials, surface dipole, charging potentials, and initial-state binding-energy shifts are cancelled, the changes determined in the Auger parameter[100] (the difference between an Auger line and a photoelectron line) should be due entirely to extra-atomic relaxation effects. A theoretical value of 10.5 eV was calculated for $\Delta E_{Auger} - \Delta E_{photo}$, compared with an experimental value of 9.9 eV, showing the change in Auger parameter to be a useful experimental parameter against which to test further developments of extra-atomic relaxation theories.

3.4.   Cadmium ($Z = 48$, $1s^2 2s^2 2p^6 3s^2 3p^6 3d^{10} 4s^2 4p^6 4d^{10} 5s^2$)

As with zinc, cadmium has a number of properties that make its study, from the point of view of comparing the results with theoretical predictions, advantageous. The element has closed subshells (this means that in the designation of the initial and final states only the interaction of the core hole and the continuum vacancy—to be filled by the Auger electron—and between the two final-state holes need to be considered) and has a monatomic vapor (the interpretation of the spectrum is not complicated by having to allow for molecular relaxation effects), and the target vapor is easily produced experimentally.

The $M_{4,5}N_{4,5}N_{4,5}$ and the $M_{4,5}N_{5,4}O_1$ spectra have been obtained by Aksela and Aksela.[96] The purpose of this study was to match the theoretically calculated spectra with the experimentally obtained ones. The method used for the theoretical calculations is the same as for xenon (see Hagmann et al.[38]). based on intermediate coupling and configuration interaction (between final-state configurations $4s^{-2}$, $4p^{-2}$, $4s^{-1}4d^{-1}$, $4d^{-2}$, $5s^{-2}$, $4s^{-1}5s^{-1}$, and $4d^{-1}5s^{-1}$).

The experimental spectra were obtained with a $127°$ cylindrical electro-static analyzer with a resolution of 0.13%. The pressure of cadmium vapor in the target region of the oven was approximately 5×10^{-4} torr. The primary electron beam was 400 μA and 2 keV.

The results for the $M_{4,5}N_{4,5}N_{4,5}$ spectrum are not presented here because they are very similar to those obtained for xenon. The effects of configuration interaction were found to be small and good agreement between theory and experiment obtained. The M_4 and M_5 binding energies were calculated, from the measured Auger kinetic energies and from optically known energies of the final states, to be $418.7 \pm .5$ and $411.9 \pm .5$ eV, respectively. These values were about 8 eV different from those calculated for solid cadmium.[50] The intensity ratio of the M_4 to M_5 groups was measured as 0.71, somewhat greater than the statistically expected value of 0.67 (on the basis of the ratio of the values of $2J + 1$), however, this ratio is sensitive to the way in which the background has been subtracted, which in this case was considered to be a continuum built up of a satellite transitions following Coster–Kronig and shake-off processes. The probability of an $M_{1,2,3}$ ionization is 30% of the $M_{4,5}$ ionization and 80% of the $M_{1,2,3}$ vacancies are filled by Coster–Kronig transitions giving rise to the $M_{4,5}X - N_{4,5}N_{4,5}X$ ($X = N_i$, O_1) satellite lines with an overall intensity of about 24% relative to the main diagram lines.

The $M_{4,5}N_{4,5}O_1$ spectrum is shown in Fig. 4.17 separated into its components. The agreement between theory and experiment may be seen from Table 4.20. The determination of the relative intensities of the $^1D_2$, $^3D_1$ and $^3D_2$, $^3D_3$ lines of the spectrum is inaccurate because of the very small energy separation. The separation into the different components was made assuming a convolution of the experimentally determined spectrometer line-broadening function and a Lorentz profile as the natural Auger line shape. The width of the Lorentzian was varied to give the best fit to the

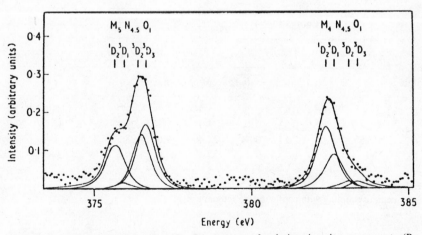

Figure 4.17. Separation of the $M_{4,5}N_{4,5}O_1$ spectrum of cadmium into its components. (Reprinted with permission of The Royal Swedish Academy of Sciences.)

low-energy side of the $^1G_4$ components of $M_{4,5}N_{4,5}N_{4,5}$ spectrum (which is not overlapped by any other diagram line). The spectrometer width was 0.48 eV [full width at half-maximum (FWHM)] and the Lorentzian was determined to be 0.35 eV.

As Auger electrons excited by a directed primary-electron beam may have a nonisotopic angular distribution if the total angular momentum of the initial vacancy is $j > 1/2$,[121,122] the angular distribution was investigated at 90° and 30° but were found to give the same relative intensities for the components of the $M_{4,5}N_{4,5}N_{4,5}$ groups.

The vapor–solid shift and broadening effects for cadmium have also been studied by Aksela et al. As for zinc, the spectra of the solid and vapor have been obtained simultaneously,[106] giving a shift energy of 11.8 eV. In an earlier study,[105,107] the spectra were obtained sequentially with the same instrument giving an energy shift of 12.2 ± 0.5 eV. The results of this study for the $M_{4,5}N_{4,5}N_{4,5}$ line groups are shown in Fig. 4.18, which very clearly illustrates both the energy shift and line broadening on going from the vapor to the solid. The background has been subtracted for both of these spectra, using the already-described method for the vapor, but for the solid it was assumed that the background was the sum of a constant or slowly decreasing (on the low-energy side) part and a part from plasmon loss peaks. The

Table 4.20. $M_{4,5}N_{4,5}O_1$ **Spectrum of Cadmium**

							Relative Intensities (%)			
	Energies (eV)						Theory			
	Experiment		Theory ($M_{4,5}$)		Experiment		M_4		M_5	
Final State	M_4	M_5	a	b	M_4	M_5	a	b	a	b
$^3D_3$	1.00^c	1.00^d	0.87	1.08	5.7^e	38.2^g	2.2	2.0	45.0	43.9
$^3D_2$	0.72^c	0.75^d	0.43	0.84	5.7^e	32.7^g	29.1	5.5	13.8	37.7
$^3D_1$	0.25^c	0.32^c	0.17	0.34	31.4^f	3.6^e	28.5	27.8	0.9	0.8
$^1D_2$	0^d	0^c	0	0	57.1^f	25.5^f	40.3	64.7	40.3	17.5

[a]L-S coupling.
[b]Intermediate coupling with configuration interaction.
[c]± 0.14.
[d]± 0.07.
[e]± 2.0.
[f]± 5.0.
[g]± 4.0.

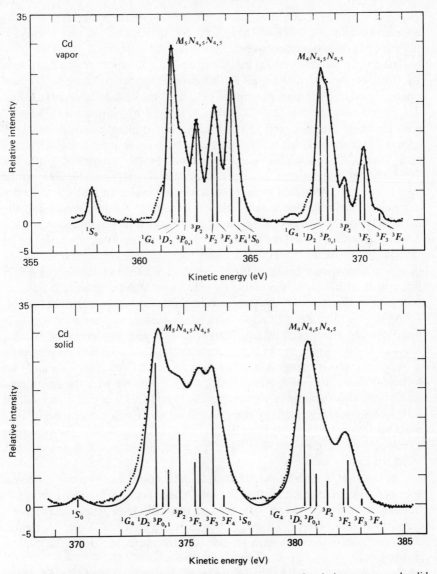

Figure 4.18.   Separation of the $M_{4,5}N_{4,5}N_{4,5}$ Auger spectra of cadmium vapor and solid into its components.[105] (Reprinted with permission of The Royal Swedish Academy of Sciences.)

spectrum was decomposed into the components by the method described earlier using the $^1S_0$ of the $L_2M_{2,3}M_{2,3}$ spectrum of argon as a standard line shape. (This is well separated from other lines so that its measured profile can be regarded as a convolution product of the instrument function and the natural Auger width.) The instrument function was found to be 0.11 ± 0.02

eV, and assuming a constant relative resolution in the range 200–400 eV, a natural width of 0.32 (FWHM) ± 0.05 and 0.79 ± 0.08 eV were found for the vapor and solid, respectively.

The line broadening arises mainly from the differences in the final-state widths. For the vapor the final-state lifetime broadening is negligible and the observed line widths are essentially those of the initial M_4 and M_5 states. The broadening process has also been considered by Nuttall and Gallon,[37] who recorded the spectrum of the solid at liquid-nitrogen temperature. A computer simulation using the energies and relative intensities of Aksela and Aksela's[96] gas-phase work produced good agreement with the experimental spectrum if a FWHM of the Auger line is taken to be 1.03 ± 0.05 eV convoluted with the spectrometer value of 0.31 eV. A better agreement was obtained if the included allowance for loss peaks and the asymmetry of the Auger peaks (Lorentzian on the high-energy side and with a non-Lorentzian tail on the low-energy side). This may cast some doubt on the validity of Aksela's method of component separation mentioned earlier in which the low-energy side of the $^1G_4$ peaks was used to establish the Auger line width, although it is possible to see that complete agreement was not obtained in Fig. 4.17. Of the three mechanisms for line broadening (initial-state lifetime, final-state lifetime, and lattice-vibrational broadening), it is final-state lifetime broadening that is considered predominant in this instance. In the vapor the $4d$ holes are long-lived since further Auger decay is impossible (their energy is less than that required to produce a $4d$ and two $5s$ holes), but in the solid, Auger processes are possible with the ejection of an electron to a vacancy in the conduction band. This mode of decay will be fast. In addition the two-hole state may decay by Coulomb repulsion onto neighboring atoms[123] or by interatomic (or cross) transitions[124] or by lattice vibrations.[125] Phonon broadening is, however, not considered a significant effect for Cd.

The fact that the spectrum of the solid can be simulated from the gas-phase spectrum is further evidence that the solid-state spectrum is quasiatomic.[126] Similarities between Auger spectra and photoelectron spectra have been noted in this respect,[127] although it is expected that the Auger spectrum, with two holes in the final state, would show the more free-atomic character.

3.5.   Mercury ($Z = 80$, $1s^2 2s^2 2p^6 3s^2 3p^6 3d^{10} 4s^2 4p^6 4d^{10} 4f^{14} 5s^2 5p^6 5d^{10} 6s^2$)

The $N_{6,7} O_{4,5} O_{4,5}$ spectrum has been obtained with 2-keV electron impact as the primary ionization, and the results compared with theoretical calculations.[108] Since the Auger energies can be readily calculated from the known core ionization potentials for $4f$ electrons[128] and known energies of the $5d^8$ configurations in Hg^{2+},[70] no special effort was made to obtain absolute energy values. The spectrum is shown in Fig. 4.19, the background subtraction and deconvolution are as already described. The $^1S_0$ peak of the N_6 group overlaps the $^1G_4$ peak of the N_7 group (to within 0.01 eV) and so these are not resolved. The experimental and calculated energies and intensities are given in Table 4.21. The calculations were performed in a mixed-coupling scheme applying j-j coupling to the initial state and intermediate coupling to the final state. No account was taken of correlation effects as they have been shown to play only a minor role in the $M_{4,5}N_{4,5}N_{4,5}$ spectra of cadmium and xenon. Angular distribution effects may need to be taken into account as the angular momentum of the initial vacancy is greater than $1/2$. Since the only decay mechanism for the final state is relatively slow atom–atom collisions (further Auger and radiative transitions being either forbidden by energy considerations or selection rules, respectively), the

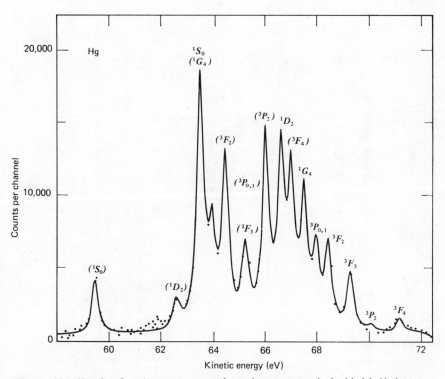

Figure 4.19. $N_{6,7}O_{4,5}O_{4,5}$ Auger spectrum of atomic mercury excited with 2-keV electrons. Labels without parentheses identify transitions originating in the N_6 shell; labels in parentheses identify those originating in the N_7 shell.[108] (Reprinted with permission of American Physical Society, New York.)

Table 4.21. $N_{6,7}O_{4,5}O_{4,5}$ Auger Spectrum of Mercury

State	Relative Energies (eV) Experiment N_6	Experiment N_7	opt[a]	Theory L-S[b]	Theory IC[c]	Relative Intensities (%) Experiment N_6	Experiment N_7	N_6 L-S	N_6 IC	N_7 L-S	N_7 IC
$^3F_4$	1.83	1.79	1.78	1.49	1.65	2.5	15.9	6.1	3.6	18.7	20.6
$^3P_2$	0.76	0.78	0.77	0.98	0.49	0.8	21.4	16.7	0.5	22.5	23.6
$^3F_3$	0.00	0.00	0.00	0.00	0.00	8.8	8.4	12.2	12.2	9.0	9.0
$^3F_2$	−0.84	−0.78	−0.84	−1.12	−1.05	12.9	18.8	12.7	15.1	3.4	13.4
$^3P_0$	−1.31	−1.30	−1.29	−2.10	−1.30	11.4	9.7	5.3	1.2	3.0	6.8
$^3P_1$			−1.30	−1.73	−1.73			14.0	14.0	10.5	10.5
$^1G_4$	−1.77	−1.75	−1.77	−1.95	−2.10	20.2	16.4[d]	13.0	15.5	13.0	11.1
$^1D_2$	−2.67	−2.67	−2.67	−1.14	−2.68	26.9	3.1	13.6	27.4	13.6	2.5
$^1S_0$	−5.80	−5.79	−5.79	−5.45	−6.25	16.7[d]	6.2	6.4	10.5	6.4	2.6

[a]Opt: values from Moore.[70]

[b]L-S: L-S coupling plus diagonal contributions from spin-orbit interaction.

[c]IC: intermediate coupling for final state.

[d]These peaks not resolved; observed intensity divided in proportion to the theoretical intensities.

widths of the Auger lines observed are essentially those of the N_6 and N_7 initial subshell vacancies. A value of 0.24 eV was found by the best fit of a Lorentzian profile to the $^1S_0$ peak of the N_7 group. This is considerably larger than the theoretical value of 0.13 eV calculated by McGuire.[129] It is somewhat surprising that, considering the experimental simplicity of obtaining mercury vapor and that the vapor is monatomic and Hg has closed subshells, the spectrum was not reported in the literature before 1977. This may be due in part to the fact that until McGuire[129] made available calculations on the radial integrals for the N-shell Auger transitions, it was not possible to calculate the intensities of the line components.

3.6. Other Free-Metal Atoms

Melhorn et al.[109] have measured core-level binding energies of Li, Na, Mg, K, Ca, Zn, Sr, and Ba by Auger spectroscopy and of Mg, Zn, Sr, and Cd by x-ray photoelectron spectroscopy. The metal-vapor pressure was about 10^{-3} torr and for the x-ray photoelectron experiments the open entrance to the oven was closed with a heated aluminium window to avoid contamination of the exit window of the x-ray tube. Since configuration-interaction satellites are observed in the $M_{4,5}NN$ Auger spectrum of krypton, the neighboring element Sr was examined to see if electron-correlation effects were also present. It was found that strong configuration-interaction effects were present in both the $3d$ spectrum of Sr and the $4d$ spectrum of Ba. The authors also discuss Coster–Kronig and super Coster–Kronig (all in the same shell) transitions. The width of the $4p$ level for $45 < Z < 70$ varying from 2 to 20 eV is mostly due to the super Coster–Kronig (sCK) transition $4p$-$4d4d$ (i.e., $N_{2,3}N_{4,5}N_{4,5}$), on the other hand, the sCK transition $3p$-$3d3d$ ($M_{2,3}$-$M_{4,5}M_{4,5}$) leads only to level widths of 1–5 eV. This is dramatically illustrated by the $3p$ and $4p$ photoelectron lines of Zn and Cd, respectively, for the solid metals. As the $3d$ state of the solid involved in the $3p$-$3d3d$ transition is better described by the $3d$ wavefunction of a neutral atom, rather than one with a vacancy in the $3p$ level, and as this effect would be absent for the vapor atom, it was of interest to find that after fitting Lorentzian line shapes to the individual components the best fit was for a level width of 2.1 ± 0.02 eV, in close agreement with the solid-state value.

Bisgard et al.[101] measured the K-shell binding energies of Be, B, and C by measuring the Auger spectra obtained when fast ion beams of the elements were excited by collision with CH_4 and He. The ion beams (Be^+, B^+, or C^+) were produced in a Nielson-type universal ion source[130] and accelerated to approximately 100 keV. The electrons ejected after collision were observed at a small angle ($\sim 6.4°$). An electrostatic parallel-plate analyzer was used with a resolution of approximately 0.25%. To correct for the Doppler shifts

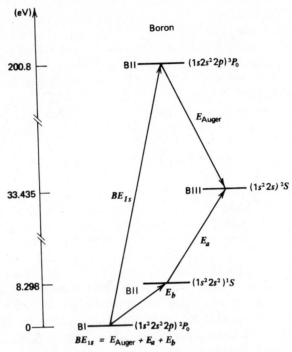

Figure 4.20.   Energy-level diagram for establishing the B K-shell binding energy.[101] (Reprinted with permission of The Royal Swedish Academy of Sciences.)

the electron energies were measured at forward and backward angles. The energy-level diagram for calculating the binding energy is shown in Fig. 4.20 for boron. The energies E_a and E_b are taken from the work of Bashkin and Stoner.[131] A value of 200.8 ± 0.5 eV was found. This compares favorably with the theoretical free-atom value of 201.1 eV[132] but is about 10 eV (as are the values for Be and C) higher than previously determined values where solid or molecular samples[11,50] have been used.

3.7. Free-Molecular metal Species

As part of a systematic investigation of the $M_{4,5}N_{4,5}N_{4,5}$ Auger spectra of the elements $Z = 47$ (Ag) to $Z = 54$ (Xe), Aksela and co-workers[39,111,112] have measured the spectra of $InCl_3$, $SnCl_2$, Sb_4, Te_2, and I_2. The solid spectra of the metals are all very similar to the $M_{4,5}N_{4,5}N_{4,5}$ spectrum of solid Cd shown in Fig. 4.18, and the spectra of the Sb$_4$, Te$_2$, and I$_2$ vapors are very similar to each other and bear a strong resemblance to the free Cd spectrum also shown in Fig. 4.18 (there is also a close resemblance to the Xe $M_{4,5}N_{4,5}N_{4,5}$ spectrum). The spectra of InCl$_3$ and SnCl$_2$ are not shown. As before, the line widths of the solid are broadened in the solids (by about 0.5–0.9 eV compared with the estimated widths of the $3d$ level), and the spectra are shifted toward higher energies by some 7–12 eV while the relative energies and intensities of the component lines remain approximately the same for the vapor- and solid-phase spectra of the same element. Molecular-broadening effects are also observed associated with the excitation of a number of vibrational states in the initial and final states of the Auger process. Since all the main features of the spectra of both the vapor and the solid can be interpreted as multiplet splitting of the two-hole final-state configuration, it appears that quasiatomic, rather than bandlike, descriptions are appropriate for the final states of these transitions. The extra-atomic relaxation energies are calculated by a simplified model in which screening electrons are assumed to occupy the first unfilled atomic orbitals. The $M_{4,5}N_{2,3}N_{4,5}$ and $M_{4,5}N_{4,5}O$ spectra of iodine are also presented[39] and discussed. The former are of particular interest because they involve a hole in the $4p$ ($N_{2,3}$) shell. For singly ionized I$_2$ and Xe these states cannot be accurately described by an independent-particle model because of considerable broadening due to Coster–Kronig transitions and the extensive configuration interaction with the $4d^8 4f$ configuration. It is found that the Xe $M_{4,5}N_{2,3}N_{4,5}$ Auger spectrum is described fairly well by the independent-particle model, whereas the I$_2$ spectrum is not. It is suggested that the reason for this may be that the presence of a $4d$ hole in the final state for Xe prevents the $4p$ hole from disappearing via super-Coster–Kronig transitions (the process responsible for the anomolous photoelectron spectrum), but that this is not so for I$_2$, and most of the $p^5 d^9$ states are broadened beyond recognition.

4. FREE MOLECULES

As was mentioned toward the end of the previous section on free-metal atoms, the Auger electron spectra of certain molecular species had been considered to be a good representation of the free-atom spectrum. The reason for this being that of the various subshells involved in the transitions none was involved in the chemical bonding so that the orbitals retained their essentially atomic character, the only effect considered was a broadening due to vibrational effects. Moreover, it was discovered that even when the metals were condensed in the solid phase, the spectral features could be accounted for in terms of an atomic-like theory (see, for example, the spectrum of solid cadmium simulated from gas-phase results), the energy shift being accounted for mainly by extra-atomic relaxation effects and the line broadening by the decreased lifetimes of the final states.

However, when the subshells taking part in the Auger transition are also involved in the molecular bonding, that is, are no longer atomic orbitals but molecular orbitals, then the interpretation of the Auger spectrum (and satellites) cannot be made on the basis of an atomic model and molecular-orbital theory must be introduced. Of course, there is considerable similarity between the basic processes involved for molecules and the processes for atoms, and transitions summarized in Fig 4.1 apply just as much to molecules as they do to atoms, although the notation will need to be changed since the molecular orbitals cannot be labeled as L, M, N, and so on (though the initial vacancy may well be in an essentially atomic orbital), hence the notation W (for weakly bound valence electron) and S (strongly bound valence electron) used by Moddeman et al.[6] Thus the interpretation of the Auger electron spectra of free molecules provides experimental results against which molecular-orbital theory can be tested. At a fundamental level, the Auger spectra provide convincing experimental evidence of the existance of molecular orbitals in much the same way as does photoelectron spectroscopy (both UV and x-ray), although as with PES, AES of free molecules arrived too late on the scene to do anything other than provide confirmatory evidence. This may in part be due to the time scale of the development of instrumentation. Because of the more-complex nature of the Auger (and satellite) processes, compared with the photoelectric effect, the interpretation of the Auger spectra of free molecules provides a more stringent test of the molecular-orbital theory.

Although the first molecular spectra were reported in 1960 (O_2, N_2, and CH_4),[10] it was not until the late 1960s and early 1970s that significant numbers of papers began to appear in the literature and, as can be seen from Tables 4.1, 4.18, and 4.22, the study of free molecular species parallels the study of the rare gases and slightly predates the study of free-metal atoms. Most of the experimental groups involved in the study of rare gases and free metals have also been involved with free molecular species and so there has been considerable overlap between the development of the study of free molecules and that of rare gases and free metals.

An outline of the literature on the Auger electron spectroscopy of free molecules is given in Table 4.22. As can be seen from the table the Auger spectra of some 50 or so volatile molecular species have been obtained. The results have been mainly discussed in terms of molecular-orbital theory from the point of view of *ab initio* calculation of the spectra or of correlation effects.

Table 4.22. Auger Electron Spectroscopy of Free Molecules

Compound	Research Group	Subject of Study	Year	Reference
CH_4	Karlsruhe	*KLL* spectrum	1960	10
	Uppsala	*K* Auger spectrum	1967/9	11, 191
	Uppsala	Study of Br-substituted methanes and some hydrocarbons (C_2H_6 and C_6H_6)	1970	133
	San Jose	Theoretical analysis, claimed to be first for molecule	1976	134
	Manchester	Configuration interaction in calculation (also HF, H_2O, CO)	1975	142
	Bergen	Roothaan's symmetry-restricted open-shell SCF method (CH_4 used as example)	1976	146
	Albuquerque	Chemical information, line-shape analysis, calculation of molecular and cluster spectra	1978 1979 1980	137, 176 138 139
C_2H_2	Loughborough	Empirical calculations (also C_2H_4, H_2S, NH_3, SO_2)	1976 1979	135, 136
	Albuquerque	References 139 and 176 above		
C_2H_4	As for References 135, 137, 138, and 139 above			
C_2H_6	As for References 11, 133, 135, and 137 above			
C_3H_8	As for Reference 138 above			
C_6H_6	As for References 11 and 133 above			
CO	Oak Ridge	*KLL* spectra, assignment of lines (also N_2, O_2, NO, H_2O, and CO_2) in C and O regions	1971	6
	Uppsala	Reference 11 above		
	CSIRO (Australia)	Discussion of vibrational structure (*C* Auger)	1971	140
	Loughborough	Analytical aspects	1979	136, 141
	Manchester	See Reference 142 above		
CO_2	As for References 6, 136, and 141 above			
C_3O_2	Uppsala	Assignment of spectrum (C and O), comparison with CO	1974	143
CH_3OH	As for References 137, 139, and 176 above			

169

Table 4.22 (Continued)

Compound	Research Group	Subject of Study	Year	Reference
$(CH_3)_2O$	As for References 137 and 176 above			
CF_4	Uppsala	C and F K-Auger spectra	1967	191
CH_3Cl, CH_2Cl_2 $CHCl_3$, CCl_4	Loughborough	Experimental, empirical assignment and discussion of chemical shift C and Cl Auger	1978	151
CH_3Br, CH_2Br_2 $CHBr_3$, CBr_4	Uppsala	Br $L_{2,3}MM$ and $M_{4,5}N_{2,3}$ spectra, C K-Auger spectra; discussion of line widths, shifts, and dissociation	1970	133
C_6H_5Cl, C_2H_5Cl	Eindhoven	Ionization cross-section determination	1973	145
NH_3	Loughborough	Empirical assignment and analytical applications	1976	135
	Bergen	Calculation of spectrum by method of Reference 146	1976	147
	Albuquerque	Experimental spectrum; comparison with Reference 147	1977	148
	Oregon	Discussion in context of isoelectric series Ne, HF, H_2O, NH_3, and CH_4	1977	149
	Rome	Experimental and assignment	1977	179
	Austin	Submonolayer on aluminium; comparison with gas- and solid- phase NH_3	1980	150
N_2	Karlsruhe	Low-resolution spectrum— KLL Auger	1960	10
	Munster	Study of excited states of N_2^{2+}	1969 1972	185 144
	Uppsala	K-Auger spectrum	1969	11
	Oak Ridge	Identification of high-energy lines	1970	152
	Oak Ridge	Assignment of spectrum (refers to Reference 152)	1971	6
NO	See Reference 6 above; N and O K-Auger			
N_2O	Manchester	O and N spectra, comparison with calculations	1976	153

Table 4.22 (Continued)

Compound	Research Group	Subject of Study	Year	Reference
NO_2	Loughborough	Experimental only, quantitative analysis (also HO, CO, CO_2, H_2S, SO_2, Ar, Ne, C_2H_2, N_2O)	1978	154
H_2O	Oak Ridge	See Reference 6 above		
	Uppsala	K-Auger spectrum and assignment; SCF and CI	1975	155, 156
	Uppsala	Semiinternal correlation effects	1980	157
	Manchester	Experimental and CI calculations	1976	142, 153
	Albuquerque	See References 137, 138, and 176 above		
CO	See References 6, 11, 131, 136, 140, 141, and 142 above			
CO_2	See References 6, 136, and 141 above			
C_3O_2	See Reference 143 above			
CH_3OH, $(CH_3)_2O$	See References 137 and 139 above			
NO	See Reference 6 above			
N_2O	See Reference 153 above			
NO_2	See Reference 154 above			
O_2	Karlsruhe	Low-resolution K-Auger spectrum	1960	10
	Oak Ridge	Experimental and assignment	1971	11
SO_2	Loughborough	Empirical assignment and quantitative analytical possibilities	1976 1978	135 154
	QEC London/ Toronto	Frozen orbital approximation	1978	178
HF	Oregon	Experimental, assignment, and ionization potentials	1975	158
	Manchester	Configuration-interaction calculations	1976	142
	Oslo	RHF calculation	1977	175
	Bergen/Oslo	Correlation effects	1979	20
CF_4	See Reference 191 above			
Over 20 organo-Si compounds	Uppsala	Chemical shifts ($^1D_2$ in S KLL spectrum) and $2p_{3/2}$ binding energies for SiX_4, (X=H, F, Cl, CH_3, OCH_3, OC_2H_5, C_2H_3), $SiCl_3R$	1980	159

171

Table 4.22 (Continued)

Compound	Research Group	Subject of Study	Year	Reference
		$(R = H, CH_3, C_2H_3, C_2H_5,$ $C_4H_9, C_6H_5), SiCl_{4-n}(CH_3)_n,$ $SiCl_{4-n}(C_2H_5)_n, Si(OC_2H_5)_{4-n}$ $(CH_3)_n$ and $SiCl_{4-n}H_n$		
SiH_4	Rome	$L_{2,3}MM$ spectrum assignment by comparison with Reference 134	1980	160
H_2S	Loughborough	Empirical assignment and qualitative and quantitative analytical aspects	1976 1979	135 141
	Helsinki/ Oak Ridge	Energies and chemical shifts of 1s level and $^1D_2(KL_2L_3)$ line in H_2S, SO_2, and SF_6	1976	161
	Uppsala	Chemical shifts and electron binding energies, $^1D_2(KL_{2,3}L_{2,3})$ line; also SF_6, COS, SO_2, CS_2	1976	162
	Uppsala	Ab initio calculations with configuration interaction; also SO_2, SF_6	1976	163
	Bergen/Helsinki	S KLL spectra, calculations by ab initio LCAO SCF method; also SO_2, SF_6	1977	164
	QEC London/ Toronto	Calculation of $L_{2,3}MM$ energies	1977	165
	Athens	Calculation of KLL energies; also SO_2 and shifts in KLL lines for H_2SO, H_4SO_2, H_2SO_2 and LMM of atomic S	1979	169
	QEC London/ Toronto	Calculation of KLL energies; also SO_2	1980	166
COS	See Reference 162 above			
SO_2	See References 135, 141, 161, 162, 163, 164, 166, 168, and 169 above			
SF_6	See References 161, 162, 163, 164 above			
CS_2	See Reference 162 above			
CCl_4	Munster	KLL spectrum of Cl	1969	167

Table 4.22 (Continued)

Compound	Research Group	Subject of Study	Year	Reference
C_6H_5Cl, C_2H_5Cl	Eindhoven	Ionization cross section	1973	145
CH_3Cl, CH_2Cl_2 $CHCl_3$, CCl_4	Loughborough	High-resolution spectra and chemical shifts	1978	151
GeH_4 and seven other volatile Ge compounds	Berkeley	Static relaxation and estimation of relaxation energy differences, $L_{2,3}M_{4,5}M_{4,5}$ spectra	1973	170
CH_5Br, CH_2Br_2 $CHBr_3$, CBr_4	Uppsala	Energies, intensities, and chemical shifts, $L_{2,3}MM$ and $M_{4,5}N_{2,3}N_{2,3}$ spectra	1970	133
$InCl_3$ $SnCl_2$, Sb_4 Tl_2, I_2	See Table 4.18 and References 39, 111, and 112			

A minority of papers have discussed chemical shifts and the possible analytical usefulness of the Auger spectra of gas-phase molecular species. This latter aspect will be discussed separately in the next section. In this section the interpretation and assignment of gas-phase molecular spectra will be discussed, with reference to selected molecules.

A discussion of the details of molecular-orbital theory is considered to be beyond the scope of the present text. There are several excellent texts available on molecular-orbital theory[171,172] and on molecular symmetry and group theory.[173,174]

4.1. The Hydrides of the First Row of the *p* Block

The molecules HF, H_2O, NH_3, and CH_4 form a series isoelectronic with Ne, and the features of the Auger spectra can be understood in terms of perturbations to the energy-level structure of neon by the proton ligands. In neon the "valence" orbitals are the $2s$ and $2p$ and for the other species the valence orbitals falls into two groups, those that are principally $2s$ and those mainly $2p$ in character.

The relative energies of the transitions may be determined by the energies of the two-hole final states (with valence vacancies) of the appropriate molecule. The energies of the final states may be approximated[149] as

$$E_{ij} = I_i + I_j + \langle i|1/r_{ij}|j\rangle \qquad (4.1)$$

where i and j are the quantum numbers describing the two vacancies, I_i and I_j are the interaction energies of a hole with the remaining electrons and the the nuclei, and $\langle i|1/r_{ij}|j\rangle$ is the Coulomb interaction between the two holes in i and j.

As an approximation the values of I_i and I_j are taken to be the measured ionization potentials for removal of electrons i and j from the neutral species and the Coulomb interaction term is taken to be independent of i and j. Thus this latter term is constant and cancels in the calculation of *relative* energies of the final states. This approach does not account for relaxation energy nor for multiplet splitting, but does account for the major features of the Auger spectra that can be seen from Fig. 4.21, a comparison of the relative energies calculated by the above equation and the experimentally determined values. The major features being, first, that there are relatively few energy levels at the extremes (Ne and CH_4) and relatively many for H_2O. The complexity of the spectrum depends on the degree of degeneracy of the molecular orbitals which in turn depends on the symmetry of the molecule. Second, the overall spacing of the levels decreases in passing from Ne to CH_4, owing to the corresponding decrease in the one-electron ionization potentials.

The bonding character of the valence orbitals also changes in moving from Ne to CH_4. In the case of Ne all electrons are nonbonding, whereas in CH_4 all electrons are involved in chemical bonds; NH_3, H_2O, and HF have one, two, and three pairs of nonbonding electrons, respectively. Thus the Ne spectrum consists of narrow well-separated lines (see Fig. 4.3); however, as one moves from HF to CH_4, vibrational and dissociative broadening effects give rise to increasingly broader and thus overlapping spectral lines. Auger transitions involving bonding electrons will produce broader spectral lines than those involving nonbonding electrons. This can cause difficulties, as the states that are most interesting are those with vacancies in bonding orbitals and it is these states that are broadest and may well overlap neighboring lines. In turn this makes the complete assignment of the spectrum to individual ion states difficult (most difficult in the case of H_2O, which has the lowest symmetry of the molecules in this group). It has been suggested[149] that because of these problems in assigning spectra it is not possible to obtain detailed information concerning electron double-vacancy states from Auger spectra except for transitions involving nonbonding electrons in which the final states are well separated in energy from the remaining manifold of states.

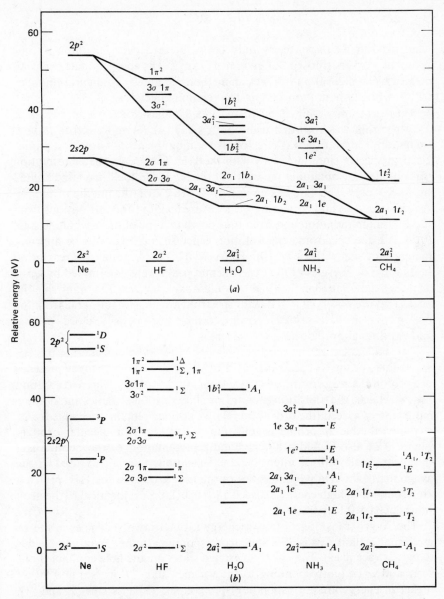

Figure 4.21. (a) Calculated relative energies of double-vacancy states in the isoelectronic series Ne, HF, H_2O, NH_3, and CH_4 using Equation (4.1) (neglecting multiplet splitting). (b) Experimental energies.[149] (Reprinted with permission of Elsevier Scientific Publishing Company, New York.)

4.1.1. HF

The spectrum has apparently only been obtained experimentally by one group of workers (those at Oregon in 1975).[158] The spectrum has been calculated by configuration-interaction methods[142] and by configuration interaction with configuration mixing (correlation effects[20]).

The spectrum obtained by Shaw and Thomas is shown in Fig. 4.22. The spectrum was recorded using both x-ray (Al K) and electron (8-keV) excitation. Calibration was effected by running the neon Auger and photoelectron spectra simultaneously with the HF. The spectrometer resolution was obtained by obtaining the width of the Ne $1s$ photoelectron line excited by Al K x-rays. This line has an energy of 616 eV (in the middle of the HF Auger spectrum) and correcting the measured width for contributions from the aluminum radiation and from the known lifetime of the Ne $1s$ hole state (0.23 eV), the spectrometer resolution function was found to be approximately Gaussian with a FWHM of 0.68 eV. This corresponds to a resolution of 0.11%. The energies of the experimental peaks were determined by least-squares fits of Gaussian or Voight profiles to the spectra. The many satellite peaks that are observed in the neon spectrum are indistinguishable from the background in the case of HF owing to their low intensity, close spacing, and vibrational or dissociative broadening.

The assignment of the spectrum may be effected by considering the HF spectrum to be the Ne spectrum with a small perturbation due to the presence of a proton at an appropriate distance from the Ne nucleus. This perturbation has two effects, the transition energies are shifted to lower values and become more closely spaced (the order and relative spacing remain unchanged) and, secondly, the Ne $2p$ orbital is split into the 3σ and 1π molecular orbitals. Figure 4.22 also includes a schematic representation of the neon Auger spectrum. Peak widths, intensities, and positions may all be used in the assignment. The relative intensities of the four groups in the HF spectrum are, from low to high energy, 0.054:0.18:0.050:1 almost identical to the neon values of 0.05:0.15:0.06:1.

The Auger transition of lowest energy is due to two vacancies in the 2σ molecular orbital, this is a $^1\Sigma$ state (corresponding to the ^1S state in Ne due to two $2s$ vacancies). The next two excited states are the hole configurations $2\sigma3\sigma$ and $2\sigma1\pi$ (corresponding to the vacancy state ^1P in Ne) and 10 eV higher in energy are corresponding spin-coupled triplet states $^3\Sigma$ and $^3\Pi$. The fourth group (around 640 eV) is derived from the Ne parent holes states ^1D and ^1S; the molecular perturbation will split the ^1D state into three levels $(1\pi^2)^1\Delta$, $^1\Sigma$, and $(1\pi3\sigma)^1\Pi$. The ^1S state becomes $(3\sigma^2)^1\Sigma$. Since the separation between the ^1D and ^1S states in neon is only 3.8 eV, it may be expected that considerable mixing between the $(1\pi^2)^1\Sigma$ and $(3\sigma^2)^1\Sigma$ states will occur.

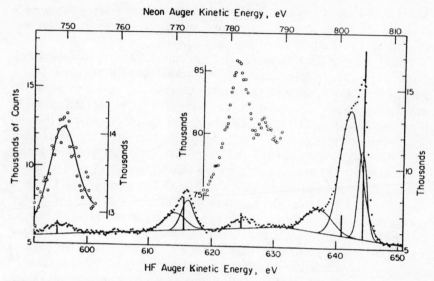

Figure 4.22. Auger spectrum of HF. Vertical bars represent the neon Auger spectrum.[158] (Reprinted with permission of American Physical Society, New York.)

Table 4.23.   Auger Transitions in HF

Final-State Configuration	Energy (eV)	Width (eV)
$(2\sigma^2)^1\Sigma$	595.6	5.13
$(2\sigma3\sigma)^1\Sigma$	614.1	4.10
$(2\sigma1\pi)^1\Pi$	616.2	2.55
$(2\sigma3\sigma)^3\Sigma$ $\}$ $(2\sigma1\pi)^3\Pi$	625.1	3.80
$(3\sigma^2)^1\Sigma$	636.9	4.5
$(3\sigma1\pi)^1\Pi$ $\}$ $(1\pi^2)^1\Sigma$	642.35	4.04
$(1\pi^2)^1\Delta$	644.28	1.62

Shaw and Thomas assign the highest-energy peak to $^1\Delta$ with the $^1\Sigma - ^1\Pi$ peaks from the $(1\pi^2)$ and $(3\sigma1\pi)$ hole configurations unresolved followed by the $^1\Sigma$ $(3\sigma^2)$ peak. The complete assignment together with peak widths is given in Table 4.23. The corresponding spin-coupled triplet states $(3\sigma1\pi)^3\Pi$ and $(1\pi^2)^3\Sigma$ were not detected.

Consideration of the line widths confirms this assignment since it can be seen that the broadest lines are those in which two bonding electrons are involved and the narrowest that in which two nonbonding electrons are involved. Transitions involving the loss of one π and one σ electron are intermediate in width.

Hillier and Kendrick's calculation[142] are almost in agreement with this assignment. Their method starts by performing an SCF calculation on a particular low-lying state of the doubly charged ion and then using the molecular orbitals calculated in this way as a basis for performing a configuration-interaction calculation to obtain the other states of the ion. The calculation produces the same order for the energies of the states apart from the $^1\Delta(1\pi^{-2})$ and $3\Pi(3\sigma\ 1\pi)$ states that, in Hillier and Kendrick's calculation, are the next lowest in energy below the $^3\Sigma(1\pi^2)$ peak, which is calculated to have the highest intensity. The absolute energy of the highest energy peak differs from the experimental value by 5 eV and the lowest-energy peak differs by almost 12 eV from the experimental value. A calculation within the restricted Hartree–Fock scheme by Faegri[175] still produced a value for this lowest-energy transition in error by about 4 eV. However, a configuration-interaction calculation using a general *ab initio* program[20] produced theoretical values 2–3 eV different from the experimental values. In this later paper, Kvalheim and Faegri also discuss transition rates.

The authors conclude that correlation effects observed in the spectrum of HF must also be important for the other first row hydrides because as the symmetry of the molecule is further reduced there will be more possibilities for configuration interaction.

4.1.2. H_2O

The spectrum of water has been reported by Moddeman et al.[6] in one of the first comprehensive reports of gas-phase Auger spectroscopy applied to free molecules, and their paper published in 1971 is something of a landmark in the development of this branch of the technique. The spectrum was excited by electron impact at 5 keV and 10 μA, and the resolution of the spectrometer was 0.06%. The energies were obtained by calibration against the neon $^1D_2$ line taken to be 804.2 eV and the $L_3M_{2,3}M_{2,3}$ $^1D_2$ line in argon taken to be 203.3 eV. The spectrum has also been reported by the Uppsala group using Mg x-rays as the exciting radiation[155] and by the Albuquerque group[176] using electron-impact ionization at 2 keV and a beam current of 4 μA. In the former report the target gas pressure was 0.1 torr, whereas the latter used a target pressure of approximately 5×10^{-5} torr. A number of research groups have discussed the assignment of the spectrum and the theoretical calculation.[142,156,157,158,177] Hillier et al.[153] report the experimental spectrum in their study of nitrous oxide obtained using electron impact at 3–5 keV with a current of 2–10 μA and an analyzer resolution of

0.015 eV. This spectrum shows possibly more detail than any other published spectrum, however, the authors only show this spectrum to demonstrate the quality of their instrument in comparison with other previously published spectra and there is no further discussion of the Auger spectrum of water.

Because of the low symmetry of the water molecule the molecular orbitals lose their degeneracy and there are considerably more transitions possible for H_2O than for HF. Thus the apparent simplicity of the spectrum as shown in Fig. 4.23 is somewhat misleading, even though the spectrum has what might be considered a normal atomlike KLL (low-Z) spectrum appearance in that there are four major peaks whose intensity increases successively on the high-energy side. The electronic configuration of water is $(1a_1)^2(2a_1)^2$-$(1b_2)^2(3a_1)^2(1b_1)^2$. The $1a_1$ orbital is pure oxygen $1s$, the $2a_1$ is almost pure oxygen $2s$, and the $1b_1$ is pure oxygen $2p$ lone pair. The hydrogens bond mainly through the $1b_2$ and $3a_1$ orbitals, which consist of 41% H $1s$ and 59%

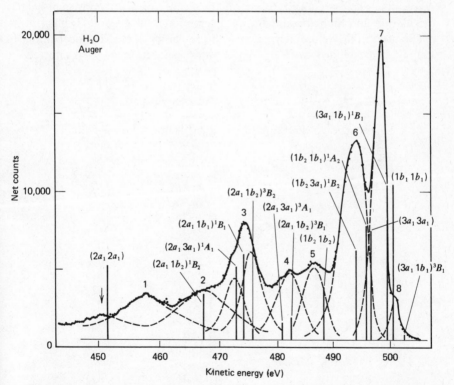

Figure 4.23. The Auger electron spectrum of water vapor excited by Mg x-rays.[155] (Reprinted with permission of North-Holland Publishing Company, Amsterdam.)

O $2p$ (approximately) and 17% H $1s$, 10% O $2s$ and 72% O $2p$ (approximately, respectively).[11] The first broad peak at the low-energy end is thus due to transitions to the final hole state $2a_1 2a_1$ (analogous to the $2s^2$ atomic hole state): as bonding electrons are removed the equilibrium bond length increases and the band is broadened due to vibrational excitation. Similarly, the highest-energy peak is due to transitions to the $1b_1 1b_1$ final hole state: as nonbonding electrons are removed the equilibrium bond length is not affected and the band is much narrower as there is less vibrational excitation. Peak number 3 contains contributions from the molecular analogs of the $2s2p$ atomic final hole state, that is, $2a_1 3a_1$, $2a_1 1b_1$, $2a_1 1b_2$, and the broad intense peak Number 6 may be viewed as the molecular analog of transitions to the $2p2p$ final atomic hole state, that is, $1b_2 1b_2$, $1b_2 3a_1$, $1b_2 1b_1$, $3a_1 3a_1$, and $3a_1 1b_1$. This simplified approach does not take into account spin coupling to produce singlet and triplet states nor does it take into account configuration interaction. The results of configuration-interaction calculations have been used to assign the peaks in the spectrum with a variety of results. The assignment based on the results of a self-consistent-field calculation and limited configuration interaction[156] is given in Table 4.24. These are in fairly close agreement with the values quoted by Siegbahn et al.,[155] but calculated by Wahlgren et al.[156] using an open-shell Hartree–

Table 4.24.   Assignment of the H_2O Auger Spectrum

Line Number	Energy (eV)	Final-State Configuration	Final-State Symmetry
1	457.7	$2a_1 2a_1$	$^1A_1$
2	467.7	$2a_1 1b_2$	$^1B_2$
3	472.4	$2a_1 3a_1$	$^1A_1$
	475.3	$\begin{cases} 2a_1 1b_1 \\ 2a_1 1b_2 \end{cases}$	$^1B_1$ $^3B_2$
4	482.1	$\begin{cases} 2a_1 3a_1 \\ 2a_1 1b_1 \end{cases}$	$^3A_1$ $^3B_1$
5	486.7	$1b_2 1b_2$	$^1B_2$
6	493.8	$\begin{cases} 3a_1 1b_2 \\ 1b_1 1b_2 \\ 3a_1 3a_1 \\ 3a_1 1b_2 \\ 1b_1 1b_2 \end{cases}$	$^1B_2$ $^1A_2$ $^1A_1$ $^3B_2$ $^3A_2$
7	498.6	$\begin{cases} 3a_1 1b_1 \\ 1b_1 1b_1 \end{cases}$	$^1B_1$ $^1A_1$
8	500.8	$3a_1 1b_1$	$^3B_1$

Fock scheme. (For some states limited configuration interaction was included.) These values are shown by the vertical bars in Figure 4.23. Agren and Siegbahn[157] have also discussed the effect of semiinternal correlation effects. Correlation effects (on intensities and energies) are normally divided into initial-state correlation and final-state correlation, which may be further divided into (a) those effects that are connected to the interaction between final-state continua and (b) those that originate from final states of the residual ion. Interactions between a primary-hole-state configuration and doubly excited configurations are expected to be particularly large for valence states in rare gases and small molecules because of the possibility of a close match in energy. If the correlation involves the excitation of one electron within the Hartree–Fock "sea" (space spanned by occupied orbitals in the reference Hartree–Fock state) and one outside the Hartree–Fock sea, the correlation is usually termed semiinternal. The effects of semiinternal correlation effects were calculated for the double-hole $2a_1 2a_1$ state. The calculated transition energy (including internal correlation effects) has been given a value of 450 eV, whereas intensity arguments would indicate that the experimental peak 1 at 457 eV is the main transition. The results of this calculation show that the transition is substantially affected, the total intensity being split among several states at least three of which have significant intensities. The three strongest peaks with relative intensities 60:12:13 correspond to significant mixing of configurations of the type $2a_1 ijk^*$, where i, j are $3a_1$, $1b_1$ and k^* is $2b_1^*$ and $4a_1^*$. The agreement between theory and experiment is now seen to be quite good as most of the $2a_1 2a_1$ intensity occurs at around 458 eV and the lower intensity peak at 450 eV can be clearly assigned to a final-ionic-state configuration-interaction satellite.

The complete assignment should include a consideration of initial- and final-state shake-up and shake-off satellites. Agren and Siegbahn's calculations indicate that the most significant contributions will occur between 455 and 460 eV due to transition to a $2a_1 1b_2$ final state from an initial shake-off state.

4.1.3. NH₃

Reports of the experimental spectrum of ammonia make an interesting collection as four independent groups each published a spectrum between August 1976 and September 1977.[135,148,149,179] The last but one of these[149] claimed that theirs was the first published spectrum (although by the time their paper was received by the journal the first[135] had been published); even more extraordinary in the last report,[179] which states "To our knowledge the experimental Auger spectrum of ammonia has not yet been published," yet by the time the journal received this paper the spectrum had been published three times, the most recent of these being seven months earlier in the same journal. It is probably fair to say that the first published spectrum, that of Thompson's group at Loughborough, could easily be overlooked by the others as it appeared in an analytical journal and was not specifically mentioned in the title (though it was in the abstract), but the failure to note the other three reports ought to cause authors and referees alike a certain amount of embarrassment.

There is little doubt that as interest in gas-phase Auger work increased the various groups recognized virtually simultaneously that the spectrum of ammonia was needed to complete the series of first row hydrides isoelectronic with neon and that having carried out the work there is a natural desire to publish.

There is close agreement between the various spectra, and the assignment appears unambiguous. This is based on Okland et al.'s symmetry-restricted open-shell self-consistent-field calculations[147] published just two months before the first experimental spectrum became available. The spectrum obtained by the Albuquerque group is shown in Fig. 4.24 together with Okland's theoretical spectrum (shifted down in energy by 2.7 eV) and a deconvolution of the experimental spectrum based on Gaussian peaks. The intensities of Okland's spectrum were calculated by analogy with the experimental intensities of neon. The spectrum is obtained by electron-impact excitation at 1.5 keV. The Oregon group obtained a virtually identical spectrum by using Mg $K\alpha$ x-rays as the exciting source and deconvoluting the spectrum according to Lorentzian functions. The assignment of the spectrum is given in Table 4.25. The electronic configuration of ammonia is $(1a_1)^2(2a_1)^2(1e)^4(3a_1)^2$ in which the $1a_1$ orbital is almost entirely N $1s$ in character and plays no part in the chemical bonding and the $3a_1$ is about 88% N $2p$ in character and is thus the nitrogen lone pair. The $2a_1$ and $1e$ orbitals are involved in the bonding. Thus, as before, it would be expected that the $3a_1 3a_1$ final hole state would be narrowest (but not as narrow as the corresponding "lone-pair" transitions in H_2O and HF) and the states involving the loss of two bonding electrons would be the broadest. The agreement between the calculated and experimental spectrum becomes progressively poorer at lower energies and has already been mentioned in the case of H_2O is most likely due to neglect of correlation effects in the calculations. The full calculation has not yet been performed.

The spectrum of solid ammonia has been obtained[180] and compared with the gas-phase spectrum; the major difference between the solid and the gas is the significant lifetime broadening of the peaks corresponding to transitions involving the $3a_1$ orbital. This "lone-pair" orbital is primarily responsible for the hydrogen bonding between the NH_3 species to form the

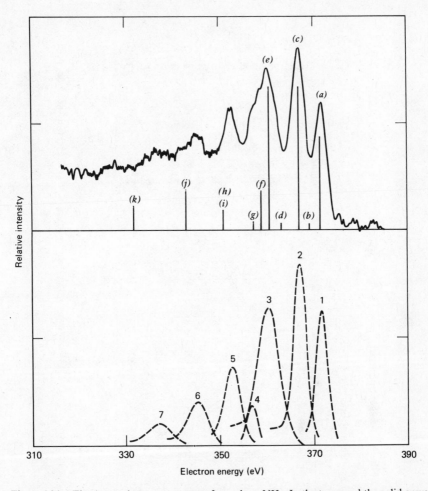

Figure 4.24. The Auger electron spectrum of gas-phase NH_3. In the top panel the solid curve and the bar graph are, respectively, the experimental and the calculated Auger spectra. The lower panel depicts decomposition of experimental spectrum.[148] (Reprinted with permission of North-Holland Publishing Company, Amsterdam.)

Table 4.25. Assignment of the NH$_3$ Auger Spectrum

Peak Number	Experimental Energy (and Width)			Theoretical Energy d	Assignment Hole Configuration	Symmetry
	a	b	c			
1	371.5 (1.8)	370.19 (2.20)	374.0 (2.6)	373.7	$3a_1 3a_1$	$^1A_1$
				370.5	$1e3a_1$	^3E
2	366.6 (2.0)	365.47 (3.00)	369.5 (3.8)	369.6	$1e3a_1$	^1E
				365.4	$1e1e$	$^3A_2$
3	360.2 (3.2)	359.47 (3.60)	363.0 (3.8)	363.1	$1e1e$	^1E
			360.0 (3.8)	360.9	$1e1e$	$^1A_1$
4	356.5 (1.5)	356.47 (3.6)	359.0 (3.8)	358.8	$2a_1 3a_1$	$^3A_1$
				352.9	$2a_1 3a_1$	$^1A_1$
5	352.4 (2.5)	350.93 (3.51)	354.5 (5)	352.8	$2a_1 1e$	^3E
6	345.1 (2.0)	343.65 (7.0)	347.5 (5)	345.1	$2a_1 1e$	^1E
7	337.0 (3.5)	334.63 (7.4)	338.8 (10)	334.0	$2a_1 2a_1$	$^1A_1$

[a] Reference 148.
[b] Reference 149.
[c] Reference 179.
[d] Calculated from data in Reference 148 by adding 2.7 eV (the downward shift shown in Fig. 4.24 to line up with the $3a_1 3a_1$ peaks).

184

molecular crystal. The spectrum of ammonia adsorbed onto molybdenum[181] and onto alumina and aluminum[150] have been obtained. In the latter study the authors conclude that on alumina NH_3 is in a similar chemical environment to the molecular crystal, but that on aluminum NH_3 is only weakly chemisorbed.

4.1.4. CH_4

The Auger spectrum of methane was one of the first to be published.[10] Tungsten x-rays were used as the exciting radiation, but the resolution of the spectrometer was only sufficient to show one peak with a low-energy tail. Since then the spectrum has been published twice, by the Uppsala group[11,14,133] and by the Albuquerque group.[176] The spectrum obtained by the latter group is shown in Fig. 4.25. The spectrum was excited using electron impact at 2 keV with a beam current of approximately 4 μA. The analyzer was a modified retarding single-pass cylindrical-mirror analyzer with a resolution of approximately 0.5%. The spectrum was signal averaged for between 2 and 5 h. Several theoretical analyses of the spectrum have been carried out[134,142,146] and and an interpretation of the spectrum given by Hartmann and Gebelein.[182] The assignment of the spectrum and the results of some of the theoretical calculations are given in Table 4.26, which is an extension of Table 2.3. The results of Faegri and Manne's calculations[146] are summarized in Fig. 4.25. In the figure the calculated spectrum has been shifted to lower energies by 4.4 eV. The widths of all the lines in the spectrum are about 6 eV, showing that all electrons are involved in bonding, that is, there are no "nonbonding" electrons which would give rise to narrower peaks. The reason that the peaks are so broad is not due to vibrational broadening, since a value of this order would suggest quite a deep potential well and therefore would imply that CH_4^{2+} is a stable entity. This is known not to be the case, because the ion is not observed by mass spectrometry and the dissociation energy is calculated[133] to be -13 eV showing that CH_4^{2+} is unstable with respect to dissociation by proton emission; also, the singly ionized entity CH_4^+ is known to have a lifetime of 3×10^{-15} sec.[183] Thus the broadening of the lines in the Auger spectrum is due entirely to the short lifetime of the upper state.

The ground-state configuration of methane is $1a_1^2 2a_1^2 1t_2^6$ where the $1a_1$ orbital is almost entirely carbon $1s$ and the other orbitals are involved in the bonding. The final hole states are $2a_1 2a_1$, $2a_1 1t_2$ and $1t_2$, $1t_2$ giving rise to terms $^1A_1$; $^3T_2$, $^1T_2$, and $^1A_1$, ^1E, $^1T_2$, $^3T_2$, respectively. The highest-energy peak is assigned to the $1t_2 1t_2$ hole state, the two middle peaks to the $2a_1 1t_2$ final state, and the low-energy peak to the $2a_1 2a_1$ final hole state. The intensi-

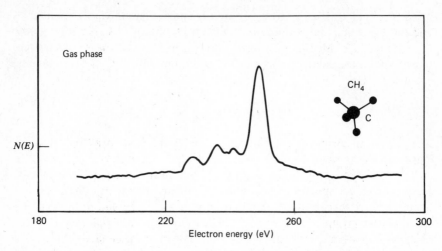

Figure 4.25. Electron-excited spectrum of CH_4.[176] The calculated spectrum is from Reference 146 and has been shifted down in enough by 4.4 eV. (Reprinted with permission of American Physical Society, New York.)

ties of the bar spectrum shown in Figure 4.25 were calculated by analogy with the neon spectrum allowing for the splitting of the ^1D level in neon into the $^1T_2$ and ^1E levels in methane.

4.2. CH_3OH, $(CH_3)_2O$

The oxygen environment in these molecules is similar to that in H_2O and the carbon environment is similar to methane. Both of these molecules have been studied by the Albuquerque group[139,176] from the point of view of comparing spectra of gas and multilayer phases as well as from calculating the theoretical spectra. The oxygen Auger spectra of H_2O, CH_3OH, and $(CH_3)_2O$ are shown in Fig. 4.26, from which the considerable similarity between the spectra can be seen. There is a slight shift in energy and a feature to high-energy side of the main peak appears in the case of CH_3OH and $(CH_3)_2O$. The carbon-region spectra are shown in Fig. 4.27, and again the high-energy features can be observed. Comparison between the gas-phase and multilayer-phase spectra shows very little difference for the carbon regions of CH_3OH and $(CH_3)_2O$, but the oxygen regions of H_2O and to a lesser extent CH_3OH and to a lesser extent still $(CH_3)_2O$ show considerable broadening reflecting the extent of the oxygen involvement to intermolecular bonding. Hydrogen bonding, of course, is particularly strong in the case of water.

Table 4.26. Assignment of the CH₄ Auger Spectrum

Energy (eV)						Assignment	
Experimental		Theoretical					
a	b	c	d	e	f	Final Hole State	Symmetry
250.0	248.9	254.71	255.1	256.11	251.9	$1t_2 1t_2$	$^3T_1$
		253.66	254.24	255.16		$1t_2 1t_2$	^1E
		251.72	252.73	253.30	249.3	$1t_2 1t_2$	$^1T_2$
		249.26	251.18	250.91		$1t_2 1t_2$	$^1A_1$
243.3	241.0	244.96	245.7	246.38	243.4	$2a_1 1t_2$	$^3T_2$
237.0	236.2	238.64	239.5	239.93	237.6	$2a_1 1t_2$	$^3T_1$
229.4	229	230.43	231.57	231.82	230.6	$2a_1 2a_1$	$^1A_1$

[a]Reference 133.
[b]Reference 176.
[c]Reference 134, self-consistent-field calculations using large basis set of contracted Gaussian-type basis functions.
[d]Reference 142, configuration-interaction calculations using medium-size Gaussian basis.
[e]Reference 146, Roothaan's symmetry-restricted open-shell SCF method.
[f]Reference 139, SCF calculation on neutral molecule.

187

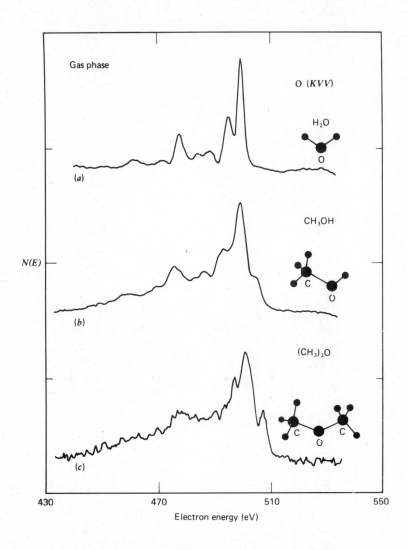

Figure 4.26. Gas-phase, electron-excited O (KVV) spectra for (*a*) H_2O, (*b*) CH_3OH, (*c*) $(CH_3)_2O$.[176] (Reprinted with permission of American Physical Society, New York.)

188

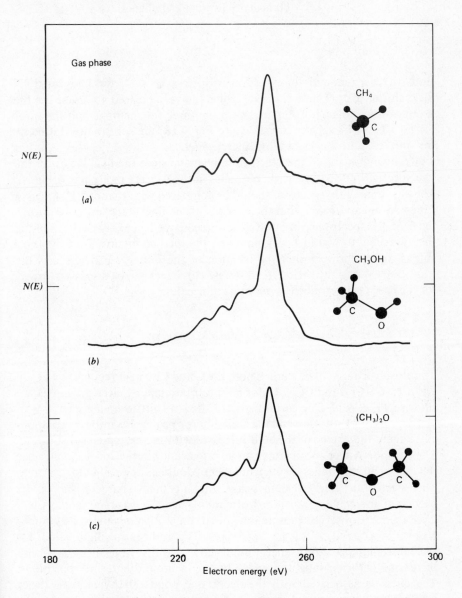

Figure 4.27. Gas-phase, electron-excited C Auger spectra for (a) CH_4 (b) CH_3OH, (c) $(CH_3)_3O$.[176] (Reprinted with permission of American Physical Society, New York.)

189

4.3.   Halogen-Substituted Methanes

4.3.1.   CF$_4$

Both the carbon and fluorine Auger spectra of CF$_4$ were reported by Siegbahn et al.[11] The molecular orbitals have increased to seven for the ground-state molecule compared with the three for methane and the three for HF. The CF$_4$ spectra are shown in Fig. 4.28; the spectra are shifted by the difference in the $1s$ binding energies, $695.2 - 301.8 = 393.4$ eV, so that a direct comparison of the spectra can be made since the final states for the two spectra are the same (two vacancies in the valence molecular orbitals). The fluorine-region spectrum should be compared with that of HF shown in Fig. 4.22. It can be seen that the energies of the final states as derived form the C and F spectra are in quite close agreement (the energy position of the lines agreeing to within 1 eV), however, the relative intensities of the transitions differ, showing that the location of the initial vacancy affects the transition rates to the various final states. This effect will be discussed further for the case of heteronuclear diatomic molecules.

4.3.2.   Chloro Methanes

Thompson et al.[151] have studied both the C and Cl Auger regions of CH$_3$Cl, CH$_2$Cl$_2$, CHCl$_3$, and CCl$_4$. Chlorine spectra excited by electron impact are shown in Fig. 4.29. The carbon spectra show a shift to higher energy with increasing Cl substitution and the most-intense peak splits into two components; again the extent of the split increases with increasing Cl substitution. The chlorine Auger spectra have an initial vacancy in the $2p$ level, and since the spin-orbit splitting is 1.6 eV, a number of doublets appear in the spectrum. The spectra are fairly rich in peaks, making them characteristic of the particular molecule. Transitions originating from a vacancy in the Cl $2s$ level were not observed presumably due to the $2s$-$2p$ conversion of Coster–Kronig processes (i.e., of the LLV type). The authors calculate values for the molecular-relaxation energy and show that its value decreases with increasing chlorine substitution for both carbon and chlorine transitions. The effect is associated with the decreasing polarizability of the valence orbitals with increasing numbers of Cl atoms. The partial charge on the carbon atom (calculated by the Sanderson procedure[184]), an inverse measure of the decreasing polarizability, correlates with the molecular-relaxation energy.

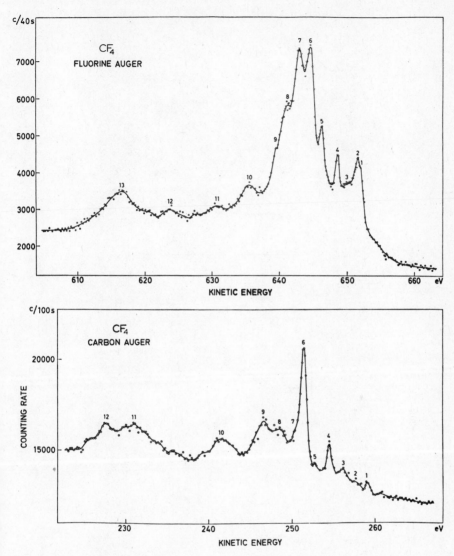

Figure 4.28. Fluorine and carbon Auger spectra of CF_4.[11] (Reprinted with permission of North-Holland Publishing Company, Amsterdam.)

191

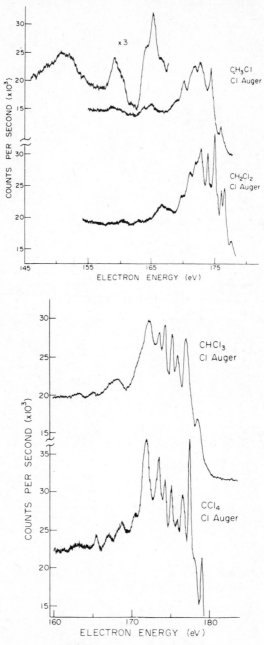

Figure 4.29.   Chlorine Auger spectra of CH_3Cl, CH_2Cl_2, $CHCl_3$, and CCl_4.[151] (Reprinted with permission of American Chemical Society, Washington, D.C.)

192

4.3.3. Bromo Methanes

Spohr et al.[133] have published the spectra of the C and Br (both $L_{2,3}MM$ and $M_{4,5}N_{2,3}N_{2,3}$) regions of CH_3Br, CH_2Br_2, $CHBr$, and CBr_4. The spectra were produced by electron-impact excitation at 5 keV and 5–20 μA with a target gas pressure of about 0.1 torr. The spectrometer resolution was about 0.6%.

The $L_{2,3}MM$ Br spectra are all very similar and in turn similar to that of Xe, but shifted to lower energy by about 80 eV. With increasing Br content in the molecule the spectrum shifts to higher energies. Since Br is more electronegative than H, the substitution of Br for H means that the binding energy of each Br atom is increased because of the more electronegative environment. The binding-energy shifts are larger for L electrons than M electrons, hence the shift to higher energy of the Auger peaks. The $M_{4,2}N_{2,3}N_{2,3}$ spectra show considerable differences and not a lot of similarity to that of Kr. The spectra are shown in Fig. 4.30, and it is easily seen that as the number Br atoms increases (a) the signal-to-background ratio increases; (b) the line at the high-energy limit of the spectrum, which corresponds to the lowest excited state of the doubly ionized molecule, shifts to higher energies; and (c) the energy interval for the whole spectrum increases

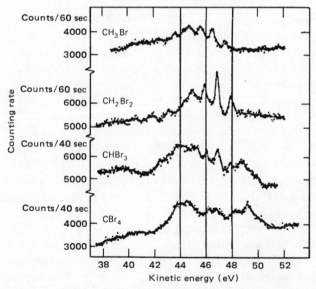

Figure 4.30.   Bromine $M_{4,5}N_{2,3}N_{2,3}$ Auger spectra obtained from CH_3Br, CH_2Br_2, $CHBr_3$, and CBr_4.[133] (Reprinted with permission of The Royal Swedish Academy of Sciences.)

from about 4 eV for CH_3Br to about 7 eV for CBr_4. The corresponding C spectra show analogous effects: (a) signal-to-background decreases; (b) shift to higher energy (by about 12 eV—this being greater than the observed shift for CCl_4 or CF_4); and (c) a *decrease* in line width from about 6 to 2 eV. As previously discussed in the case of methane, the line broadening is due almost entirely to dissociative effects, which are proton-emission processes, and thus the dissociative probability decreases as the number of H atoms in the molecule decreases. It can be seen that, overall, the Br lines are narrower than the C lines due to the fact that there is a high probability for Auger transitions to occur to final states in orbitals that derive mainly from the atomic orbitals of the atom with the primary vacancy. Thus the C Auger transitions create vacancies in the hydrogen-bonding orbitals (leading to high dissociation probabilities and broad lines), whereas the Br Auger transitions preferentially create vacancies in nonbonding orbitals (leading to more-stable final states and narrower lines). However, the $M_{4,5}N_1N_{2,3}$ and $M_{4,5}N_1N_1$, transitions are likely to be broad compared with the $M_{4,5}N_{2,3}$ transitions as the former are involved in the bonding and the removal of electrons from these orbitals leads to an increased probability of dissociation.

4.4. Other Hydrocarbons

The spectra of acetylene (C_2H_2) and ethylene (C_2H_4) have been considered by Thompson et al.[135] and ethane (C_2H_6) and benzene (C_6H_6) have been studied by the Uppsala group.[11,133] The Sandia laboratories group at Albuquerque have published the spectra of ethylene and acetylene[139,176] in the readily accessible literature, while a recent report[138] also shows the spectra of C_2H_6, C_3H_8, C_4H_{10}, C_5H_{14}, although discussion of most of these is of a very preliminary nature.

Thompson et al. make an assignment of seven peaks in the C_2H_2 spectrum and six peaks in the C_2H_4 spectrum by an empirical method, assuming the relaxation energy is constant for all transitions. This method gives surprisingly good agreement between the experimental and calculated energies.

The spectra of C_2H_6 and C_6H_6[133] are shown in Fig. 4.31 and compared with the methane spectrum. A shift to higher energy can be seen for the high-energy limit (of about 14 ev) owing to the decreasing binding energy of the outer orbitals. It is interesting to note that although ethane can be thought of as a methane derivative (one H replaced by CH_3), the spectrum of C_2H_6 does not bear such a strong resemblance to the "parent" CH_4 as do CH_3OH and $(CH_3)_2O$, whose spectra are shown in Fig. 4.27. It is also clearly seen that the line widths decrease quite sharply on going from methane to benzene.

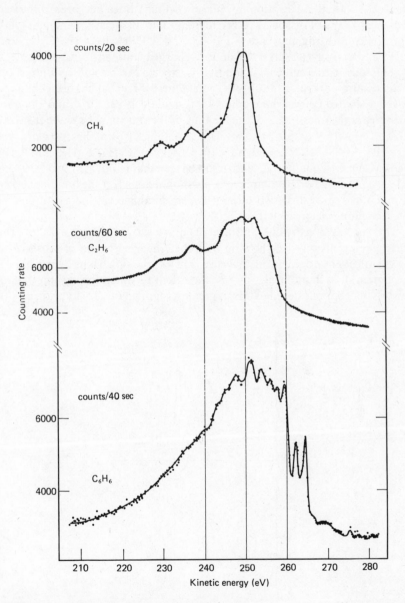

Figure 4.31. Carbon Auger spectra obtained from CH_4, C_2H_6, and C_6H_6.[133] (Reprinted with permission of The Royal Swedish Academy of Sciences.)

The reason for the exceptionally broad methane lines has been discussed already in terms of dissociative broadening. The dissociation energies (with a proton as one fragment) are -13, -8, and $+1.3$ eV for CH_4, C_2H_6, and C_6H_6, respectively (to give the doubly charged ion), and thus benzene is the only one with a molecular ion that is expected to be stable with respect to dissociation. It is also the only $+2$ ion of these three that has been observed by mass spectrometry, where detection is made between 10^{-8} and 10^{-6} sec after creation. An estimate of the life time of the state may be made from the relationship $\Delta E \times \Delta t \gtrsim h/2\pi$. The values obtained from these spectra are 3.3×10^{-16}, 1.8×10^{-16}, and 1.1×10^{-16} sec for C_6H_6, C_2H_6, and CH_4, respectively. In a qualitative sense it can be seen that removal of two electrons from a conjugated or aromatic molecule should not affect the bonding nearly so much as removal of two electrons from methane, as there will still be many bonding electrons left.

The spectra of ethylene and acetylene are shown in Fig. 4.32.[135] The spectra quite distinctive, showing that the Auger spectrum is sensitive to the chemical environment in this case. The degree of carbon hybridization is different in each molecule, that is, the extent to which the carbon s and p orbitals may be considered to mix, is greatest for methane (sp^3 hybridization),

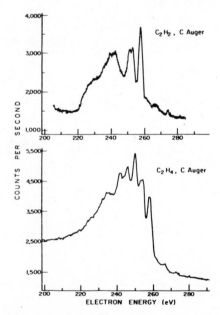

Figure 4.32. Carbon Auger spectra of C_2H_2 and C_2H_4.[135] (Reprinted with permission of American Chemical Society, Washington, D.C.)

intermediate for ethylene (sp^2 hybridization, the simplest molecule to contain a π bond), and least for acetylene (sp hybridization)ⱼ and although this approach is not particularly useful in interpreting the peaks in the spectrum, it does indicate the quite different chemical environment of the C atoms in each molecule.

4.5. Homonuclear Diatomic Molecules—N_2 and O_2

The spectra of N_2 and O_2 were first obtained as long ago as 1960,[10] but the spectra were at fairly low resolution and only two broad humps could be distinguished. An assignment of the high-resolution spectrum of N_2 was made by Stahherm et al.[185] in 1969 and by Moddeman et al. in 1971.[6] The latter research group (at Oak Ridge) discussed the identity of the high-energy satellites in the N_2 spectrum based on spectra obtained with (a) 3-keV electrons and (b) Al $K\alpha$ x-rays the previous year (1970). Moddeman et al. also discussed the spectrum of O_2 and presented results for a similar study of the high-energy satellites.

The spectra may be readily understood in terms of a molecular-orbital diagram where the ground state of N_2 is $(2\sigma_g)^2(2\sigma_u^*)^2(1\pi_u)^4(3\sigma_g)^2$ and of O_2 is $(2\sigma_g)^2(2\sigma_u^*)^2(1\pi_u)^4(3\sigma_g)^2(1\pi_g^*)^2$. The N_2 spectrum obtained by Moddeman et al. is shown in Fig. 4.33. In Moddeman's designation the Auger lines occur in regions B, C, and D, the A lines being the high-energy satellites.

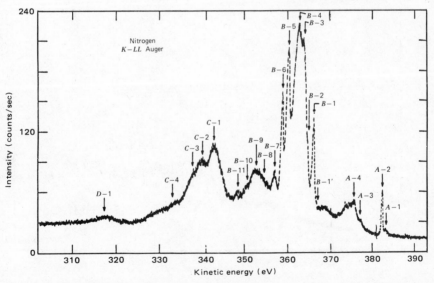

Figure 4.33. Molecular N_2 Auger electron spectrum excited by electron impact.[6] (Reprinted with permission of American Physical Society, New York.)

The molecular orbitals are designated as strongly bonding S or weakly bonding W. The S orbital is the $2\sigma_g$ and all the others are W, and the lines in region B correspond to KWW transitions, in region C to KWS, and in region D to KSS. For N_2 the $A1$ and $A2$ lines are assigned to autoionization processes $Ke\text{-}W$, whereby, initially, part of the electron-impact energy is used to promote a K electron into an excited bound orbital (designated e), when the electron returns to the K shell an electron from a W orbital is ejected. The region from 368 to 379 eV, which includes the $A3$ and $A4$ lines, is thought to include processes of the type $Ke\text{-}WWe$, that is, K-shell excitation followed by an Auger transition in which the excited electron remains a spectator. As has been discussed earlier the excitation process cannot occur with x-radiation unless a resonance condition obtains, and as this does not occur for A1 $K\alpha$ x-radiation, the satellite peaks are absent from the spectrum excited by this radiation. The peaks $B8\text{-}B11$ are assigned to $KW\text{-}WWW$ processes, that is, initial shake off (or monopole ionization) followed by a WW Auger transition.

The O_2 spectrum is shown in Fig. 4.34. Assignment of the A region is more difficult than in the case of N_2 because no K absorption data are available and there is thus an uncertainty concerning the initial Ke states formed,

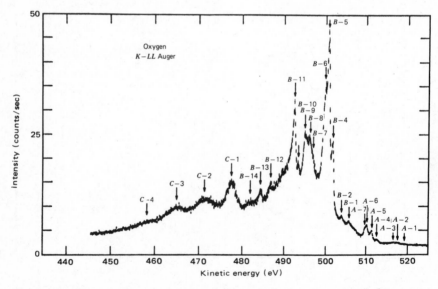

Figure 4.34.   Molecular O_2 Auger electron spectrum excited by electron impact.[6] (Reprinted with permission of American Physical Society, New York.)

also, the initial electronic states will be split by interaction of the remaining K electron with the unpaired electrons in the partially filled $1\pi_g^*$ orbital. This does not occur for N_2 because all the electrons in the molecular orbitals are spin paired. The KSS transition was not observed, and the lines $B12$ and $B13$ were assigned to KW-WWW processes. No evidence of vibrational structure was observed.

4.6. Heteronuclear Diatomic Molecules—CO and NO

The spectrum of CO has been obtained by the Uppsala group[11,143] and the Oak Ridge group.[6] The vibrational structure of the spectrum has been discussed by Hurley[140] with reference to the work of Newton and Sciamanna,[186] who had concluded that the ground state of CO^{2+} had only a very shallow local minimum in the potential-energy curve capable of supporting only one vibrational level in apparent conflict with the results of Siegbahn et al. that suggested at least five vibrational levels. Hurley suggested that the Franck–Condon factors governing the primary excitation process may lead to a preferential population of the highest vibrational levels and could explain the different findings. The spectra obtained by Moddeman et al. are shown in Fig. 4.35, in which the same notation is used to designate the lines as was used for the nitrogen and oxygen spectra. The identity of the A satellite lines and hence the highest-energy Auger line (corresponding to transition to the ground state of the molecular ion) was confirmed by exciting the spectra by Al $K\alpha$ x-rays as well as electron impact. Part of the high-resolution spectrum obtained by Siegbahn et al. (252–256 eV) is also shown in Fig. 4.35 as an inset in which the vibrational structure can be clearly seen. The oxygen and carbon spectra are shifted relative to each other by the difference in $1s$ binding energies, to aid comparison since it might be expected that the spectra would show considerable similarities considering that the final states for the Auger transitions for the C and O regions are the same, the two sets of transitions differing only in the location of the initial vacancy. However, dissimilarity between the spectra is more apparent than any similarity. The valence-bond structure of CO is $(2s\sigma^b)^2(2s\sigma^*)^2$-$(2p\pi^b)^4(2p\sigma^b)^2$ and the percentage C character of each orbital has been calculated[177] as $2p\sigma^b$, 92%; $2p\pi^b$, 33%; $2s\sigma^*$, 20%; and $2s\sigma^b$, 33%. In contrast, the percentage N character in NO, whose molecular-orbital structure is $(2s\sigma^b)^2(2s\sigma^*)^2(2p\sigma^b)^2(2p\pi^b)^4(2p\pi^*)^1$, is $2p\pi^*$, 64%; $2p\pi^b$, 33%; $2p\sigma^b$, 33%; $2p\sigma^b$, 60%; and $2s\sigma^*$, 50%. Thus in CO, when an initial vacancy is created in the C $1s$ orbital, the transitions to the $2p\sigma^b$ would be relatively more important, and this is reflected in the intensity of the $B1$–$B3$ peaks in the C spectrum while the remaining lines $B5$–$B8$ are weaker. The opposite effect may be seen in the O spectrum, in which the transition to the $2s^b2s\sigma^b$ final hole state

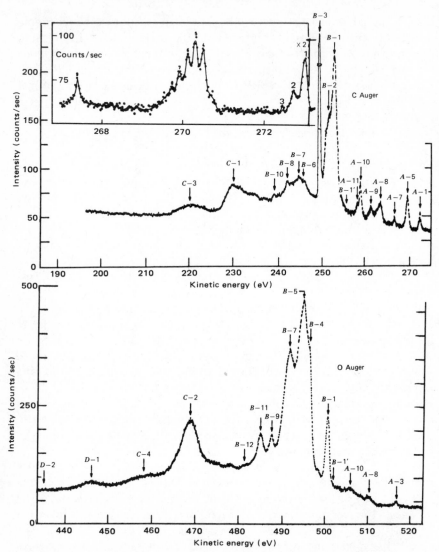

Figure 4.35. Carbon Auger spectrum of CO (upper curve); inset shows part of the spectrum at high resolution. Oxygen Auger spectrum of CO (lower curve).[6]

is observed (peak $D1$), a transition not observed in the C spectrum. The NO spectrum (both N and O regions shifted by the difference in 1s binding energies) is shown in Fig. 4.36, and it is easily seen that there is considerably greater similarity between the two spectra than between the C and O spectra of CO, reflecting the greater degree of mixing of the atomic orbitals to give

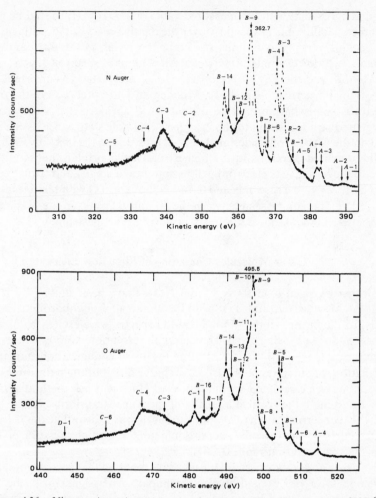

Figure 4.36.   Nitrogen Auger (upper curve) and oxygen spectrum (lower curve) of NO.[6]

the final molecular orbitals. It may also be deduced from the relative intensities of the satellite peaks in the C and O spectra that the excited orbital into which the K electron goes prior to autoionization has more C character than O character, suggesting that the excited orbital may be the π^* rather than the σ^* orbital (the percentage C and O character being reversed for the antibonding orbitals). This is confirmed from a knowledge of the C excitation energy $1s \rightarrow 2p\pi^*$ of 287.4 eV,[187] which together with the binding energy of the highest occupied molecular orbital of CO from ESCA measurements of 14.5 eV[11] gives the energy of the first autoionization peak of 272.9 eV in good agreement with the measured value of 273.0 eV). It should be noted that there is still some uncertainty over the relative energy of the NO molecular orbitals, some calculations reversing the order of the $2p\sigma^b$ and $2p\pi^b$ orbitals (see; for example, Reference 188). The photoelectron spectroscopy data[172,189] and the ESCA data[11] have been interpreted as consistent with the original ordering given above. Moddeman's interpretation of the Auger spectra is somewhat ambiguous in that the B1 peak in the N spectrum is designated as a transition to the $2p\pi^*2p\sigma^b$ final hole state, whereas the same peak in the O spectrum is designated as a transition to the $2p\pi^*2p\pi^b$. Because of the unpaired electrons in the highest molecular orbital, the initial state (due to the loss of a $1s$ electron) will be split into a singlet and triplet, separated by 1.5 eV in the case of N and 0.7 eV in the case of O with relative intensities of 1:3. This can be seen between the B1 and B2 peaks in the N spectrum and between the B4, B5, and B9, B10 peaks in the O spectrum.

4.7   Other Molecules Consisting of First-Row Elements

The other nitrogen oxides—N_2O and NO_2—have been studied by Hillier et al. at Manchester, and Tyson and Killoran at Loughborough, respectively. In the former study[153] the N and O spectra were excited by electron impact at 3–5 keV using a spectrometer with resolution 0.015%. The experimental results were compared with the results of configuration-interaction calculations of the states of N_2O^{2+}. The charge distribution in the three core-hole states was calculated by the restricted Hartree–Fock method and was used to estimate the relative intensity of the Auger transitions. In the latter study[154] concerning analytical applications, it was shown that the N and O Auger current was directly proportional to the partial pressure of NO_2 over the concentration range 0–20 mole%.

Karlsson and co-workers at Uppsala have reported the spectrum of carbon suboxide (C_3O_2)[143] excited by electron impact. An assignment of the spectrum is made, assisted by intensity and energy calculations. A comparison with the oxygen Auger spectrum of CO is made.

4.8   Molecules Containing Second-Row Elements

In general there has been more interest in the KLL spectra of the second-row elements in free molecules than in the LVV spectra, which would correspond to the spectra discussed already for first-row elements. The KLL spectra are atomlike, in that none of the electrons involved in the Auger transition are involved in the molecular-orbital scheme; however, the spectra do exhibit shifts in energy as the chemical environment of the atom in question changes, and it is this chemical shift that has been the subject of most of the investigations of the molecules in this category. Solid–vapor shifts have already been discussed earlier, and the chemical shift is discussed in more detail in Chapter 5. Often the study of the Auger shifts has been made in conjunction with a study of photoelectron (XPS) line shifts because, of course, the processes which contribute to these shifts, such as molecular relaxation, operate in both the XPS and Auger process. Indeed in the particular case of molecular relaxation, owing to the flow of negative charge toward an inner-shell-ionized atom in a molecule, this may be probed under more-severe conditions in the case of Auger spectrometry because of the double vacancy of the final state. The reasons for the shifts observed in core levels in both XPS and Auger spectra have been discussed by Wagner,[100] who proposed the use of the "Auger parameter," the difference in energy between an Auger line and a photoelectron line, as a parameter characteristic of the chemical and physical state of the element. Wagner's ideas were based primarily on results for solids, and the advantage of the Auger parameter was that terms involving static charge and work-function data cancel. For an element going from the isolated atom to a molecule the change in the Auger parameter, $\Delta\alpha$, will be due entirely to the effects of extra-atomic relaxation (or molecular relaxation). Extra-atomic relaxation energy is sometimes referred to as polarization energy.

The only systematic study of the hydrides isoelectronic with Ar (SiH_4, PH_3, H_2S, HCl) has been made from the point of view of studying the shifts in the KL_2L_3 ($^1D_2$) Auger line and the $1s$ photoelectron energies.[190] A linear correlation between the two sets of shifts was observed (though not in a 1:1 relationship). The study also included a number of phosphorus halides. The valence-shell spectra of the above molecules have been reported for H_2S[135] and SiH_4.[160] The spectrum of silane obtained by electron impact at 0.5 keV and 200 μA shows considerable similarity to the methane spectrum and covers the energy range 62–81 eV. The lines are broad, but not as broad as for methane, indicating that SiH_4^{2+} may be a more stable entity than CH_4^{2+}. Part of the H_2S spectrum obtained by Thompson et al. is shown in Fig. 4.37; this may be compared with the H_2O spectrum in Fig. 4.23. Although the molecular-orbital diagram is the same for H_2O as for H_2S, the $1b_1$ orbital is even less bonding in H_2S than in H_2O and may be regarded as pure S lone pair. The two sharp features in the spectrum are assigned to transitions to the $1b_1 1b_1$ final hole state split by the sulfur $2p$ spin-orbit coupling of the initial state. (All the transitions are assigned an initial $2p$ vacancy state.) The relative intensities of the peaks reflect the ratio of the statistical weights expected for $2p_{1/2}$ and $2p_{3/2}$ initial states. The satellite structure to the low-energy side of the peaks is attributed to vibrational structure. This is consistent with the removal of nonbonding electrons, which should leave a fairly stable final

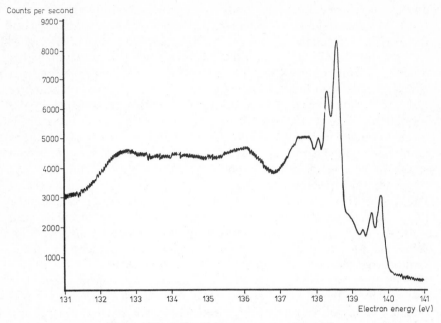

Figure 4.37. Sulfur L Auger spectrum of H_2S.[135]

state, and the excitation of the first vibrational level being the strongest, showing little change in equilibrium molecular geometry.

The $L(2p)$-MM spectrum has been calculated by multiconfiguration self-consistent-field methods,[165] and the results have been compared with Thompson's empirical assignment based on calculating a single relaxation term. As was mentioned previously surprisingly good agreement was obtained for the high energy transitions, though the agreement was poorer at the low-energy end of the spectrum. The LVV spectra of PH_3 and HCl have not been discussed in the literature.

Kelfve et al. at Uppsala[159] have made an extensive study of the silicon $KL_{2,3}L_{2,3}$ ($^1D_2$) Auger energies and the $2p_{3/2}$ binding energies for gas-phase compounds. The KLL spectrum of SiF_4 may be compared with the neon KLL spectrum. It can be seen that the ^3P peaks of the $L_{2,3}L_{2,3}$ transition have a higher relative intensity and that the satellite structure (shake off) is not nearly so pronounced in SiF_4 as it is for Ne or Ar. Semiempirical calculations were performed for most of the 20 or so compounds, and their use was evaluated in providing a satisfactory description of the valence charge densities and their changes upon core ionization.

Several papers from a number of different research groups have considered the KLL spectra of a range of sulfur compounds and, of course, have been concerned mainly with the shift effects observed. Keski-Rahkonen and co-workers studied the chemical shift of the $^1D_2$ line in H_2S, SO_2, and SF_6[161] (also the sulfur $1s$ binding energies) and concluded that molecular relaxation effects were small compared with atomic relaxation effects. They observed a shift of 6.1 eV, on passing from H_2S to SF_6, to lower energy in the $^1D_2$ Auger line and a shift of 11.6 eV to higher energies for the $1s$ photoelectron lines. In a later publication,[164] the complete KLL spectra of the three molecules are reported and the experimental results compared with *ab initio* LCAO SCF calculations. The satellite structure caused by shake off accompanying the initial K-shell ionization was also discussed, as well as the strong dependence of the lifetime broadening of final states containing an L_1 vacancy on the chemical environment. At about the same time the Uppsala group published the complete KLL spectra of the same three molecules.[163] The results were compared with *ab initio* calculations including configuration interaction. Shake-off and autoionization satellites were discussed as well as the broad lines corresponding to states with an L_1 final vacancy. One explanation for this being the occurrence of Coster–Kronig transitions rapidly converting an L_1 vacancy into $L_{2,3}M$ vacancy states. The Uppsala group has also considered the Auger shifts and S $2p$ shifts for CS_2 and COS in addition to the three molecules already mentioned.[162] Finally, the KLL spectra of several sulfur-containing molecules have been calculated[169] and the Auger chemical shifts have been considered from a theoretical point of view.[166] In this latter study, the authors show that the use of a frozen-orbital approximation is preferable to SCF methods for calculation of chemical shifts in KLL Auger energies and LL ionization potentials. A series of model compounds was considered (H_2S, H_2SO, H_4SO_2, and H_2SO_2) in which a chemical shift in the KLL Auger energy of 12 eV per formal charge on sulfur was calculated.

4.9. Other Molecules

Perry and Jolly have reported on the $L_{2,3}M_{4,5}M_{4,5}$ spectra of germane (GeH$_4$) and a series of germanium compounds.[170] The $M_{4,5}$ molecular relaxation energy was calculated to be 5.2 eV, close to the estimated free-atom value as the $4a_1$ and $3t_2$ molecular orbitals are closely related to the germanium $4s$ and $4p$ atomic orbitals.

Meyer and Vrakking[145] measured the $L_{2,3}$ ionization cross section of Cl by obtaining the Auger spectra of chlorobenzene and comparing the carbon peak with the chlorine peak using a previously published value of the ionization cross section for carbon.

As has already been mentioned, the Cl and Br spectra of the chloro and bromo methanes have been reported. The spectra of a number of volatile metal compounds have also been studied (see Chapter 4, Section 3.7).

5.   ANALYTICAL APPLICATIONS OF GAS-PHASE AUGER
ELECTRON SPECTROSCOPY

In 1975 Thomas Carlson wrote that

Good resolution, high counting rates, unique identification of molecules and sensitive
elemental identification all make Auger spectroscopy a potentially powerful tool for
gas analysis. As yet no systematic study of its analytical possibilities on gases has been
made, but it would appear to be a worthwhile undertaking.[3]

However, there has been very little development in the analytical
applications recently, despite the advantageous properties that Carlson
pointed out.

5.1.   Qualitative Analysis

The key to the qualitative analytical potential lies in the basic equation for
the energy of the Auger electron:

$$E_{XYZ} = E_X - (E_Y + E_Z) \qquad (4.2)$$

where E_X = binding energy of the initially ionized electron
E_Y = binding energy of the down electron
E_Z = binding energy of the ejected electron

As was previously pointed out, the values of E_X are so different for different
elements that qualitative analysis for elemental composition may be readily
achieved by observing the energies at which the Auger electrons occur in
the spectrum. This is clearly illustrated in Fig. 4.38, which shows the spectra
of a number of different elements. If the electrons involved in the Auger
process Y and Z are also involved in the bonding, that is, the electrons are
valence electrons, then the energy of the ejected electron will also depend on
the nature of the bonding, and thus the Auger energies can be characteristic
of the chemical environment of the atom on which the initial vacancy
occurred. Since there will probably be a number of occupied molecular
orbitals, a number of transitions will be possible whose energies and intensities will be characteristic of that particular initial vacancy, and thus
the overall Auger spectrum for any element in a molecule is likely to be
characteristic of that molecule. This is illustrated in Fig. 4.39, which shows
the C and O regions of carbon monoxide, carbon dioxide, and a mixture of
the two gases.

There is little doubt that gas-phase AES does have considerable potential
as an analytical method. The spectra of the few gas-phase species that have

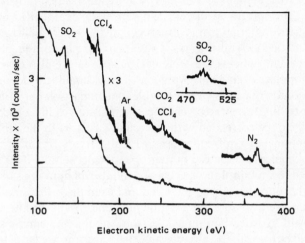

Figure 4.38.   Auger spectrum of a gas mixture.[141]

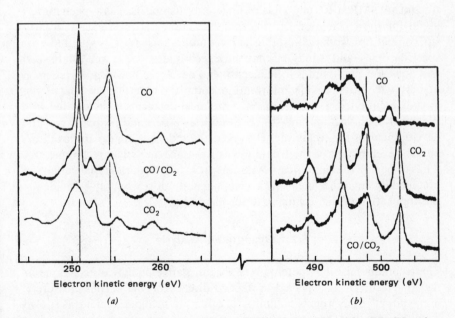

Figure 4.39.   (a) Carbon and (b) oxygen Auger spectra of CO, CO_2, and a CO–CO_2 mixture.[141]

been shown in the previous section clearly illustrate the qualitative analytical possibilities. The spectra of the elements are well separated in the energy scale (owing to the large energy difference in the binding energies of the initial ionized electron) so that, as with XPS, there is no difficulty in distinguishing which elements are present. The spectra are also characteristic of the given molecule and thus would provide a qualitative identification, if a library of known spectra were available for comparison purposes. As has been discussed and illustrated, chemical shifts are readily observed, but whether they can be correlated with molecular structure in the same way as shifts in functional group frequencies in IR absorption or shifts in nuclear magnetic resonances can be is a matter open to speculation. This aspect of AES will need to wait until the spectra of a number of homologs of a variety of functional groups have been recorded and examined. The small number of results that are available, for example, the chloro and bromo methanes and a variety of hydrocarbon molecules, indicate that systematic studies of this sort could be fruitful. It is also possible that, via a knowledge of ionization cross sections, the relative numbers of the different atoms in a molecule could be calculated, although it seems unlikely at this stage that information as to the relative numbers of atoms of an element in different environments in a molecule could be obtained, as may be obtained from the integration trace of an NMR spectrum. Thus, as far as establishing itself as a spectroscopy useful for the elucidation of molecular structure is concerned, AES faces stiff competition from well-established techniques such as IR, NMR, and mass spectroscopies, and it may be that these techniques are so well established and powerful that there is little incentive to develop a new technique which at best will only provide information complementary to the techniques already mentioned. There are also the practical problems (a) associated with the use of high vacuum and (b) cost that may also stand in the way of the development of the technique. However, these do not appear to have held up the development of the technique as a surface-analytical technique, but it has to be said that in this case AES can provide information that cannot be obtained by any other means. This claim could not reasonably be made for the application of the technique to the gas phase.

5.2. Quantitative Analysis

The technique also has potential as a quantitative analytical technique as the Auger current is expected to depend directly on the partial pressure P of the target gas as well as a number of other physical parameters, as shown from the equation for I_A, the Auger current:

$$I_A = I_p Gr\phi\gamma SP \qquad (4.3)$$

where I_p = current of the primary exciting beam

G = geometrical factor dependent on the instrumental collection efficiency and the angular distribution of the Auger electrons

r = back-scattering factor (additional contribution to Auger signal caused by scattered electrons)

ϕ = ionization cross section

γ = probability of an Auger transition occurring

S = probability of Auger electron escape without prior collision

Thus it can be seen that in quantitative terms AES can be thought of as analogous to molecular fluorescence or phosphorescence, in which electrons replace photons in the basic phenomenon. As with molecular luminescence the resultant signal depends on the intensity of the primary beam, and, because it is possible to produce high initial fluxes with commercial electron guns, there is the possibility that the Auger current is a sensitive quantitative analytical parameter.

As with a number of other analytical spectroscopies, AES would not be used to deal with mixtures of molecules containing the same elements because the problems of spectral overlap would be severe. The solution to this problem would be to effect a prior separation, by, for example, gas chromatography, which would present few difficulties in view of the existing interfaces for gas chromatographs and mass spectrometers. The Auger spectrometer would need to have a sufficiently rapid scan rate to cope with the transient nature of the gas chromatograph eluent peaks.

The linear dependence of the Auger current on the partial pressure of the target gas has been demonstrated by Killoran and Tyson[141,154] for a number of gas mixtures, namely, nitrogen and hydrogen sulfide; nitrogen and carbon monoxide; nitrogen and carbon dioxide; nitric oxide and argon; nitrogen dioxide and argon and neon; sulfur dioxide and acetylene; and nitrous oxide, acetylene, and argon. In all mixtures the Auger current was a linear function of partial pressure over the range 0–20 mole% of the target gas. As the total pressure in the collision chamber was of the order of 10^{-5} torr, the actual amount of vapor detectable is very small indeed, and these preliminary experiments show that if a suitable collecting system were devised, the Auger spectrometer could be used as a sensitive, trace analytical tool. However, as has been pointed out in the case of qualitative analysis, there are many competitive techniques in the field of trace-gas analysis. The interfacing of a gas chromatograph and an Auger spectrometer would provide a very powerful combination in much the same way as the gas chromatograph/ mass spectrometer combination has proven to be, whose success may well be a good reason for a reluctance to investigate an expensive alternative.

It would seem likely that while Auger spectroscopy is still developing as a surface-analytical method and while the study of gas-phase species is still firmly the province of physicists and theoreticians, there is unlikely to be any major developments in the applications of gas-phase AES as an analytical technique. Such developments will need to wait until analytical chemists appreciate the full potential of the technique.

REFERENCES

1. P. Kelfve, B. Blomster, H. Siegbahn, K. Siegbahn, E. Sanhueza, and O. Goscinski, *Phys. Scr.*, **21**, 75 (1980).
2. S. Hofmann, in G. Svehla (Ed.), *Wilson and Wilson Comprehensive Analytical Chemistry*, Elsevier, Amsterdam, 1979, Vol. 9, p. 89.
3. T. A. Carlson, *Photoelectron and Auger Spectroscopy*, Plenum, London, 1975.
4. M. O. Krause, T. A. Carlson, and W. E. Moddeman, *J. Phys.* **32**, C4-139 (1971).
5. W. N. Asaad and E. H. S. Burhop, *Proc. Phys. Soc. (London)*, **71**, 369 (1958).
6. W. E. Moddeman, T. A. Carlson, M. O. Krause, B. P. Pullen, W. E. Bull, and G. K. Schweitzer, *J. Chem. Phys.*, **55**, 2317 (1971).
7. L. Asplund, P. Kelfve, B. Blomster, H. Siegbahn, and K. Siegbahn, *Phys. Scr.*, **16**, 268 (1977).
8. M. O. Krause, F. A. Stevie, L. J. Lewis, T. A. Carlson, and W. E. Moddeman, *Phys. Lett. A*, **31**, 81 (1970).
9. P. Auger, *Surf. Sci.*, **48**, 1 (1975).
10. W. Mehlhorn, *Z. Phys.*, **160**, 247 (1960).
11. K. Siegbahn, C. Nordling, G. Johansson, J. Hedman, P. F. Hedén, K. Hamrin, U. Gelius, T. Bergmark, L. O. Werme, R. Manne, and Y. Baer, *ESCA Applied to free Molecules*, North Holland, Amsterdam, 1971.
12. H. Korber and W. Mehlhorn, *Z. Phys.* **191**, 217 (1966).
13. W. Mehlhorn, D. Stalherm, and H. Verbeek, *Z. Naturforsch.*, **23a**, 287 (1968).
14. L. Asplund, P. Kelfve, B. Blomster, H. Siegbahn, and K. Siegbahn, *Phys. Scr.* **16**, 268 (1977).
15. T. A. Carlson, W. E. Moddeman, and M. O. Krause, *Phys. Rev. A*, **1**, 1406 (1970).
16. M. O. Krause, *J. Phys.*, **32**, C4-67 (1971).
17. D. L. Matthews, B. M. Johnson, J. J. Mackey, and C. F. Moore, *Phys. Lett. A*, **45**, 447 (1973).
18. G. Howat, T. Aberg, and O. Goscinski, *International Conference on the Physics of x-ray Spectra*, Washington D.C., 1976, p.35.
19. H. P. Kelly, *Phys. Rev. A*, **11**, 556 (1975).
20. O. M. Kvalheim and K. Faegri, Jr., *Chem. Phys. Lett.*, **67**, 127 (1979).
21. W. Mehlhorn, *Z. Phys.*, **208**, 1 (1968).
22. W. Mehlhorn and D. Stalherm, *Z. Phys.*, **217**, 294 (1968).
23. W. N. Asaad and W. Mehlhorn, *Z. Phys.*, **217**, 304 (1968).
24. M. O. Krause, *Phys. Rev. Lett.*, **34**, 633 (1975).
25. L. O. Werme, T. Bergmark, and K. Siegbahn, *Phys. Scr.*, **8**, 149 (1973).

26. T. Kondow, T. Kawai, K. Kunimori, T. Onishi, and K. Tamaru, *J. Phys. B*, **6**, L156 (1973).

27. G. N. Ogurtsov, *Zh. Eksp. Teor. Fiz.*, **64**, 1149 (1973) [*Sov. Phys. JETP*, **37**, 584 (1973)].

28. N. Stolterfoht, D. Schneider, and P. Ziem, *Phys. Rev. A*, **10**, 81 (1974).

29. E. J. McGuire, *Phys. Rev. A*, **11**, 1880 (1975).

30. Ch. Briancon and J. P. Desclaux, *Phys. Rev. A*, **13**, 2157 (1976).

31. L. O. Werme, T. Bergmark, and K. Siegbahn, *Phys. Scr.*, **6**, 141 (1972).

32. W. Mehlhorn, *Z. Phys.*, **187**, 21 (1965).

33. W. Mehlorn, N. Schmitz, and D. Stalherm, *Z. Phys.*, **252**, 399 (1972).

34. M. O. Krause, *Phys. Lett.*, **19**, 14 (1965).

35. E. J. McGuire, *Phys. Rev. A*, **11**, 17 (1975).

36. J. D. Nuttall and T. E. Gallon, *Solid State Commun.*, **15**, 329 (1974).

37. T. E. Gallon and J. D. Nuttall, *Surf. Sci.*, **53**, 698 (1975).

38. S. Hagman, G. Hermann, and W. Mehlhorn, *Z. Phys.*, **266**, 189 (1974).

39. S. Aksela, H. Aksela, and T.D. Thomas, *Phys. Rev. A*, **19**, 721 (1979).

40. K. T. Compton, and J. C. Boyce, *J. Franklin Inst.*, **205**, 497 (1928).

41. P. G. Kruger, *Phys. Rev.*, **36**, 855 (1930).

42. R. Whiddington and H. Priestley, *Proc. Roy. Soc. (London)*, **A145**, 426 (1934).

43. U. Fano, *Phys. Rev.*, **124**, 1866 (1961).

44. J. W. Cooper, U. Fano, and F. Prats, *Phys. Rev. Lett.*, **10**, 518 (1963).

45. U. Fano and J. W. Cooper, *Phys. Rev.*, **137**, A1364 (1965).

46. F. J. Comes and H. G. Sälzer, *Phys. Rev.*, **152**, 29 (1966).

47. F. J. Comes, H. G. Sälzer, and G. Schumpe, *Z. Naturforsch.*, **23a**, 137 (1968).

48. J. Macek, *J. Phys. B*, **1**, 831 (1968).

49. R. P. Madden and K. Codling, *Phys. Rev. Lett.*, **10**, 516 (1963).

50. K. D. Sevier, *Low Energy Electron Spectrometry*, Wiley, Toronto, 1972.

51. K. Codling and R. P. Madden, *Phys. Rev. Lett.*, **12**, 106 (1964).

52. H. Körber and W. Mehlhorn, *Phys. Lett.*, **13**, 129 (1964).

53. W. N. Asaad, *Nucl. Phys.*, **66**, 494 (1965).

54. A. K. Edwards and M. E. Rudd, *Phys. Rev.*, **170**, 140 (1968).

55. M. O. Krause, F. A. Stevie, L. J. Lewis, T. A. Carlson, and W. E. Moddeman, *Phys. Lett. A*, **31**, 81 (1970).

56. N. Stolterfoht, H. Gabler, and U. Leithauser, *Phys. Lett. A*, **45**, 351 (1973).

57. Ch. Briancon, *Ann. Phys. (Paris)*, **5**, 151 (1970).

58. W. N. Asaad, R. W. Fink et al. (Eds.), *Proceedings of the International Conference on Inner-Shell Ionization Phenomena and Future Applications*, Atlanta, 1972, CONF-720404. National Technical Information Service, U.S. Dept. of Commerce, Springfield, Va., 1972, p. 463.

59. E. H. S. Burhop and W. N. Asaad, *Advances in Atomic and Molecular Physics*, Academic, New York, 1972, Vol. 8.

60. D. A. Shireley, *Phys. Rev. A*, **7**, 150 (1973).

61. C. P. Bhalla, *Phys. Lett. A*, **44**, 103 (1973).

62. R. L. Chase, H. P. Kelly, and H. S. Kohler, *Phys. Rev. A*, **3**, 1550 (1971).

63. D. Stalherm, Diplomarbeit Inst. Kernphysik, Univ. Munster, 1968.

64. T. A. Carlson, M. O. Krause, and W. E. Moddeman, *J. Phys.*, **32**, C4-76 (1971).
65. P. Stalherm, B. Cleff, H. Hittig, and W. Mehlhorn, *Z. Naturforsch.*, **24a**, 1728 (1969).
66. J. P. Desclaux, *Comput. Phys. Commun.*, **9**, 31 (1975).
67. E. Balisier, Uppsala University UUIP-922, March 1976.
68. M. H. Chen and B. Craseman, *Phys. Rev. A*, **8**, 7 (1973).
69. W. L. Jolly and D. N. Hendricksson, *J. Am. Chem. Soc.*, **92**, 1863 (1970).
70. C. E. Moore, *Atomic Energy Levels*, National Bureau of Standards Circular 467, Washington, D.C., 1949, 1952, 1958.
71. R. A. Rubenstein, Ph.D. Thesis, University of Illinois, 1955.
72. R. L. Watson and L. H. Toburen, *Phys. Rev. A*, **7**, 1853 (1973).
73. D. J. Volz and M. E. Rudd, *Phys. Rev. A*, **2**, 1395 (1970).
74. H. Hanashiro, Y. Suzuki, T. Sasaki, A. Mikuni, T. Takayanagi, K. Wakiya, H. Suzuki, A. Danjo, T. Hino, and S. Ohtani, *J. Phys. B*, **12**, L775 (1979).
75. E. J. McGuire, *Phys. Rev. A*, **3**, 587 (1971).
76. E. J. McGuire, *Phys. Rev. A*, **3**, 1801 (1971).
77. A. Rosen and I. Lindgren, *Phys. Rev.*, **176**, 114 (1968).
78. W. Lotz, *Z. Phys.*, **232**, 101 (1970).
79. M. J. Van der Wiel and G. Wiebes, *Physica*, **49**, 441 (1970).
80. J. A. Bearden and A. F. Burr, *Rev. Mod. Phys.*, **176**, 114 (1968).
81. S. Hagmann, Diplomarbeit, University of Freiburg, 1973.
82. E. J. McGuire, *Phys. Rev. A*, **5**, 1052 (1972).
83. C. W. Nestor, T. C. Tucker, T. A. Carlson, L. D. Roberts, F. B. Malik, and C. Froese, ORNL Report No. ORNL-4027 (unpublished).
84. M. Gryzinski, *Phys. Rev.*, **138**, A336 (1965).
85. J. C. Slater, *Quantum Theory of Atomic Structure*, McGraw-Hill, New York, 1960.
86. E. U. Condon and G. H. Shortley, *Theory of Atomic Spectra*, Cambridge University, Cambridge, 1970.
87. J. B. Mann, Los Alamos Scientific Laboratory Report LA-3690, 1967.
88. K.-H. Huang, M. Aoyagi, M. H. Chen, B. Craseman, and H. Mark, *At. Data Nucl. Data Tables*, **18**, 243 (1976).
89. B. W. Shore and D. H. Menzel, *Principles of Atomic Spectra*, Wiley, New York, 1968.
90. J. A. D. Matthew, *Surf. Sci.*, **20**, 183 (1970).
91. R. G. Albridge, K. Hamrin, G. Johansson, and A. Fahlman, *Z. Phys*, **209**, 419 (1968).
92. H. Hillig, B. Cleff, W. Mehlhorn, and W. Schmitz, *Z. Phys.*, **268**, 225 (1974).
93. B. Breukmann and V. Schmidt, *Z. Phys.*, **268**, 235 (1974).
94. S. Aksela and H. Aksela, *Phys. Lett. A*, **48**, 19 (1974).
95. S. Aksela, J. Väyrynen, and H. Aksela, *Phys. Rev. Lett.*, **33**, 999 (1974).
96. H. Aksela and S. Aksela, *J. Phys. B*, **7**, 1262 (1974).
97. W. J. Lotz, *J. Opt. Soc. Am.*, **58**, 915 (1968); **60**, 206 (1970).
98. D. A. Shirley, R. L. Martin, S. P. Kowalczyk, F. R. McFeely, and L. Ley, *Phys. Rev. B*, **15**, 544 (1977).

99. W. Mehlhorn, B. Breuckmann, and D. Hausamann, *Phys. Scr.*, **16**, 177 (1977).
100. C. D. Wagner, *Faraday Discuss. Chem. Soc.*, **60**, 291 (1975).
101. P. Bisgard, R. Bruch, P. Dahl, B. Fastrup, and M. Rødbro, *Phys. Scr.*, **17**, 49 (1978).
102. E. Breuckmann, B. Breuckmann, W. Mehlhorn, and W. Schmitz, *J. Phys. B*, **10**, 3135 (1977).
103. B. Breuckmann, V. Schmidt, and W. Schmitz, *J. Phys. B*, **9**, 3037 (1976).
104. R. Hoogewijs, L. Fiermans, and J. Vennik, *J. Elec. Spectrosc. Rel. Phenom.*, **11**, 171 (1977).
105. J. Väyrynen, S. Aksela, and H. Aksela, *Phys. Scr.*, **16**, 452 (1977).
106. R. Kumpula, J. Väyrynen, T. Rantala, and S. Aksela, *J. Phys. C*, **12**, L809 (1979).
107. S. Aksela, H. Aksela, M. Vuontisjärvi, J. Väyrynen, and E. Lähteenkorva, *J. Elec. Spectrosc. Rel. Phenom.*, **11**, 137 (1977).
108. H. Aksela, S. Aksela, J. S. Jen, and T. D. Thomas, *Phys. Rev. A*, **15**, 985 (1977).
109. W. Mehlhorn, B. Breuckmann, and D. Hausamann, *Phys. Scr.*, **16**, 177 (1977).
110. R. Abouaf, *J. de Phys.*, **32**, C4-128 (1971).
111. M. Pessa, A. Vuoristo, M. Vulli, S. Aksela, J. Väyrynen, T. Rantala, and H. Aksela, *Phys. Rev. B*, **20**, 3115 (1979).
112. H. Aksela, J. Väyrynen, and S. Aksela, *J. Elec. Spectrosc. Rel. Phenom.*, **16**, 339 (1979).
113. A. Fahlman, R. Nordberg, C. Nordling, and K. Siegbahn, *Z. Phys.*, **192**, 476 (1966).
114. S. Hagström and S. E. Karlsson, *Ark. Fys.*, **26**, 451 (1964).
115. R. Hoogewijs, L. Fiermans, and J. Vennik, *Chem. Phys. Lett.*, **38**, 471 (1976).
116. D. A. Shirley, *Phys. Rev. B*, **9**, 381 (1974).
117. J. Friedel, *Philos. Mag.*, **43**, 2392 (1952).
118. R. Hoogewijs, L. Fiermans, and J. Vennik, *Surf. Sci.*, **69**, 271 (1977).
119. C. D. Wagner and P. Biloen, *Surf. Sci.*, **35**, 82 (1973).
120. L. Yin, I. Adler, M. H. Chen, and B. Crasemann, *Phys. Rev. A*, **7**, 897 (1973).
121. W. Mehlhorn, *Phys. Lett. A*, **26**, 166 (1968).
122. B. Cleff and W. Mehlhorn, *Phys. Lett. A*, **37**, 3 (1971).
123. J. A. D. Matthew, *Phys. Lett. A*, **32**, 361 (1970).
124. P. H. Citrin, *Phys. Rev. Lett.*, **31**, 1164 (1973).
125. J. A. D. Matthew and M. G. Devey, *J. Phys. C*, **7**, L335 (1974).
126. L. I. Yin, I. Adler, T. Tsang, M. H. Chen, D. A. Ringers, and B. Craseman, *Phys. Rev. A*, **9**, 1070 (1974).
127. M. H. Chen, B. Crasemann, L. I. Yin, T. Tsang, and I. Adler, *Phys. Rev. A*, **13**, 1435 (1976).
128. J. S. Jen and T. D. Thomas, *Phys. Rev. B*, **13**, 5284 (1976).
129. E. J. McGuire, Sandia Research Lab. Research Report, No. San-75-0443, 1975.
130. O. Almen and K. O. Nielsen, *Nucl. Instrum. Methods*, **1**, 302 (1957).
131. S. Bashkin and J. O. Stoner, *Atomic Energy Levels and Grotrian Diagrams*, North-Holland, Amsterdam, 1975, Vol. 1.
132. U. I. Safranova and V. N. Kharitonova, *Opt. Spectrosc.*, **27**, 300 (1969).
133. R. Spohr, T. Bergmark, N. Magnusson, L. O. Werme, C. Nordling, and K. Siegbahn, *Phys. Scr.*, **2**, 31 (1970).

134. I. B. Ortenburger and P. S. Bagus, *Phys. Rev. A*, **11**, 1501 (1975).
135. M. Thompson, P. A. Hewitt, and D. S. Wooliscroft, *Anal. Chem.*, **48**, 1336 (1976).
136. G. N. Killoran and J. F. Tyson, *European Spectroscopy News*, **23**, 43 (1979).
137. R. R. Rye, J. E. Houston, D. R. Jennison, T. E. Madey, and P. H. Holloway, Report SAND-78-0141C.
138. D. R. Jennison, Report SAND-79-1240C.
139. D. R. Jennison, *Chem. Phys. Lett.*, **69**, 435 (1980).
140. A. C. Hurley, *J. Chem. Phys.*, **59**, 3656 (1971).
141. G. N. Killoran and J. F. Tyson, *Proc. Analyt. Div. Chem. Soc.*, **16**, 29 (1979).
142. I. H. Hillier and J. Kendrick, *Mol. Phys.*, **31**, 849 (1976).
143. L. Karlsson, L. O. Werme, T. Bergmark, and K. Siegbahn, *J. Elec. Spectrosc. Rel. Phenom.*, **3**, 181 (1974).
144. W. Mehlhorn, "Physics of Electronic and Atomic Collisions," *VII ICPEAC*, 1971, p. 169.
145. F. Meyer and J. J. Vrakking, *Phys. Lett. A*, **44**, 511 (1973).
146. K. Faegri, Jr. and R. Manne, *Mol. Phys.*, **31**, 1037 (1976).
147. M. T. Okland, K. Faegri, Jr., and R. Manne, *Chem. Phys. Lett.*, **40**, 185 (1976).
148. J. M. White, R. R. Rye, and J. E. Houston, *Chem. Phys. Lett.*, **46**, 146 (1977).
149. R. W. Shaw, Jr., J. S. Jen, and T. D. Thomas, *J. Elec. Spectrosc. Rel. Phenom.*, **11**, 91 (1977).
150. C. T. Cambell, J. W. Rogers, Jr., R. L. Hance, and J. M. White, *Chem. Phys. Lett.*, **69**, 430 (1980).
151. M. Thompson, P. A. Hewitt, and D. S. Wooliscroft, *Anal. Chem.*, **50**, 690 (1978).
152. T. A. Carlson, W. E. Moddeman, B. P. Pullen, and M. O. Krause, *Chem. Phys. Lett.*, **5**, 390 (1970).
153. J. A. Connor, I. H. Hillier, J. Kendrick, M. Barber, and A. Barrie, *J. Chem. Phys.*, **64**, 3325 (1976).
154. G. N. Killoran, M. Phil. Thesis, University of Loughborough, 1978.
155. H. Siegbahn, L. Asplund, and P. Kelfve, *Chem. Phys. Lett.*, **35**, 330 (1975).
156. H. Agren, S. Svensson, and U. I. Wahlgren, *Chem. Phys. Lett.*, **35**, 336 (1975).
157. H. Agren and H. Siegbahn, *Chem. Phys. Lett.*, **69**, 424 (1980).
158. R. W. Shaw, Jr., and T. D. Thomas, *Phys. Rev. A*, **11**, 1491 (1975).
159. P. Kelfve, B. Blomster, H. Siegbahn, K. Siegbahn, E. Sanhueza, and O. Goscinski, *Phys. Scr.*, **21**, 75 (1980).
160. F. Maracci, R. Platania, and R. Salomone, *J. Elec. Spectrosc. Rel. Phenom.*, **19**, 155 (1980).
161. O. Keski-Rahkonen and M. O. Krause, *J. Elec. Spectrosc. Rel. Phenom.*, **9**, 371 (1976).
162. L. Asplund, P. Kelfve, H. Siegbahn, O. Goscinski, H. Fellner-Feldegg, K. Hamrin, B. Blomster, and K. Siegbahn, *Chem. Phys. Lett.*, **40**, 353 (1976).
163. L. Asplund, P. Kelfve, B. Blomster, H. Siegbahn, K. Siegbahn, R. L. Lozes, and U. I. Wahlgren, *Phys. Scr.*, **16**, 273 (1977).
164. K. Faegri and O. Keski-Rahkonen, *J. Elec. Spectrosc. Rel. Phenom.*, **11**, 275 (1977).
165. R. H. A. Eade, M. A. Robb, G. Theordorakopoulos, and I. G. Csizmadia, *Chem. Phys. Lett.*, **52**, 526 (1977).

166. G. Theodorakopoulos, I. G. Csizmadia, and M. A. Robb, *Chem. Phys. Lett.*, **69**, 66 (1980).

167. B. Cleff and W. Mehlhorn, *Z. Phys.*, **219**, 311 (1969).

168. J. F. Tyson, *Anal. Proc.*, **18**, 120 (1981).

169. G. Theodorakopoulos, C. A. Nicolaides, and D. R. Beck, *Int. J. Quantum Chem.*, *Quantum Chem. Symp.*, **13**, 671 (1979).

170. W. B. Perry and W. L. Jolly, *Chem. Phys. Lett.*, **23**, 529 (1973).

171. C. J. Ballhausen and H. B. Gray, *Molecular Orbital Theory*, W. A. Benjamin, New York, 1965.

172. R. E. Ballard, *Photoelectron Spectroscopy and Molecular Orbital Theory*, Adam Hilger, Bristol, 1978.

173. A. Vincent, *Molecular Symmetry and Group Theory*, Wiley, London, 1977.

174. F. A. Cotton, *Chemical Applications of Group Theory*, 2nd ed., Wiley-Interscience, New York, 1971.

175. K. Faegri, *Chem. Phys. Lett.*, **46**, 541 (1977).

176. R. R. Rye, T. E. Madey, J. E. Houston, and P. H. Holloway, *J. Chem. Phys.*, **69**, 1504 (1978).

177. D. B. Neumann and J. W. Moskowitz, *J. Chem. Phys.*, **30**, 673 (1959).

178. M. A. Robb, G. Theodorakopoulos, and I. G. Csizmadia, *Chem. Phys. Lett.*, **57**, 423 (1978).

179. R. Camilloni, G. Stefani, and A. Giardini-Guidoni, *Chem. Phys. Lett.*, **50**, 213 (1977).

180. F. P. Larkins and A. Lubenfield, *J. Elec. Spectrosc. Rel. Phenom.*, **15**, 137 (1979).

181. T. Kawai, K. Kunimori, T. Kondow, T. Onishi, and K. Tamaru, *Phys. Rev. Lett.*, **33**, 535 (1974).

182. H. Hartmann and H. Gebelein, *Theor. Chim. Acta*, **22**, 39 (1971).

183. R. W. Ditchburn, *Proc. Roy. Soc. A*, **229**, 44 (1955).

184. R. T. Sanderson, *Chemical Periodicity*, Reinhold, New York, 1960.

185. D. Stalherm, B. Cleff, H. Hillig, and W. Mehlhorn, *Z. Naturforsch.*, **24a**, 1728 (1969).

186. A. S. Newton and A. F. Sciammana, *J. Chem. Phys.*, **53**, 132 (1970).

187. M. Tronc, G. C. King, R. C. Bradford, and F. H. Read, *J. Phys. B*, **9**, L555 (1976).

188. N. L. Jorgensen and L. Salem, *The Organic Chemist's Book of Orbitals*, Academic Press, New York, 1973.

189. D. W. Turner, C. Baker, A. D. Baker, and C. R. Brundle, *Molecular Photoelectron Spectroscopy*, Wiley-Interscience, London, 1970.

190. R. G. Cavell and R. Sodhi, *J. Elec. Spectrosc. Rel. Phenom.*, **15**, 145 (1979).

191. K. Siegbahn, C. Nordling, A. Fahlman, R. Nordberg, K. Hamrin, J. Medman, G. Johansson, T. Bergmark, J. E. Karlsson, I. Lindgren, and B. Lindberg, "ESCA—Atomic, Molecular and Solid State Structure Studied by Means of Electron Spectroscopy," *Nova Acta Regiae Soc. Sci. Upsaliensis Ser. IV*, Vol. 20, 1967.

CHAPTER

5

THE AUGER CHEMICAL SHIFT AND SOME APPLICATIONS OF AUGER ELECTRON SPECTROSCOPY IN CHEMISTRY

1. INTRODUCTION

The analytical usefulness of Auger electron spectroscopy has already been discussed in terms of surface atomic identification and quantification, which represent by far the greatest use to which the technique is put, and in Chapter 4 the analytical information obtainable from gas-phase spectra was outlined for the limited number of molecular species studied so far. Some speculation about the future role and possibilities of gas-phase AES was made in the previous chapter. As was indicated, however, in the discussion of the theoretical studies of free-metal-atom spectra and of certain molecular species, the motivation behind several of these studies has been the investigation of various shifts in energy observed in the Auger lines of the same element in a series of compounds, that is, chemical environments, in the case of vapor-phase molecular species; and in the case of free-metal atoms, shifts were observed between the lines from the same transition in the solid and in the vapor, although the difficulties of establishing appropriate reference energy levels for the two phases in the latter case were shown to lead to quite considerable errors. Although the energy shifts in a transition on going from the gas phase to the solid might be considered, at least by a chemist, to be a physical effect, both of these effects are known by the general term "chemical shift." The origin of the effects in both cases, of course, being the difference in the chemical environment or bonding or interaction with neighboring atoms of the atom that is core ionized at the start of the Auger process.

As was also discussed in the previous chapter, a shift in the energy of the lines is not the only manifestation of a change of chemical environment. In addition, line shapes were found to change, particularly changes in line width (mainly due to changes in the relative lifetimes of the states involved in the transitions—usually the final state). For the studies on the phase change of the metals, this phenomenon is universally observed and is referred to as "solid-state broadening." In molecular species, unresolved vibrational structure arising from long-lived final states may also contribute to changes in line shape. Changes in the spectra of molecular species owing to a change from vapor to molecular solid will be discussed later in this chapter. Changes in relative line intensities are also observed which may be due to changing roles of the various satellite processes (shake up, shake off, Coster–Kronig transitions etc.). Such effects can cause problems in quantitative surface analysis.

In addition to attracting the interest of theoretical physicists, since "chemical" effects provide a severe test for any atomic or molecular-orbital theory, the existence of these effects is of considerable interest to the analytical chemist. It is no accident that x-ray photoelectron spectroscopy has come to be known colloquially as ESCA—electron spectrometry for *chemical analysis*—since the shifts in energies of the ejected photoelectrons provide a powerful method of probing the chemical environment of mainly surface species but also of atoms in free molecules. A quick scan of any recent review of surface characterization[1,2] reveals the extent to which XPS (and AES) is used to provide information about the chemical composition of surfaces and is therefore applied to the study of a wide range of processes: chemisorption, adsorption reactions on surfaces, corrosion, contamination, lubrication, electrochemistry, etc.

The vast majority of the applications of both these techniques takes the form of following the spectra as a function of time as the surface is subjected to some perturbation (fracture, ion-etching, etc.) or as a reaction proceeds (oxidation, catalysis, etc.) and comparing the "before" and "after" situations as the basis of the extraction of the information of analytical utility. Thus the analytical chemist does not need to understand the detailed theoretical physics behind these chemical-state effects, but requires a qualitative, general term picture (not quite empirical!) of the origin of these effects to be able to apply them to solving analytical problems.

2.   THE ORIGIN OF THE AUGER CHEMICAL SHIFT

Obviously, since three orbitals are involved in the Auger transition, the overall shift in the Auger energy will be the algebraic sum of the shifts of the individual orbital energies. The first level to be considered is the initial core level. As this is the level from which the corresponding photoelectron in XPS is obtained, the chemical shift in this level gives rise to the chemical shift of the XPS spectrum. A detailed discussion of XPS chemical shifts is considered beyond the scope of this chapter. The topic has been treated on a rigorous theoretical basis (see, for example, References 3 and 4), but more readily understandable treatments are also to be found.[5]

2.1   Free-Atom–Molecule, Free-Atom–Conducting Solid Core Binding and XPS Shifts

For an isolated atom the binding energy of an electron, E_a^B, is given by

$$E_a^B = h\nu - E^{PE} \tag{5.1}$$

where $h\nu$ is the energy of the ionizing x-radiation and E^{PE} is the kinetic energy of the photoelectron. Unlike the situation for UV photoelectron spectroscopy, it is not correct to equate E_a^B with the orbital energy ε (calculated by rigorous nonrelativistic Hartree–Fock methods), an approximation known as Koopman's theorem, since the electrons in the remaining orbitals "relax" as the photoelectron leaves. This relaxation is a flow of negative charge inward under the action of the increased nuclear charge, which is experienced by the remaining electrons owing to the removal of a screening electron. Koopman's theorem is thus known as a "frozen-orbital approximation." The relaxation stabilizes the ion and thus

$$E_a^B = \varepsilon - R \tag{5.2}$$

where E_a^B is the binding energy referred to the vacuum level (the energy required to remove an electron to infinity) and R is the relaxation energy, sometimes referred to as rearrangement energy. To avoid confusion between another type of relaxation energy, which will be discussed shortly, R is known as the intra-atomic relaxation energy.

For the same atom in a gaseous molecule, two new effects manifest themselves; first, there will be a chemical shift in ε, $\Delta\varepsilon$, due to the bonding and, second, electrons from the neighboring atoms will be able to take part in the relaxation process. Assuming that the intra-atomic relaxation energy does not change, then

$$E_m^B = \varepsilon + \Delta\varepsilon - (R + R_m^{ea}) \tag{5.3}$$

where E_m^B is the binding energy of the electron in the molecule referred to the vacuum level and R_m^{ea} is the extra-atomic relaxation energy arising from the involvement of electrons from neighboring atoms. R_m^{ea} is sometimes referred to as polarization energy. Subtracting Eq. (5.2) from Eq. (5.3) gives the chemical shift in binding energy between atom and molecule:

$$\Delta E_{am}^B = \Delta\varepsilon - R_m^{ea} \tag{5.4}$$

and thus, via Eq. (5.1), the shift in the line in the photoelectron spectrum, E_{am}^{PE}, would be given by

$$\Delta E_{am}^{PE} = R_m^{ea} - \Delta\varepsilon \tag{5.5}$$

Similar equations are derived for the shift between the free atom and the conducting solid:

$$\Delta E_{ac}^{B} = \Delta\varepsilon_c - R_c^{ea} \tag{5.6}$$

$$\Delta E_{ac}^{PE} = R_c^{ea} - \Delta\varepsilon_c \tag{5.7}$$

The value of $\Delta\varepsilon$ in this case is negligible compared with the extra-atomic relaxation energy term,[6] which is given by[7]

$$E_c^{ea} = [e^2(1 - 1/k)]/2r \tag{5.8}$$

where e is the charge on the ion, r is the radius of minimum electron screening distance, and k is the dielectric constant.

2.2   Free-Atom–Molecule, –Conducting Solid Auger Shifts

Starting with Eq. (2.15) for the complete Auger process and rewriting in terms of the notation used in this section, for a KLL transition

$$E_a^A = E_a^B(K) - E_a^B(L) - E_a^B(L) + R^S + S(LL) \tag{5.9}$$

where E_a^A = kinetic energy of the electron ejected from the free atom
$E_a^B(K)$ = binding energy of the K electron in the free atom
$E_a^B(L)$ = binding energy of the L electron in the free atom
$S(LL)$ = electron–electron interaction term, the sum of the final state coupling terms due to the presence of the two holes
R^S = intra-atomic relaxation energy for the L shell, sometimes known as static relaxation energy.[8]

Substituting from Eq. (5.2) with R and ε terms for both K and L shells

$$E_a^A = \varepsilon(K) - R_k(K) - 2\varepsilon_k(L) + 2R_L(L) + R^S - S_{LL} \tag{5.10}$$

Thus the shift in energy when the atom is incorporated in a gas-phase molecule

$$E_{am}^A = \Delta\varepsilon_m - 2\Delta\varepsilon_m(L) + R_m^{ea}(L) - R_m^{ea}(K) \tag{5.11}$$

where $R_m^{ea}(L)$ is the extra-atomic relaxation energy for the two-hole final state. To a good approximation $\Delta\varepsilon_m(K) = \Delta\varepsilon_m(L)$, thus Eq. (5.11) reduces to

$$\Delta E_{am}^A = -\Delta\varepsilon_m(L) + R_m^{ea}(L) - R_m^{ea}(K) \tag{5.12}$$

As $R_m^{ea}(L)$ is likely to be several times the magnitude of $R_m^{ea}(K)$, because the final state for the Auger transition is doubly charged, it would be expected that Auger molecular shifts would be positive and would increase with increasing polarizability of the bonding, that is, decreasing electronegativity of the neighboring atoms. According to Eq. (5.5), the corresponding XPS line shifts would also be positive, but it is not possible to predict the relative magnitude of the shifts.

There does not appear to have been an experimental study of photoelectron and Auger shifts on going from the free atom to a series of gas-phase molecular species. However, a number of gas-phase molecules have been studied from the point of view of chemical shift (see Chapter 4), and the results for a study of sulfur compounds[9] is shown in Table 5.1. The Auger line used was the $^1D_2$ line of the $KL_{2,3}L_{2,3}$ transition. The photoelectron line originated from the sulfur $2p_{3/2}$ orbital. In both cases the results from Reference 9 have been recalculated with sulfur hexafluoride as the reference compound; the XPS shift is given as a photoelectron line shift rather than as a shift in binding energy. As can be seen from the table, the general conclusions reached on the basis of Eqs. (5.5) and (5.12) would appear to be valid.

It is important to note that the trends exhibited in the Auger line shifts for this series of molecules are for transitions that involve core electrons only, that is, none of the electrons participating in the transition is involved in the bonding. The situation is more complex for Auger transitions involving valence electrons and can produce trends in the opposite sense to that exhibited by the molecules listed in Table 5.1. For example, in a study of bromine-substituted methanes, Spohr et al.[10] found that the carbon KLL Auger transitions shifted to higher energy with increasing bromine substitution, although bromine is more polarizable than hydrogen.[11]

The shift in the Auger line obtained on going from the free atom to the conducting solid is given by

$$\Delta E^A_{ac} = -\Delta\varepsilon_c(L) + 3R^{ea}_c(K) \tag{5.13}$$

This is assuming that Eq. (5.8) applies with r the same for a one-hole and a two-hole state. The Auger final-state charge of $+2$ means that $R^{ea}_c(L)$ will be four times the magnitude of $R^{ea}_c(K)$.

Table 5.1. XPS and Auger Shifts (eV) For a Series of Gaseous Sulfur Compounds

Compound	ΔE^A ($^1D_2$)	ΔE^{PE} ($2p_{3/2}$)
CS_2	8.88	10.25
COS	6.72	9.51
H_2S	5.90	9.84
SO_2	2.88	5.44
SF_6	0	0

Again there is a shortage of experimental results against which to check the predictions of Eqs. (5.7) and (5.13). The results of a study on Zn and Cd vapors[12] in which the XPS and Auger spectra of the vapor and solid metals were obtained simultaneously are given in Table 5.2. Also given in the table are data compiled by Wagner[13] for some inert gases. The solid-state values were obtained by ion implantation in iron and the gas-phase values were taken from the book by Siegbahn et al.[14] The solid-phase values are greater than the vapor-phase values.

2.3. Insulating Solid Shifts

If, instead of the solid element shift considered in the previous treatment, an insulating solid compound is considered, then an equation analogous to Eq. (5.7) is derived for the XPS shifts and to Eq. (5.13) for the Auger shift. Again it would be expected that the Auger shifts would be larger than the XPS shifts. The same argument would apply if the shift on going from solid element to insulating solids were considered. The appropriate equations are derived as follows.

The shift in energy of a core binding level on going from atom to insulating solid is

$$\Delta E_{ai}^{B} = \Delta \varepsilon_i - R_i^{ea} \tag{5.14}$$

and thus the shift on going from solid element to compound is, by subtracting Eq. (5.6) from Eq. (5.14),

$$\Delta E_{ci}^{B} = \Delta \varepsilon_i - \Delta \varepsilon_c + R_c^{ea} - R_i^{ea} \tag{5.15}$$

Table 5.2.   XPS and Auger Vapor–Solid Shifts (eV)

Element	ΔE_{am}^{A}	ΔE_{am}^{PE}
Zn	13.10 (LMM)	3.25 ($2p_{1/2}$)
		3.15 ($2p_{3/2}$)
Cd	11.80 (MNN)	3.00 ($3d_{3/2}$)
		2.95 ($3d_{5/2}$)
Ne	8.9 (KLL)	2.5 ($1s$)
Ar	9.1 (LMM)	2.5 ($2p_{3/2}$)
Xe	7.8 (MNN)	1.9 ($3d_{5/2}$)

Thus the shift in the photoelectron line energy will be

$$\Delta E_{ci}^{PE} = R_i^{ea} - R_c^{ea} + \Delta\varepsilon_c - \Delta\varepsilon_i \qquad (5.16)$$

Similarly the shift in energy of an Auger line on going from free atom to insulating solid is

$$\Delta E_{ai}^{A} = -\Delta\varepsilon_c(L) + 3R_i^{ea}(K) \qquad (5.17)$$

and thus the shift on going from solid element to compound is obtained by subtracting Eq. (5.13) from Eq. (5.14),

$$\Delta E_{ci}^{A} = \Delta\varepsilon_c(L) - \Delta\varepsilon_i(L) + 3R_i^{ea}(K) - 3R_c^{ea}(K) \qquad (5.18)$$

There are considerably more data on photoelectron and Auger shifts of various compounds of the same element. The shifts for a series of sodium compounds are shown in Table 5.3.[13] The anions at the top of the table are the most polarizable and those at the bottom are the least polarizable. Examination of Eq. (5.18) shows that the trend in Auger shifts found experimentally is predicted by the theoretical approach adopted here. If the terms involving $\Delta\varepsilon$ are ignored as approximating to zero, then it can be seen that

Table 5.3. **Photoelectron and Auger Line Shifts (eV) Relative to the Solid Metal for a Series of Sodium Compounds**

Compound	ΔE^{PE} (1s)	ΔE^A (KLL $^1D_2$)
NaI	+0.3	−2.8
NaBiO$_3$	+1.1	−2.4
NaBr	+0.2	−3.4
Na$_2$CrO$_4$	+0.9	−2.8
Na$_2$MoO$_4$	+1.0	−3.0
NaCl	+0.3	−3.7
NaSCN	+0.6	−3.5
Na$_2$S$_2$O$_4$	+0.7	−3.4
Na$_2$SO$_3$	+0.6	−3.6
Na$_2$WO$_4$	−0.1	−4.4
NaNO$_2$	+0.3	−4.2
NaPO$_3$	+0.2	−4.7
NaNO$_3$	+0.5	−4.4
Na$_2$SO$_4$	+0.7	−4.2
Na$_3$AlF$_6$	+0.1	−6.0
NaF	+0.7	−5.4
NaBF$_4$	−0.8	−6.9

as the polarizability of the anion decreases, the amount of extra-atomic relaxation energy due to the anion environment becomes smaller, that is, $R_i^{ea}(K)$ becomes smaller and ΔE_{ci}^A thus becomes more negative, since the value of $R_c^{ea}(K)$ is a constant. Comparison with Eq. (5.16) would suggest that the Auger shifts will be larger than the corresponding photoelectron line shifts and this is borne out by the data given in Table 5.3. However, it would be expected that the shifts in energy would be in the same direction, but this turns out not to be the case, showing that the somewhat simplified approach adopted here is not capable of predicting accurately small shifts in core binding energies.

For a more-detailed and rigorous treatment of relaxation processes the papers by Hoogewijs[15] and Thomas[16] should be consulted.

2.4. Practical Aspects

From the previous discussion it might appear that for the characterization of the chemical state the use of Auger electron spectroscopy would be preferable to XPS because the magnitude of the shifts is much greater (see Table 5.3). In practice, however, it is XPS that is used predominantly to monitor changes in the chemical environment in surface studies, while AES is used to provide element identification and quantification, which can also be spatially resolved. The major drawback to the use of Auger chemical shifts is that most Auger instruments use an electron beam to create the initial core vacancy in the sample material rather than the x-radiation used in XPS. There is, consequently, a high background level owing to scattered secondary electrons, and to improve the Auger signal-to-background ratio, the spectra are normally obtained in first-derivative form (see Chapter 3, Section 2) making location of the exact position of the peak maximum difficult. Moreover, in order to increase the sensitivity of the spectrometer, the slits are opened to allow as large a flux of electrons to pass through as practicable with a consequent loss in resolution which means that the line shapes are rather broad and the profiles may contain several overlapping lines, again making the location of a peak maximum difficult. Thus a comparison of peak energies obtained on different instruments becomes rather difficult, since it is customary to measure the energy to the sharp peak of the downside of the first differential which corresponds to the inflection point on the high-energy side of the undifferentiated curve.

In addition to these instrumental problems, which may also include the problem of which reference level is used as the base line for the energy scale, bearing in mind that the energy actually measured is the Auger kinetic energy reduced by the difference between the work functions of the analyzer and sample, there are the problems associated with using an electron beam as the ionizing source. When the sample material is an insulator, then there are difficulties associated with sample charging and radiation damage. The first of these will distort the energy at which the peaks appear in the spectrum and the second will give rise to a change in the chemical state that is being measured. Obviously both effects makes the use of energy shifts as a means of identifying chemical states very difficult indeed.

It would appear, then, that the use of Auger spectra of solids from an instrument using an electron gun as source to identify chemical states through energy shifts is not possible. However, it may be that electron-excited Auger spectra can be used for line-shape analysis methods of investigating chemical states. This topic will be discussed later (see Section 5 and Chapter 6). Thus x-ray excitation is required to produce spectra that are useful from the viewpoint of energy shift information. With x-rays the radiation damage is much less serious than with electrons because the energy absorbed per unit time per unit weight at the surface of the sample is very much less. Obviously, charging problems are not apparent with x-ray sources. Auger instruments are not, in general, constructed with x-ray sources, but Auger lines are observed with x-ray photoelectron (XPS) instruments, displayed as is normal with *XPS* as the undifferentiated spectrum. To make use of Auger line shifts for identification of chemical states, the Auger lines observed in XPS spectra should be used. The role of Auger lines in photoelectron spectroscopy has been discussed by Wagner,[17] who earlier had proposed a combination of Auger and XPS line shifts as a means of characterizing chemical states,[13] known as the Auger parameter. This parameter is discussed in the following section.

3. THE AUGER PARAMETER

Following an early publication[18] concerning the Auger lines appearing in XPS spectra, Wagner[13,19] proposed that the energy difference between an Auger line and a photoelectron line be used as a parameter, which he called the Auger parameter, α, characteristic of chemical state. The reasoning behind this is that because α is the difference between two lines, any static charge corrections cancel and thus α can be measured more accurately than the individual absolute line energies. The parameter is normally calculated as the difference between the sharpest Auger line and the most intense photoelectron line. This idea was touched on by a number of other workers— Shirley,[20] Siegbahn and co-workers,[9] and Lawson.[21]

An expression for α for a free atom may be readily derived from the appropriate equations in Section 5.2; by definition

$$\alpha_a = E_a^A - E_a^{PE} \tag{5.19}$$

substituting for E_a^A from Eq. (5.10) and for E_a^{PE} from Eq. (5.1)

$$\alpha_a = \varepsilon(K) - R(K) - 2\varepsilon(L) + 2R(L) + R^S - S_{LL} - (h\nu - E_a^B)$$

substituting for E_a^B from Eq. (5.2),

$$\alpha_a = 2\varepsilon(K) - 2R(K) - 2\varepsilon(L) + 2R(L) + R^S - S_{LL} - h\nu \qquad (5.20)$$

When shifts to different chemical states are considered (to free molecule, solid conducting metal, or insulating solid), most of the terms in Eq. (5.20) are constant and thus cancel when $\Delta\alpha$ is calculated. Thus for the shift from free atom to free molecular environment, subtracting Eq. (5.5) from Eq. (5.11) and assuming the $\Delta\varepsilon$ terms are approximately equal,

$$\Delta\alpha_{am} = R_m^{ea}(L) - 2R_m^{ea}(K) \qquad (5.21)$$

Thus the shift in the value of α is entirely an effect of extra-atomic relaxation. The analogous expression for the change in α on going from the free atom to the atom in a conducting solid is obtained by subtracting Eq. (5.7) from Eq. (5.13),

$$\Delta\alpha_{ac} = 2R_c^{ea}(K) \qquad (5.22)$$

Finally, the expression for the shift between the element in a conducting solid and an insulating compound is

$$\Delta\alpha_{ci} = 2[R_i^{ea}(K) - R_c^{ea}(K)] \qquad (5.23)$$

This equation is derived by subtracting Eq. (5.16) from Eq. (5.18). Thus in all cases in this treatment the change in the Auger parameter with change in the chemical environment reflects the change in extra-atomic relaxation energy, that is, the difference in polarization of the media. The corresponding $\Delta\alpha$ values for the situations exemplified in Table 5.1, 5.2, and 5.3 can readily be calculated by simply subtracting the ΔE^{PE} value from the corresponding ΔE^A value. Table 5.4 gives the full data for the series of sodium compounds listed in Table 5.3 and also the corresponding data for a series of indium compounds which cover a range of formal oxidation states. The Auger line for indium was the $M_4N_{4,5}N_{4,5}$ most-intense transition, and the corresponding photoelectron line was for the $3d_{5/2}$ electron. The values of $\Delta\alpha$ and α for the series of sodium and indium compounds illustrates the drawbacks of this particular way of defining α (and hence $\Delta\alpha$) in that the value of α may be either positive or negative, and thus, although the value of α decreases for each set of compounds as the polarizability of the anion decreases in the sense at least that α becomes more negative, the trend in $\Delta\alpha$ for the set of indium values is difficult to visualize. One way of overcoming this problem was proposed by Wagner and that was to arbitrarily add 1000 or 2000 to

Table 5.4. Auger and Photoelectron Line Energies (eV) and the Corresponding Auger Parameter (the Spectra were Obtained using Al K_α x-rays at 1486.6 eV

Compound	E^A	E^{PE}	α	$\Delta\alpha$ (relative to metal)
Na	994.2	414.9	579.3	0
NaI	991.4	415.2	576.2	-3.1
$NaBiO_3$	991.8	416.0	575.8	-3.5
NaBr	990.8	415.1	575.7	-3.6
Na_2CrO_4	991.4	415.8	575.6	-3.7
Na_2MoO_4	991.2	415.9	573.3	-4.0
NaCl	990.5	415.2	575.3	-4.0
NaSCN	990.7	415.5	575.2	-4.1
$Na_2S_2O_4$	990.8	415.6	575.2	-4.1
Na_2SO_3	990.6	415.5	575.1	-4.2
Na_2WO_4	989.8	414.8	575.0	-4.3
$NaNO_2$	990.0	415.2	574.8	-4.5
$NaPO_3$	989.5	415.1	574.4	-4.9
$NaNO_3$	989.8	415.4	574.4	-4.9
Na_2SO_4	990.0	415.6	574.4	-4.9
Na_3AlF_6	988.2	415.0	573.2	-6.1
NaF	988.8	415.6	573.2	-6.1
$NaBF_4$	987.3	414.1	573.2	-6.1
In	410.3	1042.5	-632.2	0
In_2S_3	407.5	1042.1	-634.6	-2.4
InI_3	406.0	1041.0	-635.0	-2.8
$InBr_3$	405.0	1040.8	-635.8	-3.6
InCl	405.9	1042.0	-636.1	-3.9
$InCl_3$	404.8	1040.8	-636.0	-3.8
In_2O	407.0	1042.5	-635.5	-3.3
In_2O_3	406.6	1042.5	-635.9	-3.7
InF_3	403.9	1040.6	-636.7	-4.5
$(NH_4)_3InF_6$	404.3	1041.2	-636.9	-4.7

the value of α if it happened to be negative in order to turn it into a large, positive number.

A much more elegant solution to the problem was used by Gaarenstroom and Winograd[22] almost incidentally in their consideration of initial- and final-state effects in the XPS spectra of cadmium and silver oxide, which was to redefine α as the sum of the Auger line energy and the binding energy corresponding to the photoelectron line in the XPS spectrum. That is, in the notation of Section 2 above,

$$\alpha' = E^A + E^B \qquad (5.24)$$

A comparison with Eq. (5.19) shows that the new Auger parameter α' is simply the old value α with the addition of the energy of the ionizing x-rays, $h\nu$. Thus in addition to producing a positive number for the Auger parameter, the new definition makes α' independent of the energy of the x-rays used in a particular apparatus. This, of course, would not affect the values of $\Delta\alpha$ since the x-ray energy cancels in calculating the change in Auger parameter.

It can be seen from Table 5.4 that although certain trends show up well, the Auger parameter does not distinguish between the different formal oxidations of $+1$ and $+3$ of indium. Examination of the values for the anions of the sodium salts shows that fluoroanions have low polarizability, the halogens appear in the expected order; nitrate, phosphate, and bismuthate are shown to be more polarizable as the atomic number of the element bound to oxygen increases, whereas the opposite behavior is noted for the oxyanions containing metals, namely, chromate, molybdate, and tungstate. There is also good correlation with Pearson's[23] designation of anions as "hard" (least polarizable) or "soft" (most polarizable) on the basis of their behavior in solution, despite the fact that the $\Delta\alpha$ values correspond to the anions in a crystal lattice without solvation.

The Auger parameter is essentially a zero-dimensional representation of the chemical state and as can be seen from the table does not always provide a unique value characteristic of a particular environment. Better discrimination between chemical environments can be obtained from a two-dimensional display of the information. This may be conveniently done by plotting Auger kinetic energy as the ordinate and photoelectron binding energy as the abscissa of normal Cartesian coordinates; the Auger parameter would thus be characterized by a series of diagonal lines of slope -1 forming an Auger parameter grid and converting the zero dimensionality to one dimensionality. This is illustrated in Fig. 5.1 for a number of elements from the first long row of the periodic table. The data are plotted according to the modified Auger parameter, Eq. (5.24), for the $2p_{3/2}$ photoelectron and the LMV, LVV, or LMM Auger transition. A similar plot for elements in the second long row is shown in Fig. 5.2 for the $3d_{5/2}$ photoelectron and the M_4VV, M_5VV, or $M_4N_{4,5}N_{4,5}$ Auger transition. The Auger parameters are represented by the set of diagonal lines. Obviously, as would be expected from the fact that E^A and E^B are both unique functions of atomic number, the Auger parameter provides unambiguous elemental identification. However, for the possibility of distinguishing between different chemical environments of the same element it is the distribution of points within the small squares that is of interest. It is clear from Figs. 5.1 and 5.2 that the magnitude of the chemical

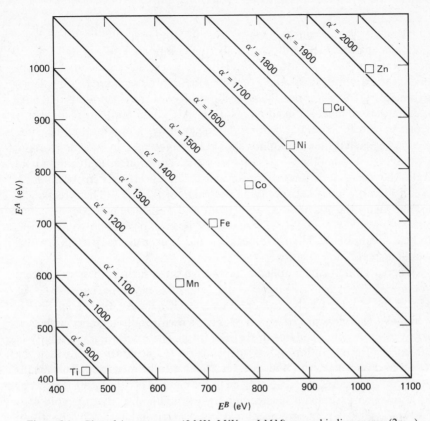

Figure 5.1. Plot of Auger energy (*LMV*, *LVV*, *or LMM*) versus binding energy ($2p_{3/2}$).

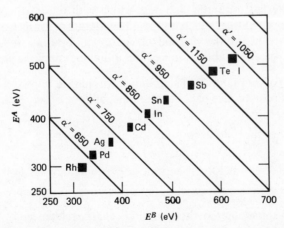

Figure 5.2. Plot of Auger energy (M_4VV, M_5VV, *or* $M_4N_{4,5}N_{4,5}$) versus binding energy ($3d_{5/2}$).

229

shift (both Auger and photoelectron) is small compared with the intraelement shift. The intraelement shift for the sodium compounds shown in Table 5.4 is depicted in Fig. 5.3. In other words Fig. 5.3 is a scale expansion of the type of data shown in Figs. 5.1 and 5.2. As can be seen by representing the Auger parameter as a line, it is possible to distinguish between compounds that have the same value of α', for example, NaF, NaAlF$_6$, and NaBF$_4$ are all clearly separated on the $\alpha' = 2059.8$ line. The uncertainty in making the measurements means that the compounds should be represented by rectangles about 1 eV in length along the α' lines (due to static charge uncertainty) and about 0.2 eV between the α' lines. In other words, the uncertainty in specifying α' is considerably less than specifying the value of E^A or E^B. These rectangles are shown for the fluoride-containing anions. Distinction between the sulfur-containing anions S$_2$O$_4$, SCN, and SO$_3$ on the basis of α' alone would appear to be difficult, but C, N, and O peaks would appear in the spectrum, which would also aid identification.

Wagner et al. have compiled a number of plots such as these in a recent volume of *Analytical Chemistry*[24] for a total of 24 elements, namely, F, Na, Mg, Si, Ti, Mn, Fe, Co, Ni, Cu, Zn, As, Se, Br, Rh, Pa, Ag, Cd, In, Sn, Sb, Te, I, and W. The plots are based on Wagner's own results together with data taken from 26 other sources in the literature. The discriminating ability of this form of presentation of data depends on the accuracy with which the Auger and photoelectron line energies are able to be measured. This means that some consideration must be given to the method of calibrating the energy scale. Wagner assumed a spectrometer work function so that the gold 4$f_{7/2}$ line was at 83.8 eV; this produced a carbon 1s line from residual hydrocarbon at 284.6 eV and this value was then used to correct the energies of lines from insulating materials for static charging, which with the instrument used in this study amounted to as much as 5 eV. The literature sources of data were critically evaluated for the method of energy calibration, and data were considered to be acceptable if any of the following methods were used:

1. The use of thin samples. The thinness was judged to be suitable if lines from the conductive backing material could be observed with reasonable intensity.
2. The use of the gold decoration method.
3. The use of residual hydrocarbon C 1s line.
4. The use of internal hydrocarbon or cyano ligand[25] C 1s lines.
5. The use of codeposition of a hydrocarbon from the vapor phase.

Data based on some other methods, notably charge referenced by a calibrant powder in admixture and use of C 1s line from a polymer film sample mount, were not included.

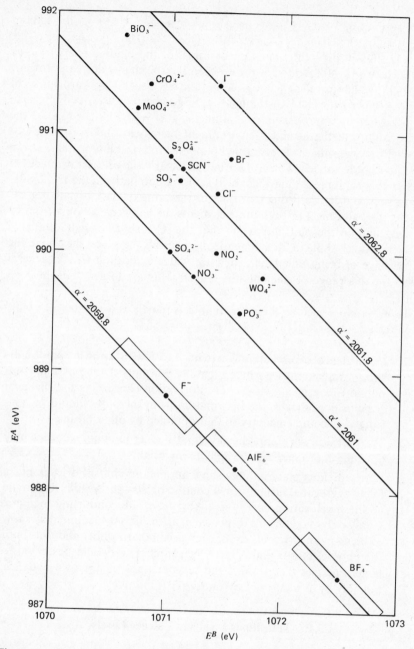

Figure 5.3. Plot of Auger energy versus binding energy for sodium compounds given in Table 5.4.

Examples of Wagner's plots are shown in Figs. 5.4 and 5.5 for sodium and indium, the elements considered in Table 5.4. The presentation of the data is slightly different from that of Fig. 5.3; the binding energy of the photoelectron is still plotted as the abscissa but the energy decreases toward the right, which has the effect of making the α' "isobars" appear as lines of slope $+1$. The net effect is as though Fig. 5.3 has been transformed by reflection through an appropriate vertical line, say $E^B = 1073$ eV. In Fig. 5.3 the polarizability of the chemical environment increases from bottom left to top right, whereas in the transformed plots the polarizability increases from bottom right to top left. The plot for indium clearly shows the advantages of this method of representing α' as the discrimination between the two oxidation states and between different salts of the same oxidation state, for example, the halogens. For both these elements the halogens fall on an almost vertical line on the plot, indicating that the XPS shift is small compared with the Auger shift. In fact, generalizing over all the plots, they have the appearance of being taller than they are wider, indicating that in general terms for this range of elements the Auger shifts are larger than the XPS shifts.

In using plots such as these for qualitative analysis a number of points should be borne in mind:

1. Coincidence of an unknown (with 1 eV) with known compounds should be taken as positive identification only if the possible compound is consistent with other information concerning the chemical composition, that is, the information from the plots should be considered as complementary to that obtained by other means.

2. The plots are not comprehensive, and a compound not recorded may have the coordinates of the analyte material.

3. Finely divided material may show anomalous shifts. For example, a finely aggregated insulator in a conductor may show shifts due to the addition relaxation energy available from the conducting medium. Conversely, a conductor dispersed in an insulating medium may show as being in an environment of decreased polarizability and thus both the photoelectrons and Auger electron energies would be shifted to lower energies (i.e., toward the bottom right of Figs. 5.4 and 5.5), possibly by as much as 1 eV and 3 eV, respectively.

3.1. Transitions Involving Valence Levels

The expressions derived for the relation between the Auger parameter and relaxation energy and hence polarizability of the environment of the analyte

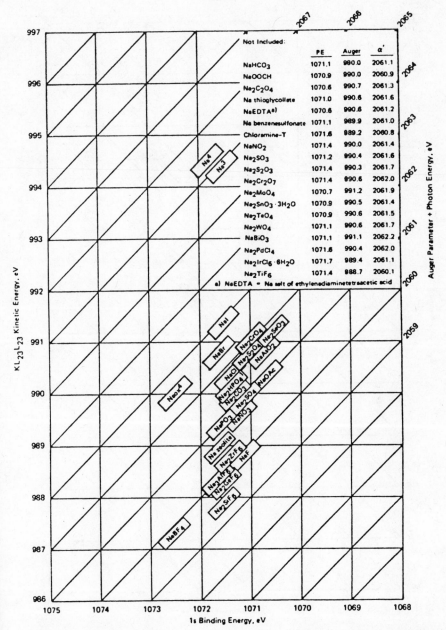

Figure 5.4. Wagner plots for sodium compounds.[24] (Reprinted with permission of American Chemical Society, Washington, D.C.)

233

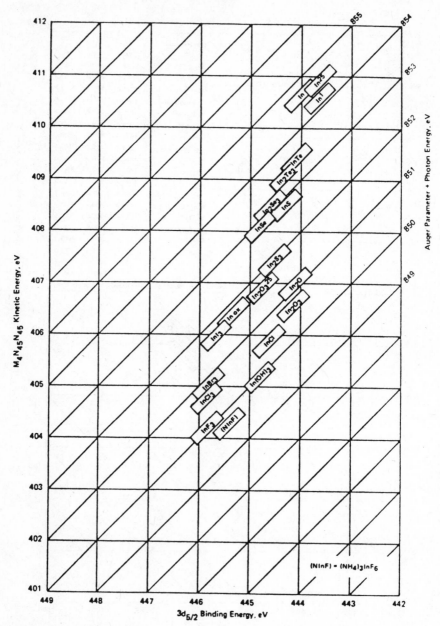

Figure 5.5. Wagner plots for indium compounds.[24] (Reprinted with permission of American Chemical Society, Washington, D.C.).

atom are based on Auger transitions involving core levels only, that is, the initial and final two-hole states are localized on one particular atom. Chemical shifts are observed for transitions of the type core–core–valence (CCV) or core–valence–valence (CVV), but the analysis of such shifts is more of a problem than for the CCC transitions because, for solids, the valence levels lie in somewhat broad energy level bands for which the energy locations and density of states can change greatly with variation in chemical state. In theory, analysis of CCV and CVV transitions should yield more information about chemical changes, but it would appear so far that the major source of information from these types of transition is from line-shape analysis (see Section 5) rather than by measuring shifts in the energies of the lines. Such shifts can be rather difficult to interpret, and the values reported in the literature often vary considerably for the same chemical system. For example, on oxidation the shift in the CCC Auger lines is always negative by as much as 8 eV in the case of germanium (element 32), which exhibits the largest shift. For the CCV and CVV transition the shift may be positive and almost as large as the negative shifts. Chromium (element 24) exhibits the largest positive shift (10 eV), but the shift is strongly dependent on the oxide bonding state. Nonetheless, these types of transitions can be used in Auger parameter plots quite successfully. Figure 5.6 shows Wagner's[24] plot for iron in which the Auger transition is the L_3VV line around 701 eV. Similarly, plots for cobalt, nickel, and copper all demonstrate good resolution of a variety of chemical environments. Plots for titanium and manganese using the $L_3M_{2,3}V$ transition also gave good discrimination. In the case of fluorine, Fig. 5.7, the KVV spectra usually exhibit $KL_{2,3}L_{2,3}$ character, and it is possible to identify regions on the plot corresponding to different categories of compounds. Thus covalent organic fluorine compounds are grouped in the lower-left-hand corner, complex fluoroanions are grouped in the middle of the plot, and binary fluorides are grouped in the upper-right-hand corner.

Finally, transitions may occur in the Auger spectrum of a compound, which have no analog in the solid metal, owing to an initial vacancy on the metal atom being filled by an electron from the anion, accompanied by ejection of the Auger electron from the anion. Such transitions are referred to as cross transitions and have been observed for a number of metal oxides[26,27] in which a single CVV peak in the solid metal spectrum apparently splits into a doublet in the oxide spectrum, one part of which has apparently moved to higher energy and the other to lower (by a much larger amount than could be accounted for in terms of spin-orbit coupling).

3.2 Correlation of Auger Parameter with Other Properties

One of the earliest reports and discussions of Auger chemical shifts[28] indicated that there was a close relationship between the shift in the Auger

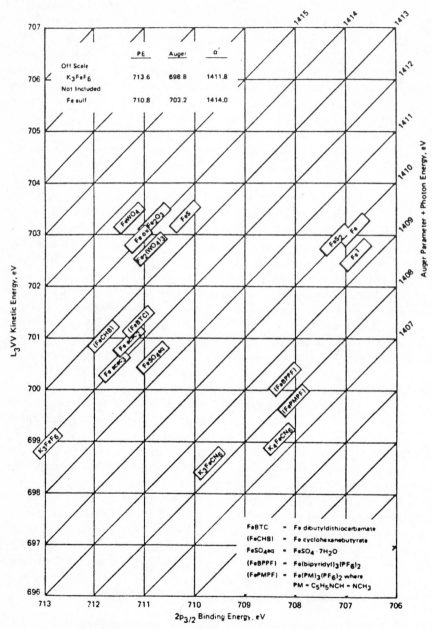

Figure 5.6. Wagner plots for iron compounds.[24] (Reprinted with permission of American Chemical Society, Washington, D.C.)

236

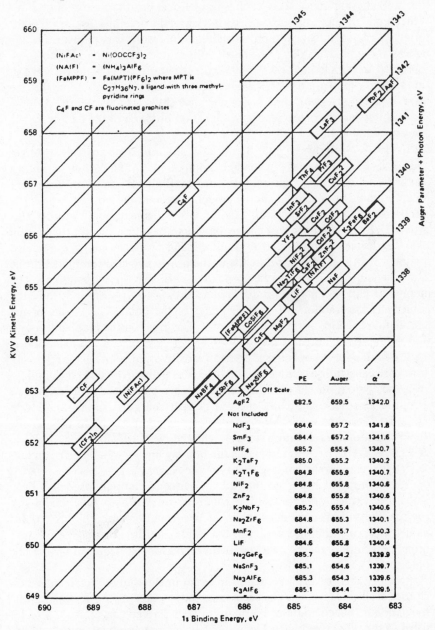

Figure 5.7. Wagner plots for fluorine compounds.[24] (Reprinted with permission of American Chemical Society, Washington, D.C.)

237

line and the difference in electronegativity between the anion and cation. This should not come as any surprise in view of the previous discussion of the origin of the chemical shift, but at the time this work was published (1972) it was thought that Auger chemical shifts would be comparable to those observed in XPS because it was thought that only what is now known as intra-atomic relaxation energy was involved. It is interesting to note that the experimental observations of larger chemical shifts, which gave rise to ideas concerning the role of extra-atomic relaxation processes, were first described as "abnormal chemical shifts."[28]

In 1974, Szalkowski and Somorjai[29] attempted to correlate Auger chemical shifts of a series of vanadium compounds with a parameter known as the bond ionicity, a concept developed by Phillips[30] and Van Vechten[31] in an attempt to develop a simple model to explain the observed dielectric properties of solid compounds. The ionicity is defined as the ratio C^2/E_g^2, where E_g is the total average energy band gap based on the approximation that the usually complicated crystalline band structure may be modeled by the isotropic bands associated with the nearly free electron model of the band structure and C is the heteropolar part of this energy band gap. The hompolar part, E_h, makes up the total according to $E_g^2 = E_h^2 + C^2$; thus, the ionicity is the fraction of ionic character of the band. The results of the correlation exercise are shown in Fig. 5.8. The break in the graph at around VO is due to a change in the involvement of electrons as the oxidation is increased. Up to VO only the $4s$ electrons are involved, which have relatively little core penetration compared with the $3d$ electrons involved in the bonding in higher oxidation states. Thus for the higher oxides the remaining core electrons experience a greater effective nuclear charge. A similar study was made by Fiermans et al.[32] for a series of zinc compounds. In this report, the extra-atomic relaxation energy change was calculated as $\Delta R_{ea} = \Delta[E(3d) - E_B(^1D)]$[33], where $E(3d)$ is the binding energy of a $3d$ electron and $E_B(^1D)$ is the apparent binding energy of the Auger peak.

Although it was observed that for the series of compounds ZnTe, ZnS, ZnSe, the extra-atomic relaxation energy decreased as a function of the ionicity, the authors concluded that a direct correlation with Phillips ionicity values was not straightforward. For the halides, it was shown that the extra-atomic relaxation energy decreases with increasing electronegativity of the halogen.

Recently, a correlation between Auger parameter and refractive index for a series of silicate minerals has been described[34] following the demonstration by Gallon[35] that the data of Fiermans et al. described earlier could also be fitted to the refractive index. The use of silicon compounds to modify inter-

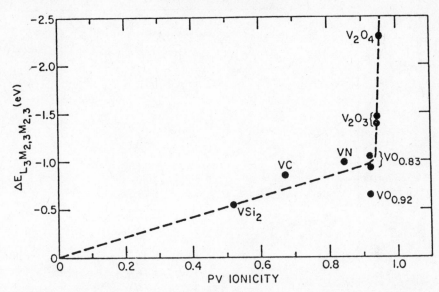

Figure 5.8. Magnitude of the observed chemical shift of the vanadium $L_3M_{2,3}M_{2,3}$ Auger transition in the vanadium compounds relative to vanadium metal plotted versus the calculated ionicity values for those compounds (Phillips–Van Vechten).[29] (Reprinted with permission of American Physical Society, New York.)

facial or surface-chemical properties is an important area of application in many technologies, not least is the development of microcircuitry. Because of the difficulty in establishing the chemical environment from the application of techniques such as XPS and SIMS, workers in this area have had to be content with compositional rather than structural information. In this study, 30 silicate materials were studied by AES using Zr $L\alpha$ x-rays to generate the initial vacancy. The Auger parameters for aluminum and silicon were compared with refractive-index data. The authors suggest that an understanding of charge redistribution in three-dimensional silicate structures could be used to interpret Auger parameters from two-dimensional networks. As has been previously discussed, the Auger parameter is a measure of lattice polarizability, and it is to be expected that there would be a reasonable correlation between them. This is shown in Fig. 5.9. The expression $(n^2 - 1)/(n^2 + 2)$ is considered to be the lattice-site polarizability. The bulk polarizability of the oxygen ions (which accounts for nearly all of the polarizability of the lattice structure) is calculated by two methods: (a) from a relationship derived by Kuroda et al.,[36] namely, $\Delta\alpha = Kp$, where K is a con-

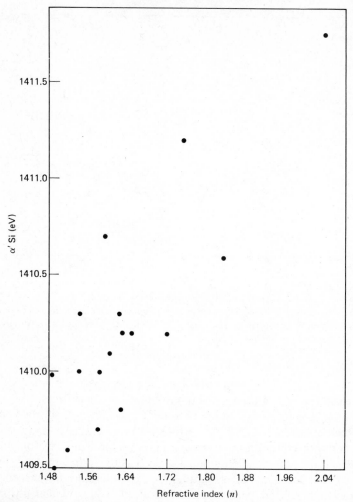

Figure 5.9. Plot of Auger parameter for silicon versus refractive index for a number of silicates (data from Reference 34).

stand dependent on the given central ion and p is the oxygen ion polarizability; and (b) from the Lorentz–Lorenz relationship, namely:

$$\frac{n^2 - 1}{n^2 - 1} = \frac{4\pi}{3} \frac{N}{V} P_B \qquad (5.25)$$

where N = Avogadro constant
V = molar volume
P_B = bulk polarizability

A very good correlation is found between the polarizabilities calculated each of these ways (slope 1.02, correlation coefficient 0.997). As the authors had previously reported that the Auger parameter was sensitive to hydration of the surface,[36] they conclude that the Auger parameter is well able to detect changes in coordination, water content, and degree of ionic bonding, and, furthermore, that the changes are readily measured for both crystalline and amorphous solids and relate to the state of the near surface in all cases.

4. AUGER LINE-SHAPE ANALYSIS (ALA)

4.1. General

Most of the discussion of the preceding sections in this chapter has been concerned with Auger transitions involving core electrons and the chemical information that can be obtained from a study of the energy shifts of the Auger lines in the spectrum. Little has been said about the shape of the lines, since for these types of transitions it is difficult to extract information concerning chemical environment from the line shape because the factors which govern the line shape are not particularly sensitive to the chemical environment. The effects that affect the lineshape in solids include lifetime broadening, phonon broadening (essentially vibrational because Auger transitions can be considered to obey the Franck–Condon principle; if the potential-energy surfaces of the initial and final states are spatially shifted, then vibrational excitation will accompany the transition), and various loss processes. These processes include the inelastic scattering of electrons and the excitation of discrete collective oscillations of the electron gas (known as plasmons). This latter effect is particularly pronounced for nearly free-electron metals (such as sodium, magnesium, and aluminum). The intensity of energy-loss peaks depends on the sample density in gas-phase studies and so are readily identified experimentally. This is not true for solids, although the use of very thin layers may be a possible approach to the problem provided the film thickness can be made homogeneous. In solids, excitations of an electron from an occupied to unoccupied orbital due to collision with an electron (giving rise to the loss in energy of the bombarding electron) will correspond to interband transitions and will therefore give rise to broad features.

In the case of Auger transitions involving valence electrons, that is, core–core–valence (CCV) or core–valence–valence (CVV), information about the chemical environment of the atom on which the initial vacancy occurs should be available from both the energy shifts of the lines and the line shape. Thus in contrast to UV photoelectron spectroscopy, which contains valence information that reflects the spatial average of the valence-electron distribution and to x-ray photoelectron spectroscopy where chemical information is extracted from the chemical shifts (as in CCC Auger transitions), Auger transitions involving coupling of a core hole to one or two valence electrons have both site specificity and valence sensitivity. Thus these transitions should provide an ideal way of probing the local chemical environment of an atom. As the Auger matrix elements are sensitive to the local electronic charge of the atom with the initial vacancy, the Auger electrons emitted from the valence region can reflect the local density of states on that atom. However, as Netzer[38] has pointed out, the feature that gives the Auger transition its potential as a tool for probing local chemical environments (the two-hole final state) makes a rigorous theoretical analysis of the line shape particularly difficult. Recently, Jennison[39] has summarized the various approaches to gaining a theoretical understanding of CVV Auger line shapes and has examined the relative importance of factors that affect the decay amplitudes (local charge density, initial-state screening, valence nonorthogonality, and configuration mixing) and the peak energies (independent final-state holes or coupled final-state holes). Although Lander suggested as long ago as 1953[40] that ALA would provide chemical information, it is only recently that this suggestion has received any serious attention, indicating possibly that it has taken 30 years for an appropriate understanding of the theory to be achieved. Considerable information, though, may be obtained from using Auger spectra as "fingerprints" of a known chemical environment and comparing a spectrum from an unknown chemical environment with the known spectra. This approach will be illustrated in later sections in this Chapter when the use of AES to examine surface adsorption, follow surface reactions etc. is discussed. The work of Rye et al.[41] concerning carbon KVV spectra for a series of gas-phase compounds discussed in the previous chapter, also illustrates the "fingerprint" made of ALA. They showed that the differences in carbon hybridization (sp, sp^2, and sp^3) gave rise to considerable differences in the Auger line shapes.

The simplest model for the solid state would predict that for a CVV transition the maximum energy possible would correspond to the "down" and "out" electrons originating from the top of the valence band, whereas for the lowest-energy transition, they would originate from the bottom of the valence band. Thus the total line width would be twice the valence band width. If the transition rate is assumed to be constant across the valence band, then the line shape should be the self-convolution of the valence-band density of states. However, when the spectra of some solids are compared with the spectra predicted on the basis of this simple model, there is little agreement and in some cases, for example, copper, not even the line width is correctly predicted by the theory.

Jennison[39,42] has started from this basis and included the necessary corrections to take account of the various many-body factors contributing to the line shape. It is thus apparent that a two-electron model isolated from the rest of the solid-state electrons is not adequate, but so far only a few approaches to this problem have been proposed.[43-47] One important factor is the correlation between the final two valence holes. Two extreme situations can be envisaged, one in which the two holes are almost uncorrelated (in which the hole motion is adequately described by canonical Bloch functions or molecular orbitals) and the other in which the holes are highly correlated through being localized on the same site or in a state in which they move together. Cini[43] and Sawatzky[44,45] have developed theories in which the appropriate criterion for predicting whether the final holes will be correlated or not is the relative magnitude of the effective hole–hole interaction energy U_{eff} and the valence band width W.

Further aspects of ALA, particularly a brief survey of theoretical methods and experimental consequences with respect to metals, are discussed in the following chapter.

4.2. Applications of ALA

4.2.1. Comparison with Band-Structure Calculations

At present there does not appear to be a great need for another valence-band spectroscopy to provide experimental evidence against which to test bulk band-structure calculations. However, theoretical calculations of the electronic properties of solid *surfaces* are not nearly so well developed as those for bulk solid properties, clusters, or molecules, so that high-resolution AES is among the spectroscopies against whose experimental findings surface theories will be tested. The information from AES is complementary to XPS or UPS valence-band data since these latter two techniques provide information on spatially averaged valence-band density of states through the direct interaction of photons with valence-band electrons, whereas AES is site specific and gives information concerning the valence electron distribution local to the atom emitting the Auger electron.

Such a comparison has recently been made by Madden,[8] who made measurements on the dihydride layer chemisorbed on Si(100)[49] and compared the results with the calculations of Appelbaum et al.[50] As there would probably be a contribution from elemental silicon below the hydride layer, the measured Si(H) spectrum was corrected by subtracting the elemental Si spectrum before the comparison was made. However, the only agreement that was evident between theory and experiment was in a negative peak on the high-energy side of each difference signal. This feature was interpreted as being due to reduced states at the top of the valence band; the chemisorbed hydrogen is considered to remove surface electron states by "tying off dangling bonds." The new electron states arising from the Si–H bonds were considered to give rise to the positive features in the difference spectrum, but the agreement with the calculated features with respect to both shape and energy were poor.

4.2.2. Study of Beam Damage and Effects of Excitation Mode

Holloway et al.[51] have studied the effects of electron bombardment on condensed layers of dimethyl ether, methanol, and water. The layers were estimated to be about 10 nm thick, sufficient to avoid substrate (Ni) emission and substrate–condensate interaction. They found that the dose equivalent to that accumulated in 1 sec from a 1-μA beam, 0.5 mm in diameter, was sufficient to cause observable changes in line shape. In order to compare spectra, the peak shapes were corrected for the secondary-electron background and then deconvoluted. This was carried out, for example, for the C region of dimethyl ether. The good agreement between the spectrum for the damaged layer and the spectrum constructed from the sum of the gas-phase ethane spectrum and the fresh undamaged methyl ether spectrum (in the ratio of $1:1.8$) was taken as evidence for the formation of new hydrocarbon products by electron bombardment. Similar results were obtained for the methanol layers.

A comparison of the spectra obtained by electron excitation with those obtained by x-ray excitation showed no significant difference in the O region, but in the C region the x-ray excited spectra showed an additional small peak about 20 eV lower than the main peak. The authors were unable to offer a satisfactory explanation for this phenomenon, though they pointed out that theoretical calculations of the methanol spectrum indicate that the peak arises from a normal transition based on known molecular-orbital structure and that it is possible that the the intensity of this feature may be due to secondary processes whose probability depends on the excitation mode. However, most evidence so far has suggested that the probability of shake-up or shake off occurring is independent of the excitation mode unless the energy of the exciting radiation is close to the threshold energy for the core-level ionization. The experiments described here used Mg or Al $K\alpha$ radiation well above the threshold of the carbon $1s$ ionization energy.

4.2.3. Study of the Molecular Surface Reaction Products

Netzer[38] has recently reviewed briefly the literature concerning ALA of gas-phase, condensed-phase, and adsorbed molecules and concludes that only for gas-phase molecules is the level of interpretation of the spectra well advanced and for most studies of adsorbed molecules the approach has been to use the Auger spectrum as a "fingerprint." This approach is greatly facilitated by a comparison with the gas-phase molecular spectrum and with the metal complex (where the adsorbate is present as a ligand). Thus, features in the spectrum arising from coupling of the molecule with the substrate surface and the formation of surface chemical bonds may be identified. For example, in the study of the adsorption of CO on certain metals (a system that has been the subject of a large number of publications), the spectrum of adsorbed CO may be compared with that of the free molecule and with the corresponding metal carbonyl. A number of metal carbonyl spectra have been reported in the literature.[52-54] A comparison of spectra for indium shows that the adsorbed CO spectrum[55] strongly resembles that of the condensed carbonyl $Ir_4(CO)_{12}$ [52] (and also the Ir_4 cluster spectrum closely resembles that of the pure metal), but with some differences in the carbon spectrum due to (a) the participation of the 5σ orbital of CO in the CO-metal bonding and (b) the participation of metal electrons that screen the ligand core hole in the Auger process. Further discussion of the interpretation of metal carbonyl spectra can be found in Reference 54.

At the other end of the interaction scale, Goodman et al.[56] were able to distinguish between graphitic carbon and carbidic carbon produced by heating a clean nickel surface in the presence of CO, by a comparison of the line shapes with those of pure graphite and nickel carbide.

4.2.4. Interfactial Characterization

There are many areas such as the study of adhesion, bonding, corrosion, fatigue, wear, and diffusion that have a requirement for interfacial characterization. As discussed elswhere in this text Auger depth profiling offers the advantages of speed, minimum specimen preparation, sensitivity to all elements apart from H and He, semiquantitation, and shallow sampling depth. Although in most applications the information provided concerns only the relative elemental composition derived from peak-to-peak height measurement, it has been shown that chemical information can be obtained when the changes in chemical state produce changes in the Auger line shape.[57] However, to utilize the changes in line shape, computerized-data-acquisition techniques are required since the entire Auger line shape must be repeatedly sampled during the etching process (as opposed to the usual method of simply following peak-to-peak or integrated intensity) and a multicomponent analysis is then required to sort out the various chemical states from the various overlapping Auger spectra. Such multicomponent analyses are well established in the areas of application of other analytical spectroscopies such as UV-visible, IR, and mass spectroscopies. The use of a technique known as target transformation was used by Gaarenstroom[58] to extract chemical-state information from a depth-profile experiment on sputter-deposited platinum on zirconia. Under certain sputtering conditions a blackened material was produced that showed a different line shape in the interfacial region from the unblackened material. The target transformation showed that the interfacial region of the blackened material contained an intermetallic or solid solution containing platinum and zirconium but not oxygen, of composition approximately Pt_2Zr.

5. CHEMISORPTION

There is an enormous literature on this topic and it continues to grow rapidly. Nearly all studies reported so far have been of gases interacting with metal surfaces. The interest in "gases-on-metals" is often academic in character, although catalysis and mechanisms thereof are always the long-term subject for study. Generally, there are two main objectives of chemisorption studies:

1. To elucidate the nature of the chemisorption bond and to quantify it in terms of geometry (orientation) and electronic and vibrational properties of the adsorbate–substrate system.
2. To gain understanding of the kinetics and mechanisms of the adsorption and desorption processes.

Experimental results prior to 1950 are of little interest because of the contamination of the surfaces studied. In fact, one important contribution of AES to this area is the ability to monitor surface cleanliness. Most fundamental studies have been with "model" systems: simple adsorbate molecules —CO, H_2, N_2—and easily cleaned ideally structured crystal surfaces, for example the low-index planes of body-centred-cubic transition metals such as W and Mo. The CO/W system has been extensively studied because clean W surfaces can be produced simply by high-temperature flashing, field- and ion-emission microscope tips can be easily fabricated, and CO is molecularly adsorbed at room temperature and below.

A wide variety of experimental techniques have been used to study chemisorption, such as the perturbation of the surface electronic properties of work function, magnetic susceptibility, and electrical conductivity as well as a host of surface analysis methods—LEED, AES, XPS, UPS, and SIMS.

A huge part of the literature in this area is taken up with studies of the chemisorption of oxygen and of oxidation processes. We consider in more detail some examples of this work in the next chapter.

In order to place an emphasis in this text on the practical implications of AES studies and to economize on space, a summary of literature for the period 1976–1982 is presented in Table 5.5. Papers concerning oxidation of the surface as compared to conventional chemisorption of oxygen are labeled Ox. Note that these data are not intended to be all-encompassing.

Table 5.5. A Partial Summary of the Literature on the Study of Chemisorption by AES and Adjunct Techniques (1976–1982)

Substrate	Adsorbate	Techniques[a]	References
Al	CO, NH_3, O_2, Ox	AES, LEED	59–65
Al–Mg	Ox	EELS, WF	66
Au	I_2, O_2	AES, LEED, EELS	67–70
Ag	Cl_2, Br_2, S, O_2	AES, XPS	71–74, 68
Ag–Rh	Ag, N_2O, O_2	AES, LEED	75
Ba	Various gases	AES, XPS	76
Be	Ox	AES	77
Bi	Ox	AES, EELS, LEED	78
Cr	Ox	AES, LEED, XPS EID, WF	79, 80
$CuIrSe_2$	Ox	AES, SIMS, XPS	81
C (graphite)	Xe	AES	82
Cd	Ox	AES, XPS	83, 89
Cu–Ru	O_2	AES	85
Co	CO, O_2, Ox	LEED, UPS, WF, EELS	86–91
CoSi, $CoSi_2$	O_2	AES	92
Co, phthalocyanine	H_2O, O_2		93
Cu	S, C_2H_4, C_2H_2, CO, C_2N_2 HNCO, O_2, N_2O, Xe, P compounds Ox	AES, XPS EELS, TDS, LEED EL	94–112
Cu–Ni	CO, hydrocarbons, O_2	EL, AES	113–116
Fe	C_2H_4, Hg, N_2, H_2, CBr_4 CCl_4, Br_2, I_2, Cl_2, CO, NH_3, O_2	AES, LEED TDS	117–128
Ni–Fe	Ox	AES, LEED	129
Fe–Pd	Ox	AES, XPS	130
Cr–Ni steel	H_2O, Ox	AES	131, 132
Ge	O_2	AES, EELS, WF, TD	133, 134
Ir	NO, N_2, D_2, CO, O_2, H C_2H_4, C_6H_6	XPS, UPS, AES LEED	135–138
In	Ox	XPS, AES	83
LaB_6	Cs, O_2	AES, WF	140, 141
Mg	O_2, Ox	WF, LEED, EELS XPS, UPS, AES	142–144

Table 5.5 (Continued)

Substrate	Adsorbate	Techniques[a]	References
Mo–C	N_2, H_2, C_2H_4, CO	AES, LEED	145
Mo	H_2, CO, $(CH_3)_3N$, N_2,		
	NH_3, O_2, Ox	AES	146–151, 91
MoS_2	Amylamine	AES	152
Ni	I_2, CO, NO, S, N_2,	AES, LEED, UPS	
	C_2H_2, C_2H_4, HCOOH,	SIMS, TDS, EL	153–180, 110
	CH_4, Cl_2, O_2, Ox		
Nb	O_2	AES, SIMS	181, 182
Pd	CO, O_2, NO, S, Cl_2, HI	AES, LEED, WF,	
	Xe, N species	TDS	183–191
Pd–Au	O_2, H_2	AES, XPS	192
PbSnTe	O_2	AES, UPS, XPS	193
Pt	S, O_2, CO, C_2H_4,	AES, LEED, TDS,	184–216
	C_2H_2, HBr, HCl, NO,	XPS, EELS, WF	95–97, 188
	HNCO, I_2, HI		189, 68
	NH_3, Br_2, C_2N_2, C_6H_6,		
	N_2, O_2		
Pu	Ox	AES, XPS	217
Rh	Cl_2, O, H_2O_2, CO, CO_2,	AES, LEED, TD	218–222
	NO		
Ru	O_2, H_2, CO, NO, H_2S,	AES, LEED	
	NH_3, H_2, N_2, CO, O_2	XPS, UPS	223–233
Re	NO, H_2O, CO, O_2	XPS, AES, SIMS	234–236
Si	H_2, H_2O, H_2S, HCl,	AES, EELS, EID	
	HBr, CO, O_2, Ox		237–244
Sn	O_2, Ox	AES, EELS	245, 83
Sb	Ox	AES, XPS	83
Ta	Cl_2, Ox	AES, LEED	246–247
Th	Ox	AES, LEED	248
Ti	Ox	AES	249, 91
U	Ox		250
V	NO, NH_3, N_2, Br_2	AES	251, 252
W	C_2H_2, NO, H_2O, H_2S,	AES, LEED, UPS	253–271
	Br_2, I_2, O_2, N_2, Cl_2	XPS, EELS	
Zr	CO, NO, N_2, O_2, D_2, I_2	WF, TDS	272, 273
Zn	Ox	AES, RHEED,	
		LEED	274

[a]AES, Auger electron spectroscopy; XPS, x-ray photoelectron spectroscopy; UPS, ultraviolet photoelectron spectroscopy; LEED, low-energy-electron diffraction; SIMS, secondary ion mass spectrometry; EELS, electron-impact energy-loss spectroscopy; EID, electron-impact desorption; EL, ellipsometry; TDS, thermal desorption spectroscopy; WF, work function; TD thermal desorption RHEED, reflection high energy electron diffraction.

6. SURFACE CHEMICAL REACTIONS AND CATALYSIS

Despite the enormous amount of AES work carried out on various aspects of gas adsorption and chemisorption, relatively few papers have been concerned with the chemical-kinetic aspects of molecular reactions on surfaces (involving, for example, mixed gases over the surface or sequential gas exposure) where the reacting entities are altered to other gas-phase species. This area, which is clearly of crucial importance in the understanding and design of catalysts, is covered in this section with a few examples of direct use of AES in the catalyst field.

6.1. Inorganic-Surface Reactions

The decomposition and reaction of CO with O_2, H_2, and NO on various metals has attracted considerable attention. In 1977, Hooker and Grant[55] pointed out that ALA can be useful in distinguishing between graphite overlayers, surface carbides, and other carbon-containing species with respect to C Auger. The C and O line shapes were measured following the room-temperature adsorption of CO on Ta, Nb, Ti, W, Co, Rh, Pd, Ir, and Pt. In the case of Ta, Nb, and Ti, CO decomposes, while it adsorbs molecularly on Pd, Ir, and Pt. Slight CO decomposition may occur with Rh, whereas significant decomposition is found for Co. Although the detection by AES of fractional monolayers of carbon on Ru is difficult because of interference between the Ru (273 eV) and C (272 eV) signals, use of an AES feature of carbidic carbon can be used to study surface carbide build up and removal in a methanation reaction $(3H_2 + CO \rightarrow CH_4 + H_2O)$.[275] In this examination of the reaction of CO with clean Ru, the point has been made that the rate of carbide formation is equal to the rate of methane formation in a CO–H_2 mixture. Bridge and Lambert[276] studied NO chemisorption and NO/CO reaction on Co(0001). NO is adsorbed dissociatively at room temperature at all but the highest coverages, and thermal desorption and AES experiments confirm that N_2 is removed. When CO is preadsorbed on the crystal followed by exposure to low doses of NO, the adsorption behavior of NO is not noticably affected by the presence of CO, indicating that removal of CO from the surface is very rapid. There is an increase in the partial pressure of gas-phase CO_2, and residual surface oxygen is only observed by AES after all CO has been removed. This suggests that the overall process is

$$NO(g) + CO(a) \rightarrow CO_2(g) + N(a)$$

rather than

$$NO(g) + CO(a) \rightarrow NO(a) + CO(g)$$

In a reverse experiment, if an NO adsorbed layer is heated to 900 K to remove N_2, and then the crystal is exposed to a larger CO dose than the initial NO dose, desorption of α_2-CO is observed, and after heating to 500 K Auger analysis shows that no residual oxygen is present. These results imply that surface oxygen produced by the higher-temperature removal of N_2 is still reactive to CO, presumably owing in some way to the presence of nitrogen.

Other examples in the inorganic area are the adsorption of molecular chlorine to Cu(111) and Ag(111), where desorption involves Cl atoms,[277] NH_3 on Fe,[278] and H_2S on Ir(110).[279]

6.2.   Organic-Surface Reactions

The study of reactions of ethylene and acetylene on various surfaces, often using an adjunct technique for analysis of the gas phase, such as mass spectrometry, represent a significant portion of published work in this area. For example, ethylene and acetylene on Ni has been looked at a number of times (see References 161, 167, and 280). The kinetics of interaction of ethylene with Ni(110) at $T < 350°C$ was studied by a combination of AES measurements of surface carbon concentration and modulated-beam mass-spectrometry analysis of scattered ethylene and desorbed hydrogen fluxes.[167] The surface carbon concentration rises rapidly and then trends off at a saturation coverage of one-half monolayer of ethylene molecules. Detection of the mass-two signal shows that H_2 is evolved after 105°C is reached. The carbonaceous layer is completely dehydrogenated above 250°C. Above 350°C a close-packed monolayer of carbon graphitic structure is achieved. After the earlier stages of chemisorption described above, a process is observed that involves nondissociatively adsorbed ethylene. The authors comment on this phase of the reaction with respect to practical catalysis. The interaction of ethylene and benzene with Ir(110) was studied by Nieuwenhuys and Somorjai[138] using LEED, AES, and flash desorption mass spectrometry, and ethylene adsorption on W has been the subject of other investigations.[265,281] Similar techniques were used in a study of the decomposition of formic acid on Ni(100).[282] HCOOD readily adsorbs on Ni(100) at temperatures from 110 to 300 K. When formic acid was adsorbed at 300 K, the only products seen to desorb were CO_2, H_2, and CO, whereas at 150 K desorption of D_2O occurred during the flash. The product distribution measured by mass spectrometry and oxidization of the Ni(100) surface as evidenced by AES are shown in Fig. 5.10. The results of the study were interpreted in terms of the following mechanism for formic acid decomposition on nickel:

15. R. Hoogewijs, L. Fiermans, and J. Vennik, *J. Elec. Spectrosc. Rel. Phenom.*, **11**, 171 (1977).

16. T. D. Thomas, *J. Elec. Spectrosc. Rel. Phenom.*, **20**, 117 (1980).

17. C. D. Wagner, "The Role of Auger Lines in Photoelectron Spectroscopy," in *Handbook of UV and X-ray Photoelectron Spectroscopy*, Heyden, London, 1977, p. 249.

18. C. D. Wagner, *Anal. Chem.*, **44**, 967 (1972).

19. C. D. Wagner, *Anal. Chem.*, **47**, 1203 (1975).

20. S. P. Kowalczyk, L. Ley, F. R. McFeely, R. A. Pollak, and D. A. Shirley, *Phys. Rev. B*, **9**, 381 (1974).

21. P. E. Larson, *J. Elec. Spectrosc. Rel. Phenom.*, **4**, 213 (1979).

22. S. W. Gaarenstroom and N. Winograd, *J. Chem. Phys.*, **67**, 3500 (1977).

23. R. G. Pearson, *J. Am. Chem. Soc.*, **85**, 3533 (1963).

24. C. D. Wagner, L. H. Gale, and R. H. Raymond, *Anal. Chem.*, **51**, 466 (1979).

25. N. G. Vannerberg, *Chem. Scr.*, **9**, 122 (1976).

26. R. Weismann, R. Koschatzky, W. Schnellhammer, and K. Huller, *Appl. Phys.*, **13**, 43 (1977).

27. D. T. Quinto and W. D. Robertson, *Surf. Sci.*, **27**, 645 (1971).

28. C. D. Wagner and P. Biloen, *Surf. Sci.*, **35**, 82 (1973).

29. F. J. Szalkowski and G. A. Somarjai, *J. Chem. Phys.*, **61**, 2064 (1974).

30. J. C. Phillips, *Rev. Mod. Phys.*, **42**, 317 (1970).

31. J. A. Van Vechten, *Phys. Rev.*, **182**, 841 (1969).

32. L. Fiermans, R. Hoogewijs, and J. Vennick, *Surf. Sci.*, **47**, 1 (1975).

33. S. P. Kowalczyk, L. Ley, F. R. McFeely, R. A. Pollack, and D. A. Shirley, *Phys. Rev. B*, **9**, 381 (1974).

34. R. H. West and J. E. Castle, *Surf. Interface Anal.*, **4**, 68 (1982).

35. T. E. Gallon, in L. Fiermans, J. Vennik, and W. Dekeyser (eds.), *NATO Advanced Study Institute on Electron and Ion Spectroscopy of Solids*, Ghent 1977, Plenum, New York, 1978.

36. H. Kuroda, T. Ohta, and Y. Sato, *J. Elec. Spectrosc. Rel. Phenom.*, **15**, 21 (1979).

37. J. E. Castle, L. B. Hozell, and R. H. West, *J. Elec. Spectrosc. Rel. Phenom.*, **16**, 97 (1979).

38. F. P. Netzer, *Appl. Surf. Sci.*, **7**, 281 (1981).

39. D. R. Jennison, *J. Vac. Sci. Technol.*, **20**, 548 (1982).

40. J. J. Lander, *Phys. Rev.*, **91**, 1382 (1953).

41. R. R. Rye, T. E. Madey, J. E. Houston, and P. H. Holloway, *J. Chem. Phys.*, **69**, 1504 (1978).

42. D. R. Jennison, *J. Vac. Sci. Technol.*, **17**, 172 (1980).

43. M. Cini, *Solid State Commun.*, **24**, 082 (1977).

44. G. A. Sawatzky, *Phys. Rev. Lett.*, **39**, 504 (1977).

45. G. A. Sawatzky and A. Lenselink, *Phys. Rev. B*, **21**, 1790 (1980).

46. O. Gunnerson and K. Schönhammer, *Phys. Rev. B*, **22**, 3710 (1980).

47. S. Abrahim-Ibrahim, B. Caroli, C. Caroli, and B. Roulet, *Phys. Rev. B*, **18**, 6702 (1978).

48. H. H. Madden, *J. Vac. Sci. Technol.*, **18**, 677 (1981).

49. H. H. Madden, *Surf. Sci.*, **105**, 129 (1981).
50. J. A. Appelbaum, G. A. Baraff, D. R. Hamann, H. D. Hagstrum, and T. Sakurai, *Surf. Sci.*, **70**, 654 (1978).
51. P. H. Holloway, T. E. Madey, C. T. Campbell, R. R. Rye, and J. E. Houston, *Surf. Sci.*, **88**, 121 (1979).
52. F. P. Netzer, E. Bertel, and J. A. D. Matthew, *J. Elec. Spectrosc. Rel. Phenom.*, **18**, 199 (1980).
53. E. W. Plummer, W. R. Salaneck, and J. S. Miller, *Phys. Rev. B*, **18**, 1673 (1978).
54. D. R. Jennison, G. D. Stucky, R. R. Rye, and J. A. Kelber, *Phys. Rev. Lett.*, **46**, 911 (1981).
55. M. P. Hooker and J. T. Grant, *Surf. Sci.*, **62**, 21 (1977).
56. D. W. Goodman, R. D. Kelley, T. E. Madey, and J. T. Yates, *J. Catal.*, **63**, 226 (1980).
57. S. W. Gaarenstroom, *Appl. Surf. Sci.*, **7**, 7 (1981).
58. S. W. Gaarenstroom, *J. Vac. Sci. Technol.*, **20**, 458 (1982).
59. S. A. Flodström and C. W. B. Martinsson, *Appl. Surf. Sci.*, **10**, 115 (1982).
60. K. Khonde, J. Darville, and J. M. Gilles, *Vacuum*, **31**, 499 (1981).
61. C. T. Campbell, J. W. Rogers, R. L. Hance, and J. M. White, *Chem. Phys. Lett.*, **69**, 430 (1980).
62. C. W. B. Martinsson and S. A. Flodström, *Surf. Sci.*, **80**, 306 (1979).
63. Y. Katayama, K. L. I. Kobayashi, and Y. Shiraki, *Surf. Sci.*, **86**, 549 (1979).
64. R. Michel, C. Jourdan, J. Castaldi, and J. Derrien, *Surf. Sci.*, **84**, L509 (1979).
65. A. M. Bradshaw, P. Hofmann, and W. Wyrobisch, *Surf. Sci.*, **68**, 269 (1977).
66. B. Goldstein and J. Drosner, *Surf. Sci.*, **71**, 15 (1978).
67. S. A. Cochran and H. H. Farrell, *Surf. Sci.*, **95**, 359 (1980).
68. C. N. R. Rao, V. P. Kamath, and S. Yashonath, *Chem. Phys. Lett.*, **88**, 13 (1982).
69. P. Légaré, L. Hilaire, M. Sotto, and G. Maire, *Surf. Sci.*, **91**, 175 (1980).
70. D. D. Eley and P. B. Moore, *Surf. Sci.*, **76**, L599 (1978).
71. Y-Y. Tu and J. M. Blakely, *Surf. Sci.*, **85**, 276 (1979).
72. P. J. Goddard, K. Schwaha, and R. M. Lambert, *Surf. Sci.*, **71**, 351 (1978).
73. G. G. Tibbetts and J. M. Burkstrand, *J. Vac. Sci. Technol.*, **15**, 497 (1978).
74. Y-Y. Tu and J. M. Blakely, *J. Vac. Sci. Technol.*, **15**, 563 (1978).
75. W. M. Daniel, Y. Kim, H. Peebles, and J. M. White, *Surf. Sci.*, **111**, 189 (1981).
76. J. Verhoeven and H. Van Doveren, *J. Vac. Sci. Technol.*, **20**, 64 (1982).
77. D. E. Fowler and J. M. Blakely, *J. Vac. Sci. Technol.*, **20**, 930 (1982).
78. C. T. Campbell and T. N. Taylor, *Surf. Sci.*, **118**, 401 (1982).
79. Y. Sakisaka, H. Kato, and M. Onchi, *Surf. Sci.*, **120**, 150 (1982).
80. J. C. Peruchetti, G. Gewinner, and A. Jaegle, *Surf. Sci.*, **88**, 479 (1979).
81. L. L. Kazmerski, O. Jamjoum, P. J. Ireland, S. K. Deb, and R. A. Mickelson, *J. Vac. Sci. Technol.*, **19**, 467 (1981).
82. M. Bienfait and J. A. Venables, *Surf. Sci.*, **64**, 425 (1977).
83. P. Sen, M. S. Hegde, and C. N. R. Rao, *Appl. Surf. Sci.*, **10**, 63 (1982).
84. R. N. Joyner, M. W. Roberts, and G. N. Salaita, *Surf. Sci.*, **84**, L505 (1979).
85. S.-K. Shi, H.-I. Lee, and J. M. White, *Surf. Sci.*, **102**, 56 (1981).
86. H. Papp, *Ber. Bunsenges. Phys. Chem.*, **86**, 555 (1982).
87. K. A. Prior, K. Schwaha, and R. M. Lambert, *Surf. Sci.*, **77**, 193 (1978).

88. M. E. Bridge, C. M. Comrie, and R. M. Lambert, *Surf. Sci.*, **67**, 393 (1977).

89. M. E. Bridge and R. M. Lambert, *Surf. Sci.*, **82**, 413 (1979).

90. T. Matsuyama and I. Ignatiev, *Surf. Sci.*, **102**, 18 (1981).

91. A. Benninghoven, O. Ganschow, and L. Wiedmann, *J. Vac. Sci. Technol.*, **15**, 506 (1978).

92. G. Castro, J. E. Hulse, J. Kuppers, and A. Rodriquez Gonzales-Elipe, *Surf. Sci.*, **117**, 621 (1982).

93. J. C. Bucholz, *Appl. Surf. Sci.*, **1**, 547 (1978).

94. A. F. Armitage and D. P. Woodruff, *Surf. Sci.*, **114**, 414 (1982).

95. M. A. Chesters and N. D. S. Canning, *Vacuum*, **31**, 695 (1981).

96. M. D. Baker, N. D. S. Canning, and M. A. Chesters, *Surf. Sci.*, **111**, 452 (1981).

97. N. D. S. Canning, M. D. Baker, and M. A. Chesters, *Surf. Sci.*, **111**, 441 (1981).

98. F. Solymosi and J. Kiss, *Surf. Sci.*, **108**, 388 (1981).

99. F. Solymosi and J. Kiss, *Surf. Sci.*, **104**, 181 (1981).

100. F. H. P. M. Habraken, C. M. A. M. Mesters, and G. A. Bootsma, *Surf. Sci.*, **97**, 264 (1980).

101. F. H. P. M. Habraken, G. A. Bootsma, P. Hofmann, S. Hochscha, and A. M. Bradshaw, *Surf. Sci.*, **88**, 285 (1979).

102. F. H. P. M. Habraken and G. Bootsma, *Surf. Sci.*, **87**, 333 (1979).

103. A. Glachant and U. Bardi, *Surf. Sci.*, **87**, 187 (1979).

104. F. H. P. M. Habraken, E. Ph. Kieffer, and G. A. Bootsma, *Surf. Sci.*, **83**, 45 (1979).

105. S. Ferrer, A. M. Baró, and J. M. Rojo, *Surf. Sci.*, **72**, 433 (1978).

106. J. Kessler and F. Thieme, *Surf. Sci.*, **67**, 405 (1977).

107. J. R. Noonan, D. M. Zehrer, and L. H. Jenkins, *Surf. Sci.*, **69**, 731 (1977).

108. A. Spitzer and H. Lüth, *Surf. Sci.*, **118**, 121 (1982).

109. C. S. McKee, L. V. Renny, and M. W. Roberts, *Surf. Sci.*, **75**, 92 (1978).

110. A. F. Armitage, H. T. Liu, and D. P. Woodruff, *Vacuum*, **31**, 519 (1981).

111. G. T. Burstein and R. C. Newman, *Appl. Surf. Sci.*, **4**, 162 (1980).

112. H. R. Pinnel, H. G. Tompkins, and D. E. Heath, *Appl. Surf. Sci.*, **2**, 558 (1979).

113. C. M. A. M. Mesters, A. F. H. Wielers, O. L. J. Gijzman, J. W. Geus, and G. C. Bootsma, *Surf. Sci.*, **115**, 237 (1982).

114. M. E. Schrader, *Thin Solid Films*, **86**, 49 (1981).

115. C. Benndorf, K. Gressmann, J. Kessler, W. Kirstein, and F. Thieme, *Surf. Sci.*, **85**, 389 (1979).

116. C. M. A. M. Mesters, A. F. H. Wielers, O. L. J. Gijzman, G. A. Bootsma, and J. W. Geus, *Surf. Sci.*, **117**, 605 (1982).

117. N. D. S. Canning and M. A. Chesters, *J. Mol. Struct.*, **79**, 191 (1982).

118. R. G. Jones and D. L. Perry, *Vacuum*, **31**, 493 (1981).

119. G. Ertl, M. Huber, S. B. Lee, Z. Paal, and M. Weiss, *Appl. Surf. Sci.*, **8**, 373 (1981).

120. W. Hartweck and H. J. Grabke, *Surf. Sci.*, **89**, 174 (1979).

121. P. A. Dowben and R. G. Jones, *Surf. Sci.*, **89**, 114 (1979).

122. R. G. Jones, *Surf. Sci.*, **88**, 367 (1979).

123. P. A. Dowben and R. G. Jones, *Surf. Sci.*, **88**, 348 (1979).

124. R. G. Jones and D. L. Perry, *Surf. Sci.*, **88**, 331 (1979).

125. P. A. Dowben and R. G. Jones, *Surf. Sci.*, **84**, 449 (1979).

126. G. Brodén, G. Gafner, and H. P. Bonzel, *Surf. Sci.*, **84**, 295 (1979).

127. M. Grunze, F. Bozso, G. Ertl, and M. Weiss, *Appl. Surf. Sci.*, **1**, 241 (1978).
128. G. Pirug, G. Brodén, and H. P. Bonzel, *Surf. Sci.*, **94**, 323 (1980).
129. S. E. Greco, J. P. Roux, and J. M. Blakely, *Surf. Sci.*, **120**, 203 (1982).
130. W-Y. Lee, M. H. Lee, and J. M. Eldridge, *J. Vac. Sci. Technol.*, **15**, 1549 (1978).
131. G. R. Conner, *Thin Solid Films*, **53**, 38 (1978).
132. K. Uhlmann, G. Haupold, H. J. Muessig, and F. Storbeck, *Cryst. Res. Technol.*, **16**, 1315 (1981).
133. L. Surnev, *Surf. Sci.*, **110**, 458 (1981).
134. L. Surnev, *Surf. Sci.*, **110**, 439 (1981).
135. D. E. Ibbotson, T. S. Wittrig, and W. H. Weinberg, *Surf. Sci.*, **110**, 294 (1981).
136. D. E. Ibbotson, T. S. Wittrig, and W. H. Weinberg, *Surf. Sci.*, **110**, 313 (1981).
137. D. E. Ibbotson, T. S. Wittrig, and W. H. Weinberg, *Surf. Sci.*, **111**, 149 (1981).
138. B. E. Nieuwenhuys and G. A. Somarjai, *Surf. Sci.*, **72**, 8 (1978).
139. J. L. Taylor, D. E. Ibbotson, and W. H. Weinberg, *Surf. Sci.*, **79**, 349 (1979).
140. S. A. Chambers, P. R. Davis, L. W. Swanson, and M. A. Gesley, *Surf. Sci.*, **118**, 75 (1982).
141. S. J. Klauser and E. B. Bas, *Appl. Surf. Sci.*, **3**, 356 (1979).
142. J. C. Fuggle, *Surf. Sci.*, **69**, 581 (1977).
143. S. A. Flodström and C. W. B. Martinsson, *Surf. Sci.*, **118**, 513 (1982).
144. H. Namba, J. Darville, and J. M. Gilles, *Surf. Sci.*, **108**, 446 (1981).
145. E. I. Ko and R. J. Madix, *Surf. Sci.*, **100**, L449 (1980).
146. E. I. Ko and R. J. Madix, *Surf. Sci.*, **109**, 221 (1981).
147. B. W. Walker and P. C. Stair, *Surf. Sci.*, **103**, 315 (1981).
148. M. Boudart, C. Egawa, S. T. Oyama, and K. Tamaru, *J. Chim. Phys. Phys.-Chim. Biol.*, **78**, 987 (1981).
149. T. R. Felter and P. J. Estrup, *Surf. Sci.*, **76**, 464 (1978).
150. E. Gillet, J. C. Chiavena, and M. Gillet, *Surf. Sci.*, **66**, 596 (1977).
151. E. Bauer and H. Poppa, *Surf. Sci.*, **88**, 31 (1979).
152. M. Matsunaga, T. Homma, and A. Tanaka, *ASLE Trans.*, **25**, 323 (1982).
153. R. G. Jones and D. P. Woodruff, *Vacuum*, **31**, 411 (1981).
154. E. G. Keim, F. Labohm, O. L. J. Gijzeman, G. A. Bootsma, and J. W. Geus, *Surf. Sci.*, **112**, 52 (1981).
155. M. Breitschafter, E. Umbach, and D. Menzel, *Surf. Sci.*, **109**, 493 (1981).
156. G. A. Sargent, G. B. Freeman, and J. L.-R. Rao, *Surf. Sci.*, **100**, 342 (1980).
157. Y-N. Fan, L-X. Tu, Y-Z. Sun, R-S. Li, and K. H. Kuo, *Surf. Sci.*, **94**, L203 (1980).
158. Y. Sakisaka, M. Miyamura, J. Tamaki, M. Nishijima, and M. Onchi, *Surf. Sci.*, **93**, 327 (1980).
159. R-S. Li and L-X. Tu, *Surf. Sci.*, **92**, L71 (1980).
160. G. L. Price and B. G. Baker, *Surf. Sci.*, **91**, 571 (1980).
161. M. G. Cattania, M. Simonetta, and M. Tescari, *Surf. Sci.*, **82**, L615 (1979).
162. W. Erley and H. Wagner, *Surf. Sci.*, **79**, 333 (1978).
163. M. Mohri, H. Kabibayashi, K. Watanabe, and T. Yamashina, *Appl. Surf. Sci.*, **1**, 170 (1978).
164. G. Casalone, M. G. Cattania, M. Simonetta, and M. Tescari, *Surf. Sci.*, **62**, 321 (1977).
165. F. C. Schouten, E. W. Kaleveld, and G. A. Bootsma, *Surf. Sci.*, **63**, 460 (1977).

166. W. Erley and H. Wagner, *Surf. Sci.*, **66**, 371 (1977).
167. R. A. Zuhr and J. B. Hudson, *Surf. Sci.*, **66**, 405 (1977).
168. J. T. Grant and M. P. Hooker, *Surf. Sci.*, **55**, 741 (1976).
169. S. Masuda, M. Nishijima, Y. Sakisaka, and M. Onchi, *Phys. Rev. B*, **25**, 863 (1982).
170. H. J. Grabke and H. Viefhaus, *Surf. Sci.*, **112**, L779 (1981).
171. P. H. Dawson and W-C. Tam, *Surf. Sci.*, **81**, 164 (1979).
172. R. Sau and J. B. Hudson, *Surf. Sci.*, **102**, 239 (1981).
173. P. Holloway and R. A. Outlaw, *Surf. Sci.*, **111**, 300 (1981).
174. C. Benndorf, B. Egert, C. Nöbl, H. Seidel, and F. Thieme, *Surf. Sci.*, **92**, 636 (1980).
175. P. K. de Bolex, F. Labohm, O. L. J. Gijzeman, G. A. Bootsma, and J. W. Geus, *Appl. Surf. Sci.*, **5**, 321 (1980).
176. L. Schlapbach, D. Shaltiel, and P. Oelhafen, *Mater. Res. Bull.*, **14**, 1235 (1979).
177. C. Benndorf, B. Egert, G. Kellar, H. Seidel, and F. Thieme, *Surf. Sci.*, **80**, 287 (1979).
178. K. H. Rieder, *Appl. Surf. Sci.*, **2**, 74 (1978).
179. M. Kiskinova, *Surf. Sci.*, **111**, 584 (1981).
180. M. Kiskinova, L. Surnev, and G. Bliznakov, *Surf. Sci.*, **104**, 240 (1981).
181. J. C. Riviere, *Mater. Res. Eng.*, **42**, 49 (1980).
182. P. H. Dawson and W-C Tam, *Surf. Sci.*, **81**, 464 (1979).
183. V. S. Sundaram, S. P. da Cunha, and R. Landers, *Surf. Sci.*, **119**, L383 (1982).
184. H. Poppa and F. Soria, *Surf. Sci.*, **115**, L105 (1982).
185. P. W. Davies and R. M. Lambert, *Surf. Sci.*, **110**, 227 (1981).
186. L. Peralta, Y. Berthier, and M. Huber, *Surf. Sci.*, **104**, 435 (1981).
187. D. L. Doering, H. Poppa, and J. T. Dickinson, *J. Vac. Sci. Technol.*, **17**, 198 (1980).
188. W. Erley, *Surf. Sci.*, **94**, 281 (1980).
189. G. A. Garwood and A. T. Hubbard, *Surf. Sci.*, **92**, 617 (1980).
190. J. Küppers, F. Nitschlké, K. Wandelt, and G. Ertl, *Surf. Sci.*, **87**, 295 (1979).
191. K. Kunimori, T. Kawai, T. Kondow, T. Onishi, and K. Tamaru, *Surf. Sci.*, **59**, 302 (1976).
192. L. Hilaire, P. Légaré, Y. Holland, and G. Maire, *Surf. Sci.*, **103**, 125 (1981).
193. T. S. Sun, S. P. Buchner, N. E. Byer, and J. M. Chen, *J. Vac. Sci. Technol.*, **10**, 1292 (1978).
194. G. A. Garwood and A. T. Hubbard, *Surf. Sci.*, **118**, 223 (1982).
195. U. Köhler and H.-W. Wassmuth, *Surf. Sci.*, **117**, 668 (1982).
196. G. E. Gdowski and R. J. Madix, *Surf. Sci.*, **115**, 524 (1982).
197. S. R. Bare, P. Hofmann, and D. A. King, *Vacuum*, **31**, 463 (1981).
198. M. R. McClellan, J. L. Gland, and F. R. McFeeley, *Surf. Sci.*, **112**, 63 (1981).
199. G. A. Garwood and A. T. Hubbard, *Surf. Sci.*, **112**, 281 (1981).
200. M. J. Mummey and L. D. Schmidt, *Surf. Sci.*, **109**, 29 (1981).
201. F. Solymosi and J. Kiss, *Surf. Sci.*, **108**, 641 (1981).
202. H. H. Farrell, *Surf. Sci.*, **100**, 613 (1980).
203. J. L. Gland and B. A. Sexton, *Surf. Sci.*, **94**, 355 (1980).
204. M. J. Mummey and L. D. Schmidt, *Surf. Sci.*, **91**, 301 (1980).
205. E. Bertel, K. Schwaha, and F. P. Netzer, *Surf. Sci.*, **83**, 439 (1979).
206. J. L. Gland, *Surf. Sci.*, **71**, 327 (1978).
207. M. E. Bridge and R. M. Lambert, *Surf. Sci.*, **63**, 315 (1977).
208. T. E. Fisher, S. R. Kelemen, and H. P. Bonzel, *Surf. Sci.*, **64**, 157 (1977).

209. K. Schwaha and E. Bechtold, *Surf. Sci.*, **66**, 383 (1977).
210. J. T. Grant and M. P. Hooker, *J. Elec. Spectrosc. Rel. Phenom.*, **9**, 93 (1976).
211. J. L. Gland and V. N. Korchak, *Surf. Sci.*, **75**, 783 (1978).
212. T. Matsushima, D. B. Almy, and J. M. White, *Surf. Sci.*, **67**, 89 (1977).
213. M. Wilf and P. T. Dawson, *Surf. Sci.*, **65**, 399 (1977).
214. K. Schwaha and E. Bechtold, *Surf. Sci.*, **65**, 277 (1977).
215. J. L. Gland, *Surf. Sci.*, **93**, 487 (1980).
216. G. Maire, P. Légaré, and G. Lindauer, *Surf. Sci.*, **80**, 238 (1979).
217. D. T. Larson, *J. Vac. Sci. Technol.*, **17**, 55 (1980).
218. M. P. Cox and R. M. Lambert, *Surf. Sci.*, **107**, 547 (1981).
219. P. A. Thiel, J. T. Yates, and W. H. Weinberg, *Surf. Sci.*, **90**, 121 (1979).
220. D. G. Castner, B. A. Sexton, and G. A. Somorjai, *Surf. Sci.*, **71**, 519 (1978).
221. R. A. Marbrow and R. M. Lambert, *Surf. Sci.*, **67**, 489 (1977).
222. P. A. Thiel, J. T. Yates, and W. H. Weinberg, *Surf. Sci.*, **82**, 22 (1979).
223. J. A. Schreifels, S.-K. Shi, and J. M. White, *Appl. Surf. Sci.*, **7**, 312 (1981).
224. H-I. Lee, G. Praline, and J. M. White, *Surf. Sci.*, **91**, 581 (1980).
225. E. Umbach, S. Kulkarni, P. Feulner, and D. Menzel, *Surf. Sci.*, **88**, 65 (1979).
226. S. R. Kelemen and T. E. Fisher, *Surf. Sci.*, **87**, 53 (1979).
227. L. R. Danielson, M. J. Dressner, E. E. Donaldson, and J. T. Dickison, *Surf. Sci.*, **71**, 599 (1978).
228. L. R. Danielson, M. J. Dressner, E. E. Donaldson, and D. R. Sandstrom, *Surf. Sci.*, **71**, 615 (1978).
229. J. C. Fuggle, E. Umbach, P. Feulner, and D. Menzel, *Surf. Sci.*, **64**, 69 (1977).
230. R. Ku, N. A. Gjostein, and H. P. Bonzel, *Surf. Sci.*, **64**, 465 (1977).
231. P. D. Reed, C. M. Combrie, and R. M. Lambert, *Surf. Sci.*, **64**, 603 (1977).
232. S.-K. Shi, J. A. Schreifels, and J. M. White, *Surf. Sci.*, **105**, 1 (1981).
233. G. Praline, B. E. Koel, H-I. Lee, and J. M. White, *Appl. Surf. Sci.*, **5**, 296 (1980).
234. M. Alnot, B. Weber, J. J. Ehrhardt, and A. Cassuto, *Appl. Surf. Sci.*, **2**, 578 (1979).
235. M. Housley, R. Ducros, G. Piquard, and A. Cassuto, *Surf. Sci.*, **68**, 277 (1977).
236. R. Pantel, M. Bujor, and J. Bardolle, *Surf. Sci.*, **83**, 228 (1979).
237. H. H. Madden, *Surf. Sci.*, **105**, 129 (1981).
238. K. Fujiwara and H. Ogata, *Surf. Sci.*, **86**, 700 (1979).
239. K. Fujiwara and H. Ogata, *Surf. Sci.*, **72**, 157 (1978).
240. M. Miyamura, Y. Sakisaka, M. Nishijima, and M. Onchi, *Surf. Sci.*, **72**, 243 (1978).
241. H. F. Dylla, J. G. King, and M. J. Cardillo, *Surf. Sci.*, **74**, 141 (1978).
242. S. Tougaard, P. Morgen, and J. Onsgaard, *Surf. Sci.*, **111**, 545 (1981).
243. M. C. Muñoz and J. L. Sacedór, *Surf. Sci.*, **68**, 347 (1977).
244, S. R. Kelemen, Y. Goldstein, and B. Abeles, *Surf. Sci.*, **116**, 488 (1982).
245. R. A. Powell, *Appl. Surf. Sci.*, **2**, 397 (1979).
246. Z. T. Stott and H. P. Hughes, *Vacuum*, **31**, 487 (1981).
247. C. Palacio and J. M. Martinez-Duart, *Thin Solid Films*, **90**, 63 (1982).
248. R. Bastasz, C. A. Colmenares, R. L. Smith, and G. A. Somorjai, *Surf. Sci.*, **67**, 45 (1977).
249. J. B. Bignolas, M. Bujar, and J. Bardolle, *Surf. Sci.*, **108**, L453 (1981).
250. J. Bloch, U. Atzmony, M. P. Daniel, M. H. Mintz, and N. Shamir, *J. Nucl. Mater.*, **105**, 196 (1982).

251. K. Domen, S. Naito, M. Soma, T. Onishi, and K. Tamaru, *J. Chem. Soc. Faraday Trans.* **78**, 1451 (1982).
252. P. W. Davies and R. M. Lambert, *Surf. Sci.*, **95**, 571 (1980).
253. S. D. Foulias, K. J. Rawlings, and B. J. Hopkins, *Surf. Sci.*, **114**, 1 (1982).
254. K. J. Rawlings, S. D. Foulias, and B. J. Hopkins, *Surf. Sci.*, **108**, 49 (1981).
255. A. K. Bhattacharya, L. J. Clarke, and L. Motales de la Garza, *J. Chem. Soc. Faraday Trans.*, **77**, 2223 (1981).
256. K. J. Rawlings, G. G. Price, and B. J. Hopkins, *Surf. Sci.*, **100**, 289 (1980).
257. K. J. Rawlings, G. G. Price, and B. J. Hopkins, *Surf. Sci.*, **95**, 245 (1980).
258. S. D. Foulias, K. J. Rawlings, and B. J. Hopkins, *Chem. Phys. Lett.*, **68**, 81 (1979).
259. K. J. Rawlings, S. D. Foulias, and B. J. Hopkins, *Solid State Commun.*, **32**, 295 (1979).
260. J. C. Fuggle and D. Menzel, *Surf. Sci.*, **79**, 1 (1979).
261. R. I. Masel, E. Umbach, J. C. Fuggle, and D. Menzel, *Surf. Sci.*, **79**, 26 (1979).
262. C. Somerton and D. A. King, *Surf. Sci.*, **89**, 391 (1979).
263. G. G. Price, K. J. Rawlings, and B. J. Hopkins, *Surf. Sci.*, **85**, 379 (1979).
264. E. Bauer, H. Poppa, P. R. Davis, and Y. Viswanath, *Surf. Sci.*, **71**, 503 (1978).
265. K. J. Fawlings, B. J. Hopkins, and S. D. Foulias, *Surf. Sci.*, **77**, 561 (1978).
266. E. Umbach, J. C. Fuggle, and D. Menzel, *J. Elec. Spectrosc. Rel. Phenom.*, **10**, 15 (1977).
267. M. Hourley and D. A. King, *Surf. Sci.*, **62**, 93 (1977).
268. M. Hourley and D. A. King, *Surf. Sci.*, **62**, 81 (1977).
269. C. Steinbruchel and R. Gomer, *Surf. Sci.*, **67**, 21 (1977).
270. M. A. Chesters, B. J. Hopkins, and R. I. Winton, *Surf. Sci.*, **59**, 46 (1976).
271. E. Bauer and T. Engel, *Surf. Sci.*, **71**, 695 (1978).
272. J. S. Foord, P. J. Goddard, and R. M. Lambert, *Surf. Sci.*, **94**, 339 (1980).
273. G. N. Krishnan, B. J. Wood, and D. Cubicciotti, *J. Electrochem. Soc.*, **127**, 2738 (1980).
274. T. C. Gainey and B. J. Hopkins, *J. Phys. C*, **14**, 5763 (1981).
275. D. W. Goodman and J. M. White, *Surf. Sci.*, **90**, 201 (1979).
276. M. E. Bridge and R. M. Lambert, *Surf. Sci.*, **94**, 469 (1980).
277. P. J. Goddard and R. M. Lambert, *Surf. Sci.*, **67**, 180 (1977).
278. M. Weiss, G. Ertl, and F. Nitschke, *Appl. Surf. Sci.*, **3**, 614 (1979).
279. E. D. Williams, C. M. Chan, and W. H. Weinberg, *Surf. Sci.*, **81**, L309 (1979).
280. P. Klimesch and M. Henzler, *Surf. Sci.*, **90**, 57 (1979).
281. M. A. Chesters, B. J. Hopkins, P. A. Taylor, and R. I. Winton, *Surf. Sci.*, **83**, 181 (1979).
282. J. B. Benziger and R. J. Madix, *Surf. Sci.*, **79**, 394 (1979).
283. J. G. McCarty and R. J. Madix, *J. Catal.*, **48**, 427 (1977).
284. M. E. Schrader, *Surf. Sci.*, **78**, L227 (1978).
285. J. Augustynski and L. Balsenc, "Applications of Auger and Photoelectron Spectroscopy to Electrochemical Problems," in B. E. Conway and J. O'Bockris (Eds.), *Modern Aspects of Electrochemistry*, Plenum, New York, 1979, Vol. 13, p. 251.
286. J. S. Hammond and N. Winograd, *J. Electrochem. Soc.*, **124**, 826 (1977).
287. J. Smith and G. Lindberg, *J. Electrochem. Soc.*, **125**, 1224 (1978).

288. M. F. Weber, H. R. Shanks, X. Berolo, and G. C. Danielson, *J. Electrochem. Soc.*, **127**, 329 (1980).
289. S. A. Chambers and L. W. Swanson, *Appl. Surf. Sci.*, **4**, 82 (1980).
290. P. N. Ross, *J. Electrochem. Soc.*, **126**, 67 (1979).
291. P. N. Ross, *J. Electrochem. Soc.*, **126**, 78 (1979).
292. K. Takayanagi, D. M. Kolb, K. Kambe, and G. Lehmpfuhl, *Surf. Sci.*, **100**, 407 (1980).

CHAPTER

6

APPLICATIONS OF AUGER ELECTRON SPECTROSCOPY IN METALLURGY AND MATERIALS SCIENCE

In the relatively short span since its introduction as a potentially useful surface-analytical technique,[1] AES has developed to the extent that it is now a universally accepted method of material characterization and problem solving in a wide spectrum of fields of application. This chapter is primarily concerned with an introduction to the enormous range of applications of AES in metallurgy and materials science. The word "introduction" should be explained at this point, because the number of applications of AES in these areas of surface science appearing in the literature has reached truly astronomic proportions. An earlier bibliography of studies involving the technique, only covering publications up to 1975, contains 2146 references, many of which are in the field of metallurgy or materials science.[2] A more up-to-date source of references may be found in the Fundamental Reviews which appear in *Analytical Chemistry*,[3] but even these publications manage to filter out only a fraction of the total number of papers that utilize AES as an analytical technique. In this chapter, reference will therefore be made, whenever possible, to recently published, more-specialized reviews of the applications of AES, where further information on particular topics may be found.

Since a comprehensive review of the applications of AES in metallurgy and materials science is beyond the scope of this work, emphasis will be placed, instead, on a large number of example studies in these fields, which will serve to illustrate the extensive nature of the analytical information that AES is capable of providing on such surfaces. The application of AES to clean, metallic surfaces (elements and alloys) is discussed in Section 2, with emphasis on the fundamental aspects of Auger emission from metals, and fundamental studies of the surface properties of binary alloys. Surface studies of bulk oxides and oxide films by AES are presented in Section 3, with some discussion on the practical aspects of chemical-shift information in the Auger spectra of oxides. The transition from metal to oxide surfaces is accompanied by a short discussion on fundamental studies of oxidation, but the reader is referred to Chapter 5 for a more-detailed treatment of this

263

topic. The applications of AES to "technological" oxide surfaces (anodized metals, passivated films, and corrosion layers) are discussed in some detail in this section.

From oxide films, the discussion proceeds to the application of AES to the characterization of films and coatings in Section 4. For convenience, this section is divided into two subsections: thin films (< 1 μm) and thick films (> 1 μm). Although the type of information supplied by AES is common to both areas of application, the experimental procedures for obtaining that information (composition-depth profiling) justify their classification under separate headings. Section 5 deals with applications in interfacial analysis, and Section 6 is concerned with examples of applications of AES in the study of semiconductors.

The single, most important feature of AES, which makes it so attractive to metallurgists and materials scientists as an analytical technique, is its ability to provide a qualitative "point" analysis, with theoretical limits on both lateral *and* depth resolution approaching atomic dimensions. The requirements and limitations of spatial resolution in AES applied to materials characterization are therefore briefly discussed in the first section.

1. THREE-DIMENSIONAL MATERIALS ANALYSIS

For many applications of AES, such as the identification of a uniform contamination layer on a specimen surface, we need to know only the average value of the chemical composition over a relatively large area. In this instance, there is no requirement for either lateral resolution (indeed, high lateral resolution could result in an incorrect analysis from the statistical point of view) or, providing the contamination layer is thick enough, depth resolution. In most cases of technological interest, however, the surface of the specimen to be analyzed is usually not homogeneous in lateral terms, and rarely homogeneous in depth terms, and, in general, we wish to derive this spatial variation of composition laterally and with depth.

As introduced in Chapter 3, a surface-sensitive imaging analytical method may be extended to perform three-dimensional analytis of a solid by combining it with a depth-profiling technique.[4] [Here, depth profiling is used in its most general sense, meaning any method by which the variation of a property of the system (e.g., composition) with depth may be determined.] Indeed, the inclusion of the time dependence of surface compositions with composition-depth profiles (as in studies of surface reactions, surface segregation, diffusion, etc.) extends this capability into four-dimensional analysis of the system.

The particular application concerned will determine the requirements of lateral, depth, and temporal resolution, and the nature of the specimen to be analyzed, as well as the details of the analytical procedure, will determine the limitations imposed on each of these.

1.1. Lateral Resolution

1.1.1. Instrumental Aspects

One of the main advantages of AES over other surface-analytical techniques is the very high lateral resolution achievable using conventional electron sources. In principle, this should approach atomic dimensions, limited only by the backscattered secondary-electron cascade volume.[5,6] This is possible only at very low primary-beam currents, however, and the requirement of a minimum primary-beam current of the order of 1 nA in order to obtain useful spectra imposes a practical limit of about 50 nm on the lateral resolution.[5,6] Such resolution is presently available with field-emission electron sources,[5] but it is unlikely that they will find widespread acceptance in AES applications owing to their inherent instability and sensitivity to contamination (beam-current instabilities as high as 10% have been reported in the most up-to-date instruments[6]). More conventional thermal-emission sources (employing tungsten on lanthanum hexaboride cathodes) are capable of operating at several tens of nanoamps beam current and submicron spatial resolution. For the lanthanum hexaboride cathode, it has been shown that the beam diameter varies approximately as the cube root of the beam current in this operating regime,[7,8] hence a 1000-fold increase in primary current (and therefore also in Auger signal) will result in only a 10-fold degradation of resolution in such a system.

1.1.2. Materials Aspects

A major limitation on lateral resolution is imposed by the stability of the specimen under study of the electron beam, in its susceptibility to electron-beam damage. This damage may arise from either electronic or thermal effects, and may manifest itself in many different forms.[9,10] On the assumption that the specimen is in good thermal/contact with the spectrometer, such that all of the beam power is dissipated as conducted thermal energy, a *minimum* temperature rise in the beam center (ΔT) is predicted[11] to be

$$\Delta T = \frac{p r_0}{\beta} \tag{6.1}$$

where p is the beam power density, r_0 is the beam radius, and β is the bulk thermal conductivity of the specimen.

Some of the reported temperature rises in the literature are too high to be consistent with the above equation [e.g., a rise of 150–250°C on glass using a 3 W cm^{-2} beam,[10] when Eq. (6.1) would predict, at most, several tens of degrees], but can be explained if poor thermal contact between sample and sample holder is assumed. In such circumstances, the specimen may be able to lose thermal energy only by radiative emission. The *maximum* temperature rise is then much greater than predicted above, and is given approximately[12] by

$$\Delta T = 300p \qquad (6.2)$$

for an emissivity of 0.35 and values of p up to 0.1 W cm^{-2}. For higher values of p, ΔT varies approximately as $p^{1/4}$. In practice, most specimen surfaces will be poor thermal conductors, and thermal contact with the substrate will be far from perfect, hence the temperature rise during AES analysis will lie somewhere between the two extremes mentioned above. Typical expected ranges in surface temperature for 1 μW and 1 mW electron beams on glass are illustrated in Fig. 6.1.

Other aspects of electron-beam damage are discussed in more detail in Chapter 3. In the present context, we are concerned mainly with the extent to which this can be minimized on technological surfaces. From the data given in the literature,[10,13] it is possible to construct a graph showing the practical

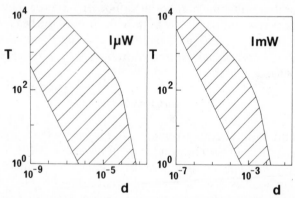

Figure 6.1.   Range of possible values for the equilibrium temperature rise (T, in degrees Kelvin) of a glass surface (bulk thermal conductivity 0.01 W cm^{-1} K^{-1}, emissivity 0.35) under static electron-beam irradiation, as a function of beam diameter (d, in meters), for 1 μW (e.g., 0.1 nA at 10 keV) and 1 mW (e.g., 1 μA at 1 keV) beam powers. The lower curves represent the bulk thermal conduction limit, the upper curves the emissivity limit (see text). In practice, the real surface temperature rise will lie between these two limits (in the shaded areas), depending on the detailed microstructure of the surface. For power densities approaching the performance limits of thermal electron sources (10^{-6} m at 1 μA and $<10^{-7}$ m at 0.1 nA, see Fig. 6.2), emissivity-limited surface temperature rises of 10,000 K are possible.

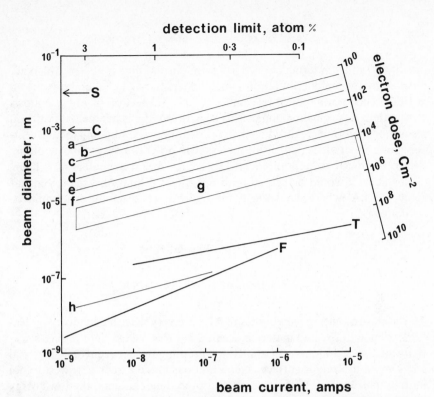

Figure 6.2. Sample degradation during AES analyses of technological surfaces. Material-dependent critical combinations of primary-beam current (abscissa) and primary-beam diameter (ordinate) at which electron-beam damage (over a 300-sec exposure) becomes significant, assuming a simple cross-sectional model for electronic damage. The electron dose scale is calculated for a 300-sec exposure, and elemental detection limits are typical values observed in the author's laboratory. Critical doses were calculated from the literature for the following materials:

Curve	Substrate	Dose (cm^{-2})	References
a	Organic films	3	13
b	Native oxides, Na_3AlF_6	10	14, 15
c	Sodium in glass	30	16, 17
d	TeO_2, K_2O	200	18, 19
e	Alkali halides, chlorine adlayers	1000	20, 21
f	CaF_2, chlorine in glass	3000	10, 22
g	SiO_2, Al_2O_3, MoO_3, Nb_2O_5, Ta_2O_5, WO_3, TiO_2, carbonaceous contamination	10^4–10^5	23–32
h	Iron oxides	$>10^9$	33

Significant damage may be avoided only by operating at a total dose below these thresholds. C and S represent typical practical upper limits on analysis diameter imposed by cylindrical-mirror and spherical-sector analyzers, respectively. T and F represent typical performance limits for thermal (hot filament) and field-emission sources, respectively.[5,7]

267

operating limits, in terms of primary-beam current and spot size, which are dictated by the stability of the specimen under study. Such a graph is shown in Fig. 6.2 for a number of relevant materials. Inspection of Fig. 6.2 shows clearly that, if specimen damage is to be avoided, the instrumental limits on resolution cannot be realized on many substrates without going to unacceptably poor sensitivities (low beam currents). In order to be certain of avoiding electron-induced reduction of silica at electron-beam diameters of ~ 1 μm, for instance, it would be necessary to operate at primary-beam currents far below 0.1 nA, where detection limits under normal conditions would be > 10 atom%.

1.2. Depth Resolution

1.2.1. Intrinsic Depth Resolution

The lateral-resolution capabilities of AES are very valuable in most applications to metallurgy and materials science, but it is the extreme surface sensitivity of the technique—its ability to discriminate between bulk material and a monomolecular surface layer—that has made it such a powerful tool in materials analysis.[34] As discussed in preceding chapters, the low-energy Auger electrons of interest to electron spectroscopists (with energies in the range 20–2000 eV) are characterized by very short inelastic mean free paths, λ, in solid materials. Typically, λ has a value of approximately 2 nm at 1000 eV, and varies as $E^{1/2}$ in most solids.[35] The electron escape depth, λ', may be defined as the component of λ normal to the sample surface, hence

$$\lambda' = \lambda \cos \theta \qquad (6.3)$$

where θ is the angle of electron emission relative to the surface normal. Ignoring attenuation of the primary beam, it can be shown[36] that the fractional contribution to the observed Auger spectrum from atoms within a depth of dnm is

$$\frac{I}{I_{\text{tot}}} = (1 - e^{-d/\lambda'}) \qquad (6.4)$$

hence, 63% of the Auger spectrum originates in a surface layer of thickness λ'. Thus, by choosing suitable values of electron energy (ε_{kin}) and emission angle (θ) for analysis, it is possible to obtain Auger spectra from the top surface with a depth resolution approaching atomic dimensions. This ability of AES has provided much invaluable information, not only of as-received surface compositions, but also of metal–gas reaction mechanisms and kinetics, substrate–adsorbate interactions and bonding, and thin-film structure and epitaxy.

1.2.2. Depth Resolution in Sputter-Depth Profiling

In many cases of technological importance, it is not only the surface mono-layer that is of interest. For example, the oxide layer on most metal and alloy surfaces will rapidly grow to a limiting thickness of <5 nm in oxidizing gases at room temperature. This is almost entirely encompassed by the sampling depth in AES; hence a detailed analysis of passivation layers may be made under high depth-resolution conditions. Such a protective layer, however, increases in thickness with increasing temperature, and the oxide film may also thicken in aqueous environments at room temperature under the action of electrochemical processes, such as anodic oxidation, or by chemical conversion of the surface. In such cases, the oxidation or corrosion layers may be several tens of nanometers in thickness, and AES will be able to probe the full extent of the film only in conjunction with a depth-profiling facility. The above argument applies to most other areas of application, such as where the ability to obtain information about the variation of atomic composition with depth, in the range of several angstroms to several microns, with high depth resolution, is usually far more valuable than a surface elemental analysis alone. For depths exceeding 5 nm, there are really only two methods of profiling that may be used in conjunction with AES analysis—inert gas ion-sputtering (etching) and mechanical sectioning. Aspects of both techniques are discussed in Chapter 3, and Section 4 of the present chapter.

The depth resolution may be defined in terms of the observed broadening of an originally atomically sharp interface at some depth Z below the surface, and is usually expressed[37] as either the observed interface width (ΔZ) or as the ratio $\Delta Z/Z$ (Fig. 6.3).

Ideally, the depth resolution ΔZ of the analysis should be constant throughout the profile, limited only by the electron escape depth discussed above. In practice, however, it is found that ΔZ deteriorates[37] markedly with increasing Z. Factors that have been shown to affect the depth resolution in AES sputter-depth profiling, for instance, include substrate roughness,[38,39] ion-induced topography,[40,41] the angle of incidence of the ion beam,[42] and secondary ion effects (such as cascade mixing and ion-induced diffusion).[38,39] Similar analyses have been made of the factors affecting depth resolution in AES combined with mechanical lapping to produce composition depth profiles.[43] The contribution of each of these factors is considered in a little more detail below.

Ion-induced knock-on effects, cascade mixing, and diffusion have all been the subject of much theoretical interest in recent years.[44] Although the subject does not lend itself to a straightforward theoretical treatment, it is generally assumed[45] that these effects will act to produce a fixed contribution to ΔZ, termed ΔZ_{mix}, which will be of the order of the primary ion range, R_p, at depths exceeding R_p.

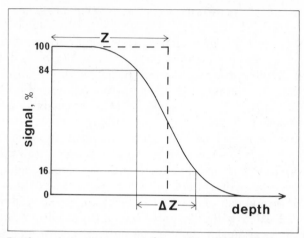

Figure 6.3.   Depth resolution in sputter-depth profiling. The normalized Auger electron intensity from a thin, uniform overlayer (true profile indicated by broken line) while sputtering through into the substrate. The measured profile is often approximated by an error (or Gaussian integral) function, and the depth resolution, ΔZ, at depth Z defined as twice the standard deviation (σ) of the Gaussian. ΔZ therefore corresponds to the depth separation between the points on the profile, which correspond to 84% and 16% of the original overlayer intensity. In the example, $\Delta Z/Z \approx 0.5$.

Early theoretical treatments of the variation of ΔZ with Z considered only the statistical nature of sputter removal (the sequential layer sputter or SLS model), and concluded a $\Delta Z_{SLS} \propto Z^{1/2}$ dependence,[46] or, in an extension of this model, $\Delta Z_{SLS} \propto \{Z(1+S)\}^{1/2}$, where S is the sputter yield.[47] On the other hand, a recent treatment by Seah et al.,[48] taking into account the variation of sputtering probability with atomic environment, concludes that the statistical contribution to ΔZ at depths greater than 10 nm is approximately constant, and will generally be in the range 1 nm $< \Delta Z_{SLS} < 2$ nm, instead of increasing with depth.

With modifications, the SLS model may also be used to describe ion-induced topographical changes, and an empirical dependence of $\Delta Z_{TOP} \propto (ZE_p)^{1/2}$, where E_p is the primary ion energy, has been proposed.[49] The formation of surface topography during ion bombardment is a complex area, however, and it does not lend itself to straightforward theoretical interpretation.[40] The exact nature of the substrate is expected to have a marked effect on ΔZ_{TOP} in most cases.[50]

Contributions to ΔZ that are expected to result in a significant deterioration of depth resolution with depth are nonuniform ion etching over the area of analysis (which should lead to $\Delta Z_{NU} \propto Z$) and the *original* surface roughness. This latter contribution (ΔZ_{ROU}) has only recently been treated rigorously

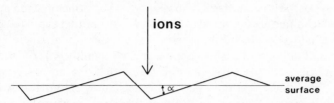

Figure 6.4.   Roughness-limited depth resolution in sputter-depth profiling. For an ion beam incident normal to the average surface, the roughness contribution to the depth resolution in sputter depth profiling is calculated to be $\Delta Z \propto \alpha_0^2 Z$, where α_0 is the mean value of the angular deviations (α) of the local surface from the average surface plane (see Reference 39).

from a theoretical point of view,[39] although its effects have been recognized for some time.[38] ΔZ_{ROU} is also expected to be strongly dependent on the ion-beam angle of incidence, and it has been proposed[39] that $\Delta Z_{\text{ROU}} \propto \alpha^2 Z$ for normal ion incidence. The variable α angle characterizes the roughness, and is defined as the angle made by the microscopically flat regions of the surface to the plane of the overall surface (Fig. 6.4). Effects due to interface (as opposed to surface) roughness have also been considered.[39,43] The overall depth resolution in sputter-depth profiling will be given by[37]

$$\Delta Z_{\text{TOT}} = (\Delta Z_{\text{MIX}}^2 + \Delta Z_{\text{SLS}}^2 + \Delta Z_{\text{TOP}}^2 + \Delta Z_{\text{NU}}^2 + \Delta Z_{\text{ROU}}^2)^{1/2} \qquad (6.5)$$

Following the arguments presented above, Z_{TOT} is expected to vary as Z^0 or $Z^{1/2}$ in ideal systems (atomically smooth substrate, perfectly uniform ion beam), but as Z in many technologically relevant systems. Most experimental measurements of the variation of ΔZ with Z report a power dependence[39,51–53] in this range. Absolute values of ΔZ range from 2 nm, obtained in the "best" profiles,[52] to $(0.16Z + 36)$ nm, obtained in typical materials science applications.[53]

Providing the contributions to ΔZ_{TOT} are known, it is possible to remove much of the broadening effects, associated with finite-depth resolution, from AES sputter-depth profiles. Such a deconvolution method has been applied to profiles through Cu/Ni and Ag/Au interfaces to facilitate retrieval of the real "profile" from the measured profile with considerable success.[38]

1.2.3.   Depth Resolution in Mechanical Sectioning

The alternative to ion sputtering for composition-depth profiling in AES is mechanical sectioning. Although direct diamond turning has been proposed as a means of producing highly accurate sections for surface analysis application,[43] it is not considered further. Taper sectioning by angle lapping, however, is a well-established technique and has been applied to depth profiling by AES with considerable success.[43,54] The principles and application of the technique are discussed in Section 4. The limiting depth resolution, assuming a perfectly flat finish, is given by $\Delta Z \approx 2r \tan \alpha$, where r is the electron-beam radius and α is the lapping angle (generally $< 1°$).[50] Although lapped surfaces with local roughnesses of less than 0.5 nm have been rereported,[55] it is likely that most surfaces of technological interest will have finishes in the range 10–50 nm. In addition, there may be contributions from smearing phenomena, and the need to remove surface contamination by ion sputtering prior to AES analysis also introduces extra contributions to ΔZ. In practice, the depth resolution using angle lapping is generally of the order of 100 nm and is independent of depth.[43,56] This compares very favorably with ion sputtering for depths exceeding 400 nm (see above). In addition, the original surface roughness should have no effect on the depth resolution using mechanical sectioning, in contrast to ion sputtering. The depth resolution in both methods of profiling, however, is limited by interface roughness.[43]

In principle, higher depth resolution ($\Delta Z < 100$ nm) may be obtained by using a variation of the angle-lapping technique—ball cratering. By lapping with a ball bearing, generally several centimeters in diameter, a spherical crater is produced in the sample surface. The advantages of this technique over angle lapping are discussed in Section 4—the only point that concerns us here is the fact that at the base of the crater the angle α becomes vanishingly small, affording the possibility of *no* interface broadening due to the taper angle.[57] The effective taper angle at the point of analysis is given by

$$\alpha \approx \tan^{-1} \left\{ \frac{2(d-Z)}{R} \right\}^{1/2} \tag{6.6}$$

where d is the total crater depth, R is the ball radius, and Z is the depth of analysis, for small values of $(d-Z)$. The quantity $(d-Z)$ represents the "overshoot" of the crater beyond the depth of interest.[43] From Eq. (6.6), the depth resolution becomes[50]

$$Z = 2r \left\{ \frac{2(d-Z)}{R} \right\}^{1/2} \tag{6.7}$$

$$= \frac{rD_1}{R} \tag{6.8}$$

where D_1 is the observed diameter of the crater at the depth Z (Fig. 6.5). Again, there will be contributions from smearing and ion-induced broadening during surface cleaning, but it has been shown that depth resolution as good as 40 nm (at a total depth of 5000 nm) may be obtained using this technique.[43] Under ideal circumstances (large ball, small electron-beam

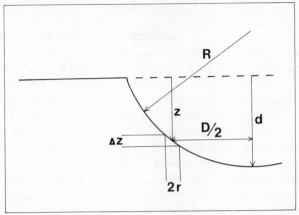

Figure 6.5.   The depth resolution in ball-cratered composition-depth profiles. The depth resolution, ΔZ, at depth Z below the original surface, assuming negligible contribution from surface roughness, is given by $\Delta Z = rD/R$, where r is the primary beam radius, D is the crater diameter at depth Z, and R is the ball radius. Alternatively, $\Delta Z = 2r\{2(d-z)/R\}^{1/2}$, where d is the total crater depth.

diameter, and $Z \approx d$) the depth resolution will be limited only by the quality of the lapping.[43] For a typical ball radius of 1 cm, ball cratering will give a better depth resolution than angle lapping at a 0.5° taper angle if the "overshoot" $(d-Z)$ does not exceed 380 nm (corresponding to a value for D of 174 μm). In this case, the contribution of the crater geometry to the depth resolution will be less than 1% of the electron-beam diameter.

The various contributions to depth resolution in profiling by AES using ion-sputtering and mechanical-sectioning techniques have been analyzed by Lea and Seah[43] in an attempt to determine criteria for choosing one profiling method in preference to the other. For the general case, when the surface and interface roughness are not correlated, and providing a number of assumptions are made, it can be shown[43] that mechanical lapping will provide superior depth resolution over ion sputtering for depths greater than $(\Delta Z_{LAP}^2 - \Delta Z_{SR}^2)/4a$, where ΔZ_{LAP} is the depth resolution of the mechanical lapping technique (determined by r and α as described above), ΔZ_{SR} is a quantity characterizing the surface roughness, and a is constant of the order of the monolayer thickness. In general, ΔZ_{LAP} will be of the order of 50 nm, ΔZ_{SR} will be between 0 and 1 μm, and a will be ~ 0.25 nm. Mechanical lapping will therefore be preferable to sputtering for *all* depths in the worst case $(\Delta Z_{SR} > \Delta Z_{LAP})$, but only for depths exceeding 2.5 μm in the best case (perfectly smooth specimen). In practice, experience and the application of "trial and error" will probably be necessary in order to determine which method will yield the better results.

1.3. Conclusions

Materials characterization in three dimensions is now a physical reality with the aid of AES combined with a depth-profiling facility.

The lateral resolution of the technique is ultimately limited by the technical aspects of electron optics. Although 50-nm resolution is feasible using sophisticated field-emission electron sources and electron optics in UHV systems, a typical state-of-the-art AES spectrometer for materials analysis will operate in the 500-nm–5-μm regime. In many cases, lateral resolution is dictated more by the susceptibility of the material under study to electron-induced damage than by instrumental limitation.

The depth resolution of the technique is often as high as one monolayer (0.25 nm) in ideal circumstances, when only the surface monolayer is of interest. When used in conjunction with a depth-profiling facility (such as ion sputtering or mechanical sectioning), the depth resolution generally deteriorates with depth, although it may be as low as 2 nm (independent of depth) on ideal systems. The highest depth resolution on perfectly flat substrates is obtained with sputter-depth profiling, but a combination of mechanical sectioning and sputter-depth profiling is preferable on rough surfaces (the general case in many technological situations) for analyses at depths exceeding several hundred nanometers. The depth resolution using mechanical sectioning may be as low as 50 nm, limited only by the electron-beam diameter and the quality of the sectioning technique.

The ultimate limits to spatial resolution in materials analysis by AES are thus generally dictated by the material under investigation. They range from several nanometers and tens of nanometers for depth and lateral resolution, respectively, in the best case (a perfectly flat, metallic substrate) to several microns and hundreds of microns, respectively, in the worst case (a rough, beam-sensitive substrate). Specific examples of the application of the techniques described in this section will be discussed in detail in Sections 2–6.

2. METALLIC SURFACES

Metallic surfaces are an area of application where problems of electron-beam-induced damage are very rare. Being excellent conductors of both heat and electricity, metals are capable of sustaining high electron current densities (of the order of several hundred A cm^{-2}) without suffering significant changes in surface composition in UHV systems. A typical temperature rise on a metal surface under such conditions would be $\sim 1°$C, for a 10-keV, 10-nA beam with a spot size of 500 nm on copper. Unfortunately, however, clean metallic surfaces are also the most difficult to work on, since most metals are highly reactive and, therefore, require very good UHV conditions (10^{-8} Pa or less) if they are to be prepared in a clean condition and remain clean for a significant period of time. Metals such as titanium, manganese, iron, zinc, zirconium, the alkali metals, alkaline earths, and rare earths rapidly form oxide layers, even in high vacuum conditions, with thicknesses comparable to the sampling depth of AES.[58]

In this section, we shall be concerned with the applications of AES to fundamental studies of clean metal surfaces (elements and alloys), since this field represents a very large body of scientific literature.

2.1. Studies of Clean Metal Surfaces

2.1.1. Theoretical Aspects

Although the theoretical basis for Auger emission has already been discussed (see Chapter 2), a few extra words are relevant at this point, if only as an introduction to AES studies of elemental metal surfaces. As pointed out in the previous chapter, it is generally the Auger line shape (fine structure in the Auger spectrum) that is of interest to electron spectroscopists.[59] Absolute Auger intensities from clean metals, although important,[60] are generally of secondary interest and, since they have already been discussed in previous chapters, they will not be considered further.

As introduced in the previous chapter, if the Auger transition under consideration involves core levels only (e.g., KLL as opposed to KVV), its shape might be expected to be relatively simple. This is, however, generally not the case. Contributions from multiplet splitting, plasmon loses, shake-up losses, Coster–Kronig transitions, doubly ionized initial states, and life-times all combine to influence the intensities of the various components to the overall Auger line shape.[34] If we then consider core–valence–valence (CVV) Auger transitions, additional and more-complex structure may be introduced into the Auger line shape through convolution of the valence band kine shape with the core level line shape. This is particularly useful in AES studies of metals, where KVV, LVV, and MVV transitions can, in principle, supply information regarding the valence-band density of states (DoS)[60] (see Section 2.1.3). Although the initial results of such investigations were not particularly encouraging,[61,62] our understanding of the theoretical principles involved in CVV transitions has developed significantly over the past few years. For metals with wide valence bandwidths, Γ (such as the s and p band elements), the line shape can usually be successfully interpreted[62] as the self-convolution of the valence-band DoS, with inclusions from matrix-element effects[63] and initial-state screening[64] in some cases. In some narrow

d-band elements, however, the valence band behaves essentially as a core level and the CVV Auger transitions have a pronounced atomic-like character. This arises when the hole–hole interaction energy in the final state (Coulomb repulsion energy, U_{eff}) is greater than Γ, thereby trapping the two holes in the same atom.[34,39] These two situations, where the two final-state holes are either trapped on the same atom (sometimes referred to as a correlated final state[59]) or delocalized (uncorrelated), represent the two extremes of the Cini–Sawatzky model,[65,66] which describes the general case. According to the model, Auger spectra from metals with $U_{eff} > 2\Gamma$ cannot yield information regarding the valence-band DoS. U_{eff} may be determined empirically from the relationship[67]

$$U_{eff} = E_b(C) - E_b(V1) - E_b(V2) - E_A \qquad (6.9)$$

where $E_b(C)$, $E_b(V1)$, and $E_b(V2)$ are the binding energies of electrons in the C, $V1$, and $V2$ levels, respectively, and E_A is the Auger electron energy. Typical values of Γ are ~ 4 eV for silver and ~ 2 eV for cadmium, indium, and tin.[60] Even in the quasiatomic spectra from high U_{eff} metals (such as copper[68]), weak band-line components, shifted to $\sim U_{eff}$ higher in kinetic energy, are expected[66] and found.[68] More subtle treatments have been made, and the reader is referred to Weightman[60] and Trégha et al.[69] for further discussion on this topic.

CVV Auger spectra from metals are also expected to be very sensitive to the width of the unoccupied portion of the valence band (density of unoccupied states, DUS), and it has been shown that even a relatively small hole density may produce qualitative changes in the spectra.[70] Increasing the DUS is expected to lead to a broadening and loss of intensity in the CVV spectrum and to a development of extra, broad structure to the low-energy side of the main peaks. Such structure is expected to be particularly useful in AES studies of metal alloys, where the DUS may be varied from 0 almost to unity by variations in alloy composition. Little experimental verification of the theory exists at present,[70] and it is likely that several refinements (such as the inclusion of interband mixing) will be necessary.

2.1.2. Experimental Line Shape and Intensities

One of the earliest experimental studies of Auger line shape and intensities from a number of clean metal surfaces was made by Palmberg and Rhodin.[71] Using a retarding-grid analyzer on a conventional LEED system, they were able to obtain spectra of reasonable intensity, but with poor energy resolution. Nevertheless, they were able to observe and identify much of the fine structure in their derivative spectra. They were also able to obtain information on the growth modes of evaporated silver films on gold, gold on silver, and gold on copper. In trying to extract useful information from the Auger intensities, it was recognized that physical similarities existed between AES and electron-probe microanalysis (EPMA).[72] An important factor to emerge from this comparison was the concept of backscattering effects, and the backscattering contribution[73] to elemental Auger intensities has recently been reviewed.[74]

High-energy-resolution LMM spectra from the first row transition metals were obtained by Allen et al.[75] using a hemispherical analyzer. By recording spectra in both the $N(E)$ and $dN(E)/dE$ modes, these workers were able to demonstrate, for the first time, the complex nature of the fine structure in the AES spectra from these metals. These workers noted the change in relative intensities of the LMM peaks across the transition series, the L_3MM transitions exhibiting intensities in the sequence $L_3M_{2,3}M_{2,3} > L_3M_{2,3}M_{4,5} > L_3M_{4,5}M_{4,5}$ for titanium, but in the reverse sequence for copper (Fig. 6.5). Their results are consistent with a model of Auger transition probabilities based on the population of the valence shell, and good agreement with theory is found. No rigorous interpretation of the fine structure was made, although it was noted that a conventional CMA analyzer (with energy resolution of 0.5%) would be unable to resolve this structure in most of the metals studied. Their inability to interpret the fine structure may be related to the quasi-atomic nature of the spectrum expected from those elements (see preceding section).

More success has been obtained in the interpretation of CVV spectra from metals with sp bands. Good agreement with theory has been obtained by assuming that the spectra are a self-convolution of the valence band, with a matrix element providing far-enhanced contribution from p^2, as opposed to sp, final states.[76,77] For CCV spectra of these elements, similar agreement with theory may be obtained if screening of the initial core hole is incorporated.[78] In this case, a difference in the extent of screening by s and p states in sodium, magnesium, and aluminum is found.

Interest in the low-energy Auger spectra of heavy metals (tungsten, gold, platinum, and iridium) has centered around the contribution of Coster–Kronig and super-Coster–Kronig transitions to these spectra. The latter (involving the $N_{4,5}N_{6,7}N_{6,7}$ levels) dominate in all of the elements studied, becoming relatively more intense as the number of $O_{4,5}$ electrons in the metal decreases.[79,80] Very-low-energy $O_{4,5}P_{2,3}P_{2,3}$ Auger emission (at 16.9 and 18.6 eV) has been reported from lead surfaces excited with UV radiation.[81] $N_{6,7}O_{4,5}O_{4,5}$ Auger spectra from thallium, lead, and bismuth have been observed under high-energy resolution, and the probable domination by surface contributions noted.[82] Good agreement with theory, including the observation of NNN and NNO Coster–Kronig and super-Coster–Kronig transitions have also been observed in manganese.[83,84]

Recent high-energy-resolution studies of the $M_{4,5}M_{4,5}N_{4,5}$ spectrum from palladium[85] may be interpreted in terms of the self-convolution of the valence band (implying an uncorrelated final state) discussed in Section 2.1.1. Confirmation of this has been obtained in an independent study,[60] which also demonstrated that the palladium $M_{4,5}N_{4,5}N_{4,5}$ transition retains its bandlike character in dilute alloys with silver.

Detailed studies of the fine structure in AES spectra from a number of metals have been made by Weightman and co-workers,[70,86-94] which illustrate the rapid development that has taken place in our understanding of Auger line shapes from metal surfaces in the past few years. Theoretical interpretation of early results, for example, was often restricted to peak assignments[88,91] or estimates of the relaxation energies.[86,89] In instances when transition probabilities were calculated, in order to provide theoretical line shapes, good fits to experiment were not always obtained.[87,90] The incorporation of quasiatomic line shapes (see preceding section) has enabled a much more complete interpretation of MNN spectra from silver, cadmium, indium, and tin to be made.[93] Values of U_{eff} and Γ for these elements are presented in Table 6.1. The quasiatomic model explains why the gas-phase and solid-state Auger spectra of these elements are almost identical in shape.[95,96] Quasiatomic and bandlike contributions have been identified in the copper MVV spectrum,[97] contributions from the DUS have been observed in the LMM spectrum from nickel.[94] The energies of L_3VV Auger features in copper and nickel have been found to be dependent on primary-beam energy, for beam energies just above the L_3 binding energy.[98] Plasmon gain peaks have been identified in sodium, magnesium, and aluminum,[99,100] and internal photoemission following core–hole ionization and decay by x-ray emission has been reported for aluminum.[100] On the assumption that the CVV spectra of aluminum and magnesium are bandlike, an attempt has been made to deconvolute the line shapes into the original valence-band line shapes.[101] Although much work still needs to be done, this may turn out

Table 6.1. Experimental Values of Γ and U_{eff} from Auger Spectra (See Text)[93]

Element	Γ	U_{eff}	U_{eff}/Γ^a
Ag	3.6 (± 0.2)	5.1 (± 0.7)	1.4
Cd	1.9 (± 0.1)	6.0 (± 0.6)	3.2
In	1.9 (± 0.1)	6.8 (± 0.7)	3.6
Sn	2.0 (± 0.1)	6.9 (± 0.6)	3.5

$^a U_{eff}/\Gamma \gg 1$ indicates quasi-atomic behavior (Cd, In, Sn); $U_{eff}/\Gamma \approx 1$ indicates band-like behavior (Ag).

to be a promising method of obtaining surface valence-band structures from Auger spectra. Impressive results have been obtained with silicon,[102] where contributions to the LVV spectrum from surface states were claimed. In principle, there is no reason why such information should not be obtainable from metal surfaces which obey the criterion $U_{eff} < \Gamma$.

2.1.3. Angular Anisotropy of Emission

Anisotropy of Auger electron emission can be observed experimentally on all surfaces, and may arise from one or more of several mechanisms.[104–110] These are listed in Table 6.2. Many of the effects listed are peculiar to AES. In XPS for instance, effects 2, 3, and 8 do not operate (effect 5 operates on polycrystalline materials[103]), since backscattering does not take place, and the exciting radiation generally penetrates a volume that is very large in relation to that being analyzed. Angular anisotropy of emission is therefore much more common in AES, may carry a lot of information, and is generally very complex. A comprehensive treatment of angle-resolved AES (ARAES) is

Table 6.2. Factors Affecting Angular Anisotropy of Auger Emission[a]

A. On All Substrates

1. Presence of contamination layer or surface roughness (attenuates signal at high θ_e[104]) and intrinsic $\cos \theta_e$ dependence if analyzed area is larger than irradiated area (true for most spectrometers).[72]
2. Variation of backscattering factor with θ_i (generally decreases with increasing θ_i at high θ_i, but may exhibit maximum at $\theta_i \approx 70°$).[74]
3. Variation of primary-beam penetration depth with θ_i ($\cos \theta_i$ dependence, except at grazing angles).[104]
4. Refraction of the Auger electrons at the solid–vacuum, or an inernal solid–solid, interface.[104]

B. On Single-Crystal Substrates

5. Intrinsic anisotropy of Auger emission.[105,106]
6. Crystallographic diffraction of the primary beam[107] and/or emitted Auger electrons.[108]

C. On Layered Substrates

7. Attenuation of substrate by overlayer.[109]
8. Backscattering differences between substrate and overlayer.[110]
9. If overlayer is epitaxial, crystallographic effects.[109]

[a] θ_i = angle of incidence of primary beam relative to surface; θ_e = angle of emission of Auger electrons relative to normal.

beyond the scope of this chapter, but the following references will allow the interested reader to pursue the topic further.

An intrinsic anisotropic contribution to Auger emission is expected from well-ordered (single-crystal) solids in the case of *CCV* or *CVV* emission, depending on the valence-state wave function.[106,111,112] In addition, there will be an extra contribution from diffraction effects, both of the emitted Auger electrons (Kikuchi effect[108]) and the primary beam (inverse Kikuchi effect[107]). Such effects have been invoked to explain the angular variation of Auger emission on iron,[113] nickel,[114], copper,[115] aluminum,[116] and silver[105] surfaces. It has been suggested that the intrinsic contribution is important only for low-energy (45–70 eV) emission from these surfaces[105,116] and the problem has been well treated from a theoretical point of view.[106,116] In many models of Auger emission, isotropy of the emitter is assumed, and a recent "cluster model" postulates that anisotropic effects may be explained by consideration of nearest neighbors only.[117] Angle-resolved results from single-crystal nickel surfaces have been interpreted in terms of diffraction effects, which were found to dominate for Auger energies above 200 eV.[108]

The effect of overlayers on the anisotropy of emission has also been investigated.[104,116,118,119] For copper and gold monolayers in tungsten, the angular distribution of Auger electrons from the overlayer is interpreted in terms of interference between direct emitted waves and waves scattered off neighboring atoms in the surface and subsurface region.[118] Conflicting results are reported for sulfur adlayers on nickel, however,[104,116] which show both the presence and absence of diffraction effects. Hydrogen adsorption on silica has been shown to enhance the anisotropy of emission, whereas oxygen on copper appears to reduce it.[119] These results are interpreted in terms of diffraction and elastic scattering, respectively, in the overlayers. Overlayer diffraction effects have also been postulated as a reason for the observation of anomalous growth modes in thin films by AES.[109] Applications of AES to thin films are discussed in detail in Section 4.

2.1.4. X-ray and Ion-Induced Auger emission

A number of AES studies are carried out with x-rays (XAES) as the primary source of exciting radiation, as opposed to electrons. This is particularly true of fundamental studies on large (several square millimeters), laterally homogeneous samples, where the major limitation of XAES (poor spatial resolution) is no longer a disadvantage. An excellent review of XAES has recently been published,[40] and there are several XPS texts that cover the subject,[120,121] therefore only a few relevant points will be covered here.

As was mentioned in Chapter 3, the advantages that x-ray excitation provides are reduced possibility of beam damage, no sample-charging problems, the possibility of using XPS in conjunction with XAES (thereby enabling use of the Auger parameter to be made; see Chapter 5),[121] and, in general, a much lower inelastic background intensity in the spectrum, since fewer secondaries are generated by the x-ray flux than by the electron flux. Against these advantages are the major disadvantages of poor spatial resolution (x-rays cannot be focused to any appreciable extent) and low Auger signals (the total x-ray power input from a typical 1-kW anode is much less than that from a conventional electron source). Both of these factors are of extreme importance in most metallurgical and materials science applications; XAES therefore finds little use in these fields. An interesting, and potentially very useful, difference between AES and XAES is the fact

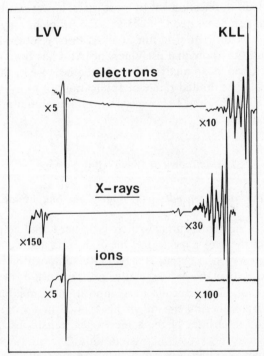

Figure 6.6. Aluminum *LVV* and *KLL* first-derivative Auger spectra under electron (5 keV), x-ray (1254 eV), and ion (5 keV argon) irradiation. The x-ray-induced spectrum is displaced slightly on the energy axis to avoid confusion. The main features to note are: (i) Relatively weak x-ray-induced Auger emission. The *KLL* features are excited, not by the 1254-eV radiation, but by the background bremsstrahlung component of the x-ray spectrum. The *LVV* features are very weak due to the low photoionization cross section of the aluminum *L* levels at 1254 eV. (ii) The absence of *KLL* Auger features in the ion-induced spectrum. The ion-induced *LVV* feature, however, is very intense and different in structure to the corresponding electron-induced feature.

that the ionization cross section, for any given electron shell, increases with increasing atomic number for x-ray excitation, but decreases with increasing atomic number for electron excitation.[122] In addition, the ionization cross section, for any given element, generally increases with increasing ionization energy (binding energy) for x-ray excitation (up to the x-ray energy), but decreases with increasing ionization energy for electron excitation. This is illustrated in Fig. 6.6, which shows both the x-ray- and electron-excited Auger spectra from aluminum.

The phenomenon of ion-induced Auger electron emission (tentatively IAES) is generally more of a hindrance than a help in AES analyses. Because Auger electrons can be excited during ion bombardment,[123] the interpretation of sputter-depth profiles, when the Auger spectra are collected during sputtering, is open to question.[124] In most cases, however, only low-energy Auger electrons are generated,[123-125] and the problem may also be overcome by sequential ion sputtering and analysis, or by electron-beam modulation.[126] Although the potential usefulness of IAES has been recognized,[123] it has not developed as an analytical technique because of the problems of interpretation and its limited range of application.

Typical AES, XAES, and IAES spectra from aluminum are presented in Fig. 6.6.

2.2. Studies of Clean Alloy Surfaces

2.2.1. The Matrix Dependence of Auger Line Shapes and Intensities

In AES studies of elemental surfaces, there is no need to attempt any form of quantification of the Auger spectrum. One of the first questions asked about alloy surfaces, however, is "What is the surface composition?" In principle, the information is available in the relative peak intensities in the spectrum.

A first approximation to the atomic composition of an alloy surface may be obtained by considering the Auger peak intensities from the alloy, as compared to the intensities of the same Auger transitions from the pure elements, analyzed under the same conditions. The atomic concentration (C_i^x) of any element i in the alloy x may then be expressed as

$$C_i^x = P_i^x / P_i^i \tag{6.10}$$

where P_i^x and P_i^i are the absolute intensities (usually measured as peak-to-peak heights, pph, in the derivative spectrum) of the Auger transition of interest in the alloy and pure i, respectively. P_i^i is usually referred to as the atomic sensitivity factor for element i. In practice, however, it is extremely difficult to reproduce absolute Auger intensities over a long period of time (due to, among other things, gain variations in the electron detector), and differences in surface roughness[127] between the elemental standard and unknown alloy will markedly influence the value obtained for C_i^x. A more accurate means of quantifying Auger spectra is therefore to normalize the total composition to unity, thus

$$C_i^x = \frac{P_i^x/P_i^i}{\sum_n P_n^x/P_n^n} \tag{6.11}$$

Even this equation may give seriously wrong results in many cases, owing to differences in backscattering yields,[74] inelastic mean free paths,[35] and atomic density between the different components of the alloy. P_i^i, for instance, is determined by the backscattering yield in pure i, whereas P_i^x is determined by the backscattering yield in the alloy, which, in the case of a dilute alloy consisting mainly of a matrix of element j, will be the backscattering yield in pure j. These so-called matrix effects may have a significant effect on the measured peak intensities and, therefore, must be taken into account.

For a dilute alloy of i in j, Hall and Morabito have shown[128] that the surface composition is given by

$$C_i^j = \frac{F_i^j \cdot P_i^j/P_i^i}{\sum_n F_n^j \cdot P_n^j/P_n^n} \tag{6.12}$$

where F_i^j is the matrix correction factor for element i in matrix j. By considering the backscattering, atomic density, and inelastic mean-free-path dependence of F_i^j, they arrived at

$$F_i^j = \left(\frac{N^j}{N^i}\right)^{1/2} \frac{(1+r^j)}{(1+r^i)} \tag{6.13}$$

where N^j and N^i are the atomic densities in pure j and i, respectively, and $(1+r^j)$ and $(1+r^i)$ are the backscattering factors for pure j and i, respectively. These latter terms are dependent on both the primary-beam energy and the threshold excitation energies for the transitions of interest, as well as the substrate atomic number.[129] For $F_i^j = F_n^j = 1$, Eq. (6.12) simplifies to Eq. (6.11). In practice, F_i^j may deviate significantly from unity. The values derived by Hall and Morabito[128] for most of the elements and transitions of interest show a standard deviation about unity of 0.5, with extreme values of F_i^j of 0.2 (for i = rubidium and j = boron) and 5 (for i = cobalt and j = cesium). Typical values of F_i^j and P_i^i are given in Table 6.3 for a range of elements and matrices. Although the matrix correction factors derived in this way are strictly valid only in the case of dilute alloys, they should also be applicable to nondilute systems, providing the absolute value of F_i^j does not deviate too greatly from unity.[128]

Table 6.3. Atomic Sensitivity Factors (P_j^i) and Matrix Correction Factors (F_i^j) for Various Elements in Aluminum^a

Element	Auger Transition	P_i^i	F_i^j
B	KLL	0.11	1.17
C	KLL	0.14	1.13
Na	KLL	0.25	0.62
Mg	KLL	0.14	0.83
Al	LMM	0.19	1.00
Si	LMM	0.28	0.93
K	LMM	0.9	0.53
Ca	LMM	0.4	0.71
Sc	LMM	0.27	0.86
Ti	LMM	0.33	1.14
V	LMM	0.37	1.29
Cr	LMM	0.3	1.41
Mn	LMM	0.24	1.41
Fe	LMM	0.21	1.45
Co	LMM	0.23	1.52
Ni	LMM	0.26	1.54

^{a}5-keV primary-beam energy from Reference 128 and the *PHI Handbook*.

All of the above discussion assumes, of course, that the alloy composition is uniform throughout the sampling depth (about 10–50 Å). In practice, this is seldom the case. Interpretative models for overlayer systems are discussed in Section 4.

Despite the relative simplicity and ease of application of Eq. (6.12), a glance through the current literature shows that it is seldom used in practice,[130-135] most authors being content to use the more approximate form of Eq. (6.11). For many systems, comparison of the Auger spectrum from the unknown with spectra from a well-characterized standard specimen provides an adequate means of interpretation and quantification,[136,137] although serious errors in interpretation may have been made through inadequate calibration[130,131] or the neglect of backscattering effects[132] in AES studies of alloy surfaces.

The matrix dependence of Auger line shapes is generally not taken into account in quantification of alloy spectra. There are two main reasons for this: First, the variation of line shape with composition is poorly understood, and little experimental work has been carried out in this area. Second, such variations are expected to make only a minor contribution to the observed intensity in most work, especially when the total energy resolution in the spectrum is greater than 1 or 2 eV. In cases where high-energy-resolution spectra of CVV levels in metallic alloys have been obtained, it has been shown that the detailed line shape is very dependent on the alloy bandwidth.[138–141] The silver $M_{4,5}N_{4,5}N_{4,5}$ line shape, for instance, has been shown to deviate significantly, from an atomic-like profile in the pure element to a more band-like profile in alloys with magnesium and aluminum, due to changes in the valence-band line shape.[138,139] In contrast, the palladium $M_{4,5}N_{4,5}N_{4,5}$ line shape changes from a broad, poorly resolved, bandlike profile in elemental palladium, to a much sharper, well-resolved, atomic-like profile in alloys with aluminum and magnesium (Fig. 6.7). These changes can be related to simultaneous changes in the valence bandwidth, which is large in pure palladium (~ 5 eV), but smaller in its alloys with aluminum (3.8 eV) and magnesium (2.6 eV).[140] These changes are consistent with the Cini model of

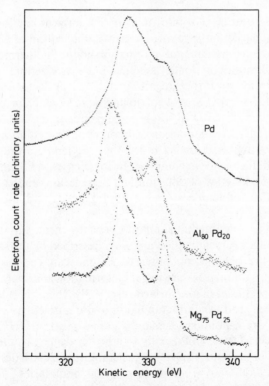

Figure 6.7.   Pd $M_{4,5}N_{4,5}N_{4,5}$ Auger spectra from Pd, $Al_{0.8}Pd_{0.2}$, and $Mg_{0.75}Pd_{0.25}$, showing bandlike structure in the element, and atomic-like structure in the alloys. (Reproduced with permission from Reference 140, © Institute of Physics.)

CVV Auger emission introduced in Section 2.1.1.[65] For alloys of palladium with silver, however, the palladium $M_{4,5}N_{4,5}N_{4,5}$ line shape does not become atomic-like, even at very high dilution, although the palladium d bandwidth is reduced to 2.1 eV in this case. Instead, new structure develops in the spectrum, which may be explained in terms of participation of silver $4d$ states in Auger emission from palladium.[141,142] The reduction in palladium $4d$ bandwidth in alloys with aluminum, magnesium, and silver is due to the reduced interatomic overlap between palladium $4d$ states on dilution—in effect, the palladium $4d$ levels become atomiclike in the alloys.[140–142]

It has also been proposed that relative Fermi energies in alloys may be measured directly from peak shifts in x-ray-excited AES, although the precision of such measurements (± 0.15 eV) appears to be very poor in comparison to the shifts expected ($\ll 1$ eV).[143]

2.2.2. Theoretical Aspects of Alloy Surface Composition

The study of segregation of solute atoms to free surfaces in alloys or almost-pure metals is a particularly important area of application of AES. Not only is it used to provide fundamental information about the thermodynamics of solid surfaces in multicomponent systems, but it is essential for studies of impurity adsorption and/or segregation which may affect surface reactions (oxidation, adhesion, etc.), and of solute interactions which promote (accelerate) or retard certain types of solute segregation which, in turn, may lead to, for example, embrittlement in steels. In practice, true equilibrium segregation is often not achieved, owing to the time (or temperature) being insufficient for the diffusion to be complete. In these cases, AES can provide information on the kinetics of diffusion and segregation, and can be used to monitor the effect of temperature on both kinetics and the equilibrium segregation.

Because of the fundamental similarity between them, surface and interfacial (grain-boundary) segregation are often described by the same theories, and many of the phenomena observed in one can be observed in the other.[144–151] It is often necessary to gain information about grain-boundary segregation from studies of free surfaces, especially whenever it is impractical or impossible to produce intergranular fracture surfaces (so that surface analysis at the grain boundaries may be carried out) in UHV. In general, segregation occurs faster, and usually to a higher degree, to the free surface than at grain boundaries, and the structural characteristics of the free surface are very different from those of grain boundaries. There is little reason, therefore, to believe that AES studies of surface segregation will ever allow us to make quantitative predictions about grain-boundary segregation. Both phenomena, however, remain thermodynamically similar, and the continuing similarity between the results of AES studies of the free surface, and of intergranular fracture surfaces, reinforce our belief in the similarity of the underlying processes.

Several excellent reviews of our present understanding of alloy surface compositions have recently been published,[144-151] therefore only a very brief summary will be given here.

The microchemistry of alloy surfaces may differ from that in the bulk by virtue of the following two processes. Sites at the surface exist with an energy different from those in the lattice. Partition between these two types of sites by solutes is in accordance with the statistics of thermodynamics in order to minimize the free energy of the system. When the overall free energy has reached a minimum, the process is termed equilibrium segregation, and may result in either an enrichment or depletion of the solute at the surface. The second process occurs by the action of kinetics. Here surface concentration is altered as a result of solute pileup at a moving interface, or by solutes coupling to vacancies, in which case nonequilibrium segregation is established.[151]

The simplest theory of equilibrium segregation in a binary solid solution is that developed by McLean.[152] According to this theory, the segregation of a solute 2 in a matrix 1 is characterized by an enrichment factor, χ, given by

$$\chi = \frac{X_2^S/[X_2^S(\text{sat}) - X_2^S]}{X_2^B/(1 - X_2^B)} \tag{6.14}$$

$$\approx \frac{X_2^S/[X_2^S(\text{sat}) - X_2^S]}{X_2^B} \quad \text{(for dilute systems)}$$

where X_2 refers to the molar concentration (mole fraction) of the solute, superscripts S and B refer to the surface and bulk, respectively, and (sat) denotes the saturation value. $X_2^S(\text{sat})$ is usually assumed to represent one monolayer,[153] in which case χ may be expressed as[147]

$$\chi = \frac{X_2^S/(1 - X_2^S)}{X_2^B/(1 - X_2^B)} \tag{6.15}$$

The enrichment factor is determined by the free energy of segregation (or adsorption), ΔG, according to[147,149-152]

$$\chi = \exp\left(\frac{-\Delta G}{RT}\right) \tag{6.16}$$

This result is similar to the Langmuir theory of gas adsorption.

A similar relation, based on the truncated Brunauer, Emmett, and Teller (BET) adsorption theory,[154] has been derived by Seah and Hondros:[149-151]

$$\frac{X_2^S/[X_2^S(\text{sat}) - X_2^S]}{X_2^B/X_2^B(\text{sat})} = \exp\left(\frac{-\Delta G'}{RT}\right) \tag{6.17}$$

where $X_2^B(\text{sat})$ is the solid solubility limit of 2 in 1, and G' is an energy term related to ΔG by

$$\Delta G' = \Delta G - \Delta G_L \qquad (6.18)$$

where ΔG_L is the free energy of condensation (or precipitation) of successive layers on top of the first monolayer. Experimental verification for this theory has been obtained in a number of binary alloys,[149-151] where the enrichment ratio shows a correlation with the atomic solid solubility (see Section 2.2.3). In the full BET theory, each segregant atom in the first monolayer creates a new adsorption (segregation) site with an associated free energy ΔG_L, and the segregation equation is written[149,150]

$$\frac{X_2^S(\text{sat})}{X_2^S} \cdot \frac{X_2^B}{X_2^B(\text{sat}) - X_2^B} = \frac{1}{K} + \frac{K-1}{K} \cdot \frac{X_2^B}{X_2^B(\text{sat})} \qquad (6.19)$$

where $K = \exp(-\Delta G'/RT)$. In the dilute limit, this equation approaches the McLean or truncated BET formulas. More-complex theories are available, which allow for interaction between the solute atoms in both binary[155] and ternary[156] systems.

The problem of estimating the free energy terms in all of the above equations is a considerable one, and several approximations are considered below.

If the bond strength between two atoms i, j is represented by ε_{ij}, where $i = 1$ for matrix atoms and $i = 2$ for solute atoms, then the bond strength ratio (solute:matrix) is given by

$$\varepsilon^* = \varepsilon_{22}/\varepsilon_{11} \qquad (6.20)$$

According to the bond-breaking theory of surface segregation, and assuming an ideal solution, it can be shown that the free energy of segregation (or adsorption), ΔG_ε, is given by[148]

$$\Delta G_\varepsilon = \tfrac{1}{2} \Delta z \varepsilon_{11}(\varepsilon^* - 1) \qquad (6.21)$$

where Δz is the difference between the effective coordination number of bulk and surface sites in the matrix ($= 5.67$ for the 100 surface of a fcc metal).[148] Equation (6.21) indicates that the component with the lower bond strength will segregate to the surface, and such segregation will be more pronounced on loosely packed surfaces (higher Δz) and for high matrix bond strengths.

Alternatively, the continuum elasticity theory considers only the effect of differences in atomic dimensions in the bulk and ignores the effect of ε^*. According to this theory, the elastic free energy of segregation, ΔG_σ, is equal to the negative of the bulk strain energy, E_{EL}, given by[148]

$$\Delta G_\sigma = -E_{\text{EL}} = -168.24\varepsilon_{11}(\sigma^* - 1)^2 \qquad (6.22)$$

where σ^* is the atomic size ratio, r_2/σ_1. Since E_{EL} is proportional to $(\sigma^* - 1)^2$, the continuum elasticity theory predicts that ΔG_σ will always be negative, that is, segregation will always occur. As shown below, this is not always the case, hence a cutoff strain energy, ΔG_{min}, is often assumed, such that segregation only occurs if $\Delta G_\sigma < \Delta G_{min}$.[157]

Another estimate of the free energy of segregation may be obtained by considering the surface tension rates, γ^*, given by

$$\gamma^* = \gamma_2/\gamma_1 \tag{6.23}$$

From the Gibbs adsorption isotherm,[147] it is possible to derive an expression for ΔG_γ, providing several assumptions are made about atomic volumes:[148]

$$\Delta G_\gamma = a_{12}\gamma_1(\gamma^* - 1) \tag{6.24}$$

where a_{12} is the effective surface area of a solute atom at the surface of the matrix.

In the absence of values of γ for the components under consideration, an approximation to γ^* may be made by[148]

$$\gamma^* = \varepsilon^*/(\sigma^*)^2 \tag{6.25}$$

In a comparison with available experimental data, Abraham and Brundle[148] found the following rule to apply:

$\gamma^* < 1$ implies solute segregation

$\gamma^* > 1$ implies solvent segregation

of the 45 sets of experimental data considered, only five exceptions were found.

It has been suggested that the total driving force for segregation may be considered as the linear sum of the bond-breaking and elastic driving forces,[158] thus

$$\Delta G = \Delta G_\varepsilon + \Delta G_\sigma \tag{6.26}$$

or, using the surface-tension approach instead of the bond-ratio approach,

$$\Delta G = \Delta G_\gamma + \Delta G_\sigma \tag{6.27}$$

An important aspect of this theory is that the prediction of segregation ($\Delta G/\varepsilon_{11} < 0$ implies solute segregation, $\Delta G/\varepsilon_{11} > 0$ implies solvent segregation) depends only on relative, and not absolute, properties of the system; that is, ε^*, γ^*, and σ^*. It is therefore possible to use theoretical estimates of ε, γ, and σ, providing it is known that theoretical values are over or underestimated by a constant, but unknown, factor.[159]

Using a combination of the bond-breaking and elasticity theories, Abraham and Brundle[148] were able to show that all of the experimental results obtained on segregation studies on binary alloys, with only one exception (iron solute in zirconium matrix) could be explained on the basis that

$\gamma^* \gtrsim 1$ and $\sigma^* \lesssim 1$ implies no solute segregation

$\gamma^* \lesssim 1$ and $\sigma^* \gtrsim 1$ implies solute segregation

Of the experimental results considered, only those in a platinum matrix fall outside the two categories defined above.[148] In general, there is very little experimental data for the categories $\gamma^* \gtrsim 1$ and $\sigma^* \gtrsim 1$, and $\gamma^* \lesssim 1$ and $\sigma^* \lesssim 1$, although some possible systems (e.g., platinum in nickel) are suggested by Abraham and Brundle.[148]

Seah[151] has proposed an alternative expression for ΔG, taking into account the solute–matrix interaction energy, Ω, which has the form

$$\Delta G = 0.64(\gamma_2 V_2 - \gamma_1 V_1) - 1.86\Omega \text{ kJ mol}^{-1} \quad \text{(for } r_1 > r_2)\qquad (6.28)$$

and

$$\Delta G = 0.64(\gamma_2 V_2 - \gamma_1 V_1) - 1.86\Omega - 4.64 \times 10^7 r_1(r_2 - r_1)^2 \text{ kJ mol}^{-1} \quad \text{(for } r_2 > r_1)$$
$$(6.29)$$

where r_1 and r_2 are in nanometers, and V is the molar volume (γV is the surface free energy). Comparison of the above equations with experiment has shown very good agreement (a standard deviation of only 8 kJ mol^{-1} over the range -20 to 80 kJ mol^{-1}) with the published results from 32 binary systems.[151]

Although derived for binary solid solutions, the above equations should be applicable to multicomponent systems, providing no competing effects (site competition, synergistic segregation, or solute–solute interaction) exist. In the case of a ternary system, with solutes 2 and 3 in a matrix 1, Guttmann and McLean have proposed[160] that the free energies of segregation are given by

$$\Delta G_2 = \Delta G_2^0 + \frac{(\varepsilon_{23} - \varepsilon_{12}) \cdot X_3^S}{X_2^B(\text{sat})}\qquad (6.30)$$

and

$$\Delta G_3 = \Delta G_3^0 + \frac{(\varepsilon_{23} - \varepsilon_{13}) \cdot X_2^S}{X_3^B(\text{sat})}\qquad (6.31)$$

where ΔG^0 refers to the free energy of segregation in a binary system (absence of competing effects).

An attempt has also been made to model the composition-depth profile through a binary, regular alloy surface, by constraining only the fifth and subsequent atomic layers to have the bulk composition, thereby allowing the four surface layers to attain equilibrium composition.[161] When applied, for example, to the lead–indium system, a monotonic decrease in lead concentration from the surface ($X^S - X^B \approx 0.3$) to the bulk is predicted. In the case of the silver–gold system, however, the silver is predicted to be alternately enriched and depleted in successive layers (i.e., enriched in the surface monolayer, depleted in the second, enriched in the third, and depleted in the fourth).[162] Experimental studies of surface segregation in binary and multicomponent alloys by AES are discussed in the following section.

2.2.3. Determination of Alloy Surface Composition

Because of the extreme surface sensitivity and elemental specificity of the technique, AES has become one of the standard methods of determining alloy surface compositions. There is already a very large body of scientific literature to which reference can be made,[144–216] and only a few specific examples will be discussed here.

Experiments to determine the extent of surface segregation in binary or multicomponent systems by AES generally follow the following procedure:

1. The sample is prepared with the desired bulk composition, generally by melting together the required quantities of pure materials (often 99.999% pure[206,211]) under vacuum or in an inert atmosphere. Melting is generally achieved in either a high-frequency induction furnace[197,211,215] or by arc-melting.[136,203] In many cases, bulk specimens of the desired composition and purity may be purchased "off-the-shelf" from commercial sources.[198,206,214] In the case of single-crystal alloy substrates, this is much the preferred route, unless highly specialized crystal-growing facilities are available.[195] Less popular, but alternative, methods of sample preparation include in situ coevaporation of the pure elements[133,134,216] and coprecipitation, and subsequent reduction, of a mixture of inorganic salts of the elements of interest.[132] Following preparation, it is generally advisable to check the final alloy composition, and crystallinity if applicable, by bulk analysis and diffraction techniques.

2. Sample preparation prior to experimentation is usually limited to cutting (by mechanical or arc-machining) and polishing to the desired size, shape, and surface finish. Satisfactory surface finishes have been achieved by polishing with alumina or diamond paste down to 1-μm size[195,203,211] and, in one case, with grit 600A emery paper.[206] Final cleaning before insertion into the spectrometer is usually accomplished by ultrasonic cleaning in a suitable solvent. In order to reduce outgassing problems, facilitate in situ cleaning procedures, and minimize the time required for specimen cooling following annealing, etc., specimens are generally made as small as possible (~ 1 cm square), with thicknesses between 100 μm and 1 mm.[198,203,211]

3. Sample mounting in the vacuum system is generally dictated by the nature of the experiment and constraints of the spectrometer. The most usual method is to spot-weld or clamp the specimen to a refractory holder (often tantalum), through which the specimen may be resistively heated and its temperature monitored (generally via a chromel–alumel thermocouple).[198,203,206,211] A heating stage which eliminates magnetic interference with the Auger signal has been described.[217] A less-popular method of sample heating is by radiative emission from a tungsten filament at a negative potential with respect to the substrate. This method is necessary when high temperatures ($\sim 1000°C$) need to be produced.[214]

4. *In situ* cleaning is usually necessary prior to performing experiments, since the specimen will be covered with an oxide layer and carbonaceous contamination. Cleaning is usually accomplished by sequential cycles of inert-gas ion etching (up to several keV argon ions, at 10–100 μA cm^{-2}) and annealing. Since the annealing step accelerates impurity segregation to the surface, such cycles may have to be repeated many times before a clean, annealed surface is produced. The principal contaminants on annealed metal surfaces are carbon, oxygen, and sulfur,[203,214] the last of these being extremely persistent in some systems. Sulfur segregation has been shown to be inhibited by the inclusion of trace amounts of manganese in ferrous alloys[218] and silver in copper alloys.[206] In addition to impurity segregation, high-temperature annealing may also cause preferential loss of one of the components by evaporation. Although this effect has been noted in some systems (e.g., Cu from Cu–Ni[214] and Mg from aluminium alloys[219]), it has not been studied in detail until quite recently. In order to maintain the conditions of cleanliness for some time following the etching–annealing process, it is generally necessary to carry out the experiments and AES analysis in the same vacuum system, with base pressures of at least 10^{-9} torr (preferably $< 10^{-10}$ torr).[198,203,211,214] This is necessary to prevent any possible adsorbate-induced segregation from taking place[203] (see following section).

5. Finally, the experiment may be carried out. This generally takes the form of monitoring the equilibrium surface composition of the ion-cleaned surface, as a function of time and temperature, with AES. The detailed problems of interpretation and quantification of the Auger spectrum have already been discussed (Chapters 2 and 3, and References 128, 150, 220, and 221). In some cases, it has been possible to derive the depth variation of composition from the variation of Auger intensities with kinetic energy,[214] although depth profiles through the segregated layer are more usually performed by sequential ion sputtering and AES analysis.[150,222] In this way, the depth distribution and thickness of the segregated layer may be estimated. The profiles for most segregants, characterized by a rapid exponential decay with depth etched, are compatible with a single atom layer of segregant

atoms at the surface.[150,220,222,223] The decay of the Auger electron intensity, I^A, for the sputtering of atoms at the surface is described by the relation

$$I^A = I_0^A \exp(-SJAt) \qquad (6.32)$$

where I_0^A is the initial intensity, S is the sputter yield of the segregant, J is the flux density of ions, A is the segregant atomic area, and t is the time elapsed. Measurements for B, Bi, Co, Cu, Mo, P, S, Sb, Sn, and Te in iron or copper all show such profiles.[150,220,222,223]

A consideration that should be included in the deductions for alloy elements is the differential sputtering of the individual components. This effect may explain the apparent observation of chromium and molybdenum segregation on steel, if the respective carbides are assumed to undergo preferential sputtering, leaving a carbon-rich surface.[150] The problem of preferential sputtering in alloys is dealt with in Section 2.1.5.

Much of the published work on AES studies of surface segregation in alloys is summarized in Table 6.4. Of the 72 systems included, only four (Co in Cu, Fe in Zr, Pt in Ni, and Fe in Ni) are observed to violate the rule-of-thumb method (described in the preceding section) that predicts surface enrichment of the component with the lower surface tension. For one of these systems (Fe in Ni), there is actually conflicting AES evidence, both in favor of enrichment[165] and depletion[173] of iron. In the case of Co in Cu, it is likely that the cosegregation of silicon influenced the AES results,[205] thereby rendering the surface tension approach invalid. For the zirconium–iron system, it is possible that the extremely different atomic sizes of these two elements has an effect on the segregation behavior.[148] Similar considerations may be applied to the nickel–platinum system.[195]

Several of the systems included in Table 6.4 represent results obtained from AES studies of intergranular fracture surfaces, and are therefore results of grain-boundary rather than surface segregation. The fact that these results also obey the rule-of-thumb method of predicting whether or not a solute will be enriched, reinforces the view that the thermodynamic processes underlying surface and grain-boundary segregation are fundamentally similar.[144–151,213]

One of the earliest applications of AES to the study of surface segregation in ferrous alloys was made by Seah and Lea,[218] who studied the Fe–Sn system with and without cosegregation of sulfur. As well as showing that the sulfur impurity interference could be suppressed by the addition of 0.04% manganese to the alloy, they were able to show that the AES results were consistent with a saturation tin coverage of two monolayers on the iron surface. Saturation coverage was achieved at all temperatures for bulk tin concentrations in excess of 1%, and coverage showed the expected $e^{1/T}$ dependence [Eq. (6.16)] for bulk tin concentrations less than 1%. Seah and

Table 6.4. Surface Segregation in Alloys Studied by AES[a]

Solvent (Matrix)	γ^b	Surface Enrichment of		Surface Depletion of		
Pb	0.47			In (162)		
In	0.56	Pb (162)				
Al	0.87		Sn (198)	Cu (212)		
Ag	0.92			Cu (165)	Au (162)	Pd (202)
Au	1.13	Ag (162)	In (167)	Cu (165)	Ni (158)	
		Ca (166)	Sn (168)	Pd (169)	Pt (199)	
Cu	1.30	Ag (182)	Au (165)			
		In (171)	Sn (171, 200)	Ni (170)		
		Al (130)	S (206)			
		Co (205)	Si (205)			
		Bi^c (192)	S^c (210)			
Zr	1.45	S (183)	O (183)			
		Fe (175)				
Pd	1.50	Ag (163, 196)	Au (169)	Ni (175)	Pt (199)	
Ti	1.55	S (184)	Cl (21, 184)			
Ni	1.75	Au (158)	Cu (170)	Fe (173)		
		Fe (165)	Pd (177, 197)			
		Pt (195)	Sn (131)			
		S^c (194)	C (187)			
		Th (203)	S (207)			
Fe	1.8	Cu^c (208)	Ni (173)	Co (132)		
		Zr (175)				
		Al (131)	Cr^c (172)			
		B (181, 201)	S^c (181, 209)			
		P (186)	C (188)			
		Sb (189)	Ti (211)			
		Te^c (191, 208)	Mn (204)			
Pt	1.9	Au (179)	Pd (199)	Ir (176)		
Nb	2.0	O (185)				
Mo	2.1			W (136)		
Ir	2.3	Pt (176)				
W	2.4	P^c (190)	Mo (136)			
		K^c (193)				
Os	2.5	Pt (178)				

[a]References to literature are in parentheses.
[b]Surface tension of solvent at the elemental melting point, from Reference 164.
[c]Segregation to grain boundaries.

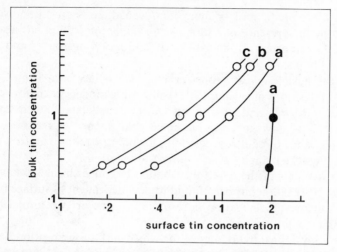

Figure 6.8. Segregation isotherms for tin to the free surface (●) and grain boundaries (○) of iron, at (a) 550°C, (b) 650°C, and (c) 750°C, as determined by AES. Surface tin concentration in monolayers of tin (absolute values subject to calibration errors) and bulk tin concentration in atomic percent. Data from Reference 149.

Hondros[153] demonstrated similar segregation behavior at the grain boundaries in iron–tin alloys, although the extent of segregation was much reduced, compared to the free-surface case, under the same conditions of bulk composition and temperature. The segregation isotherms for tin in iron are reproduced in Fig. 6.8. More-recent results for multicomponent systems have thrown this interpretation into doubt, and it has been suggested that only one monolayer of tin segregates at saturation, and not two as suggested by Seah et al.[213] The segregation of carbon to the surface of iron has also been shown to follow Eq. (6.16), for temperatures in the range 550–850°C and carbon concentrations in the range 10–90 ppm.[224] Similarly, phosphorus segregation to Ni–Cr steel follows McLean-type behavior.[217] Seah and Hondros[153] have demonstrated that the extent of grain-boundary segregation in ferrous alloys is related to the atomic solid solubilities of the segregating components. The segregation of metalloids and transition metals at the free surface of iron has been studied for a number of binary and multicomponent systems.[211] The metalloids antimony, phosphorus, carbon, nitrogen, and sulfur and the metals nickel and titanium segregate alone, whereas chromium, molybdenum, and vanadium do not. In the case of phosphorus and antimony, saturation coverage was readily obtained in the binary alloys. In the case of nickel and titanium, segregation was enhanced by the presence of antimony; in the case of molybdenum and titanium, by phosphorus; in the case of chromium, molybdenum, and vanadium, by nitrogen. Chromium, molybdenum, and vanadium segregation was not enhanced by the presence of antimony.[211] Solute interaction has also been shown to be important in the case of tin and nickel in iron[213] and silicon and carbon in iron.[213]

Over the past few years considerable interest has been shown in the surface compositions of various Cu–Ni alloys as a model system for segregation studies ($\sigma^* \approx 1$) and also in an effort to understand their catalytic properties. It is now generally accepted that copper segregation occurs over a wide range of bulk compositions, for polycrystalline surfaces and for many close-packed single-crystal surface orientations.[170,174,199,214] Absolute measurements of the depth distribution of copper in nickel have been made, and segregation has been found to be independent of surface orientation or roughness. Segregation is restricted to one or two monolayers of copper on the alloy surface.[214]

Other alloys of catalytic interest for which surface segregation has been shown to occur are Pt–Pd, Pt–Au,[199] Ag–Au,[162] Pd–Ni,[197] and Pt–Ni.[195] In the case of Pt–Pd, the catalytic activity of the alloy in benzene hydrogenation was shown to be closely related to the platinum surface concentration.[199] The addition of gold to platinum, however, has been shown to result in a slowing down of C–C bond fission reactions, C–C bond rearrangements, and a change in the contribution of various mechanisms of isomerization.[199] In this respect, gold is entirely inactive, in contrast to the copper–nickel case, where both alloying components are catalytically active, but to different extents.[199] The surface of nickel–thorium catalysts (used in methanation reactions—forming CH_4 from CO and H_2) have been shown to be enriched in thorium, the extent of segregation indicating a segregation energy of -18.8 kJ mole^{-1} in the temperature range 700–900°C.[203] AES studies of palladium segregation in nickel have shown a significant heat of segregation (-7.2 kcal mole^{-1}). Kinetic experiments at temperatures between 550 and 700°C have yielded an apparent diffusivity of Pd through the alloy described by

$$D = 2 \times 10^{-6} \exp[(37 \pm 5 \text{ kcal mole}^{-1})/RT] \qquad (6.33)$$

which is suggestive of grain-boundary transport-controlled diffusion.[197] In the case of silver–gold alloys, the surface composition is observed to differ only marginally from the bulk (a slight silver enrichment is observed for a range of bulk compositions), where the regular solution model predicts a very significant segregation of silver.[132,162]

A detailed study has been carried out on tin segregation in aluminum.[198] Using AES, it has been shown that MacLean's model describes the segregation behavior very well, giving a segregation energy of -77 kJ mole^{-1} in the temperature range 623–893 K. As with most segregation experiments, the diffusion coefficient of the segregating species (in this case tin) is too small at lower temperatures to enable meaningful data to be obtained. Although less suited to the task than radiotracer analysis, AES has also been used to obtain a value for the activation energy for tin diffusion in aluminum.[198] The ternary alloy Cu–Co–Si has also been studied by AES. In this case, CoSi clusters and the joint surface enrichment of cobalt and silicon were observed and interpreted in terms of the formation of cobalt silicide precipitates.[205] In this particular study, however, it is likely that the surface composition was significantly influenced by the presence of oxygen and oxides. The effect of gas adsorption and reaction on the surface composition of alloys, and its study by AES, is discussed in the following section.

2.2.4. The Effect of Adsorbates on Surface Composition

If gas-phase molecules are selectively chemisorbed by one of the components of an alloy, this will act to lower the surface free energy of that component in the alloy when exposed to the gas atmosphere. In this way, chemisorption can provide a driving force to enrich the surface with the element forming the strongest adsorption bands, an effect known as chemisorption-induced surface segregation.[225] The phenomenon is significantly more complex, as regards theoretical description, than that of surface segregation in the absence of adsorbates (preceding section), but the fundamental driving force, one of lowering the total free energy of the system, is the same. In addition to the multicomponent nature of chemisorption-induced surface segregation, the behavior of many systems is complicated by the fact that segregation often proceeds beyond the monolayer stage, to yield a system which may be greatly enriched in one component, to depths exceeding the Auger electron escape depth. The reason for this is clear if we consider the free energy chnge in the system for multilayer segregation. In the case of a binary alloy without chemisorption-induced segregation, the driving force for formation of the first monolayer of segregant X is given by the free energy change on exchanging an atom of X in the bulk with an atom of Y at the surface (preceding section) and is generally in the range -80 to 80 kJ mole^{-1}.[151] Subsequent (multilayer) segregation, however, will result only in an effective exchange of an atom X in the bulk with an atom Y at the X monolayer–bulk interface, an exchange which will be associated with a relatively small free energy change. In the case of a binary alloy with chemisorption-induced segregation, however, the free energy change due to the formation of X–adatom bonds will be much the same for monolayer and multilayer segregation, and the extent of segregation will, in general, only be limited by the bulk diffusivity of X, or the diffusivity of X (or the chemisorbed species) through the segregated layers.[219]

One particularly important area of chemisorption-induced segregation that has received much attention from electron spectroscopists is in the field of catalysis. It is well established that the adsorption of certain species from the gas phase can "poison" an interface-controlled reaction, either by adsorbing at active sites on the surface, thereby blocking their catalytic activity, or by inducing surface segregation of a catalytically inactive component. The phenomenon of selective chemisorption can actually be utilized in an analytical sense, and has been applied successfully to a number of binary alloys in order to determine their surface compositions,[147] but it is important to be aware of the possibility of chemisorption-induced segregation in such measurements. For a long time it was puzzling why the results of CO chemisorption experiments on copper–nickel alloys indicated a surface composition identical to the bulk,[199] whereas AES studies indicated a surface enrichment of copper.[170,174,214] The discrepancy has since been explained in terms of CO-induced segregation of nickel (for which it has a high affinity) to the (copper-rich) alloy surface, thereby resulting in a surface composition that is not very different from the bulk.[199] CO-induced segregation in Ag–Pd alloys has also been shown to lead to a surface enrichment of palladium.[147] In contrast, the surface of Pt–Pd alloys, which are normally enriched in palladium for all bulk compositions, have been shown to be *less* enriched in this element after treatment with CO.[226] The results of oxygen chemisorption on Ag–Pd alloys has recently been interpreted with the aid of AES studies of the alloy surface before and often oxygen exposure.[202] AES studies of Ni–Th alloys have demonstrated that exposure to as little as 20 langmuirs (10^{-6} torr for 20 sec) of oxygen can result in chemisorption-induced segregation of thorium to the alloy surface.[203] This observation explains the distinct behavior of nickel–thorium catalysts in methanation reactions, for which the dissociation of CO into C and O has been identified as one of the crucial steps.[227] Under methanation conditions, therefore, it is likely that the catalyst surface consists mainly of a mixture of thorium oxide and nickel, thereby influencing the surface chemistry of the adsorbate–catalyst system.[203]

Chemisorption-induced segregation is also an important phenomenon in metallurgical applications, although very few fundamental AES studies have been carried out in this area. It has been shown, for instance, that the loss of magnesium from magnesium-containing aluminum alloys (a well-known metallurgical problem) is determined by the diffusion of Mg in the oxide layer for air-exposed samples, but by bulk diffusion of Mg in clean samples examined in UHV.[219] In addition, it was shown that, although magnesium-containing aluminium alloys form a very thin, passive oxide layer in the same way that pure aluminium does, this oxide is greatly enriched in MgO.[219] Reactively-enhanced segregation of copper in aluminum alloys has been suggested as a mechanism for the observed variations in oxide film thickness, formed by anodization in phosphoric acid, between pure aluminum and copper-containing alloys. AES has shown that there is a significant enrichment of copper at the oxide–metal interface on the alloys, due to copper segregation from the bulk during anodization, and that this may determine the oxide growth mechanism, morphology, and thickness.[228] Similarly, the oxidation behavior of copper-base alloys (containing Zn, Sn, and Ni) during pickling and annealing has been shown to be the determining factor in the surface composition of the final product. AES studies have shown that acid pickling of the air-exposed alloys results in a surface that is depleted in Zn, Sn, and Ni, since all of these metals have higher affinities for oxygen than does copper, and are therefore removed as oxides in the pickling bath. Subsequent annealing, however, results in the oxygen-induced surface segregation of one of these, zinc, thereby resulting in a zinc-rich oxide surface.[229]

AES studies of oxidation of nickel-base alloys have shown a similar dependence of oxide composition and growth on preferential oxygen affinity for one or more of the alloying components.[230,231] In the case of an alloy containing 24% Al and 0.3% Zr, the high-temperature oxide layer, which was 2–3 μm thick, was found to have a thin (300 nm) interfacial region of aluminum oxide, while the underlying metal just beyond this interface was depleted in aluminum but enriched in zirconium.[230] For Ni-20 Cr alloys, high-temperature oxidation has been shown by AES to result in a phase separation of nickel and chromium oxides, to form a 500-nm NiO layer, in top of a 400-nm Cr_2O_3 layer, on top of the alloy. Subsequent reduction in hydrogen was shown to result in complete reduction of the NiO layer to Ni, but no reduction of the Cr_2O_3 layer was observed.[231]

AES observations of enhanced surface segregation in ferrous alloys have shown that the increase in crevice corrosion resistance with increasing chromium content of ferritic stainless steel is due to oxygen-induced chromium segregation to the steel surface, forming a chromic oxide-rich passive film of only 5 nm in thickness on the most resistant steels.[232] The oxide layers formed by oxidizing clean ferritic steel in very low partial pressures of oxygen ($<10^{-8}$ torr) have been shown to be enriched in chromium, and the oxidation kinetics for oxide films up to 10 nm in thickness have been shown to be determined by bulk chromium diffusion in the steel matrix.[233] Heat treatment of AISI 314 stainless steels in a carbonizing atmosphere containing oxygen was shown to result in a 50-nm-thick chromic oxide layer, containing no iron or nickel, as determined by AES.[234] All of these results are in complete contrast to the results obtained from AES studies of segregation in clean steels in UHV condition, when chromium segregation to the surface of iron was observed to occur only when nitrogen was present in the alloy.[211] An implication from these studies is that anomalous data on the equilibrium surface compositions of binary and multicomponent systems may easily be obtained if the vacuum conditions, and surface cleanliness of the specimen, are inadequate. The role of grain-boundary segregation of impurities in intergranular stress-corrosion cracking of carbon and low-alloy steels is discussed in detail by Lea and Hondros.[235] The presence of segregating elements, and their influence on the stress-corrosion cracking behavior of these materials, is shown to correlate with the equilibrium oxidation potentials of the elements concerned, and a model based on this correlation is found to predict stress-corrosion-cracking phenomena very well. Since the oxidation potentials and oxygen affinities of metallic elements are related, an alternative approach to the interpretation of stress-corrosion-cracking behavior could be based on oxygen- (or sulfur, or chlorine, etc.) induced surface segregation of solute species to the grain boundaries.

2.2.5.   AES Studies of Preferential Sputtering on Alloy Surfaces

A large proportion of AES studies of alloy surfaces, and of metallurgical and materials science applications in general, employ inert-gas ion sputtering to effect controlled erosion of surface layers, thereby producing a composition-depth profile of the specimen under study as introduced in Chapter 3. However, in addition to the topographical and atomic-mixing artifacts discussed in Section 1.3, ion sputtering causes compositional changes in the surface region of multicomponent materials through the phenomenon of preferential sputtering. A complete treatment of the topic is beyond the scope of the present work, but, because of its direct relevance to metallurgical and materials science applications of AES, and the fact that AES is often used as an analytical tool in the investigation of sputtering phenomena, and the importance of ion-induced compositional changes in the first-wall components of fusion reactors during exposure to plasma radiation, preferential sputtering does deserve a few words of comment.

The subject has recently been reviewed by Betz[236] and Furman,[237] and there is a large body of scientific literature pertaining to AES studies of preferential sputtering on binary and multicomponent alloy surfaces.[238–253]

Preferential sputtering is the rule whenever a multicomponent system is subjected to ion bombardment. Owing to differential partial sputtering yields* of the individual components, a surface layer of altered elemental composition will be established. When steady-state conditions (sputter equilibrium) are reached, the target will be sputtered stoichiometrically,

---

*The partial sputtering yield of component A is defined here as the sputter yield of component A, in atoms per incident ion, divided by the surface concentration of component A, in area fraction.[221]

simply from conservation of matter. Only in the rare cases when the partial sputter yields of all of the components in a multicomponent material are of equal magnitude will the sputtered surface have the same composition as the bulk of the material.

While sputtering of elements is quite well understood and is generally easily predicted from theory,[254] no comprehensive theory of alloy or compound sputtering has been established. The problem is complicated by the fact that the partial sputtering yields in multicomponent materials are generally not simply related to the pure, elemental sputtering yields, although this is often a very useful first approximation in the interpretation of AES sputter-depth profile results.[236] An alternative approach, which has been shown to provide a reasonable qualitative prediction of preferential sputtering phenomena in binary alloys, is one based on surface binding energies and atomic masses. Briefly, the sputtering theory of Sigmund[255] predicts the sputtering yield to be inversely proportional to the surface binding energy (sublimation energy), and the recoil energy density (amount of primary ion energy that goes into recoil atoms) to increase with increasing atomic mass of the matrix. These results have been used to show that, for alloys with components of similar masses, partial sputtering yields are determined by the surface binding energies of the components, and, in many cases, are close to the pure-element sputtering yields.[236] For alloys with components of very different mass, however (particularly for atomic masses <100), collision cascades dominate and enrichment of the heavier mass component is predicted.[256]

A more-detailed approach has been taken by Lam et al.,[238] who consider the effects of radiation-enhanced segregation in addition to direct preferential sputtering, and arrived at a kinetic model that predicts the variation of surface composition with ion dose as a function of bulk composition, ion flux, and temperature. These authors have successfully applied their model to a range of nickel-base alloys.[238] The role of segregation (at both room and elevated temperatures) during ion sputtering is rarely taken into account in the interpretation of sputter-depth profiles, even though there is evidence to suggest that it may be important in such alloys as Cu–Ni,[215,238,243] Ag–Au,[215] Au–Pd,[215] Sn–Pb,[246] Ni–Mo, and Ni–Si.[238] Segregation is expected to occur not only through thermal diffusion of components to the surface (see Section 2.2.3.), but also through ion-enhanced diffusion, via the creation and subsequent migration of ion-induced point defects.[238] In the case of Sn–Pb alloys, preferential sputtering of lead (owing to the lower sublimation enthalpy of this component) takes place during sputtering, but thermal segregation of lead from the bulk to the surface, which is enhanced by the presence of ion-induced interstitials and vacancies, occurs as soon as ion bombardment is stopped, and the alloy surface becomes enriched in lead. The effect of ion-induced point defects on the interdiffusion kinetics of tin and lead is so profound that the AES results from the sputtered alloy suggest an activation energy (from log I versus T^{-1}) for the process of only 7 kJ mole^{-1}, whereas the bulk value is 107 kJ mole^{-1}.[246]

The results of a number of other studies of preferential sputtering in alloys are summarized in Table 6.5. The behavior of the majority of systems studied may be explained in terms of relative pure element (S_E) sputter yields alone, although sputtering behavior in alloys with at least one light element component (e.g., silicides) is best described in terms of preferential sputtering of the lighter component. It is of interest to note that most oxides (see Section 3.1.3) fall into this latter category, although the underlying mechanisms for ion-induced oxygen loss from oxide surfaces are unlikely to be so straightforward.

Table 6.5. Preferential Sputtering Behavior of Alloy Surfaces[a]

A. Exhibiting No Preferential Sputtering

As–Ga	Ni–Fe	In–Sb	Cr–Fe
Cu–Au	Cd–Te	Mn–Pd	

B. Exhibiting Elemental Sputter-Yield Dependence
(Preferentially Sputtered Component Listed First)

Ag–Au	Au–In	Pb–In	Gd–Co
Ag–Co	Au–Pd	Pd–Fe	Gd–Fe
Ag–Cu	Cu–Mo	Pd–Sn	(Cu–Gd)–Co
Ag–Ni	Cu–Pd	Pd–Ni	Fe–Cr–Mo
Ag–Pd	Cu–Pt	Te–Ge	Ni–(Cr–Fe)
Al–Si	Cu–Ni	U–Nb	Te–(As–Ge)
Au–Cr	Ni–Mo	Sn–Pt	Ag–Au–Cu
Au–Ni	Ni–Cr	Zn–Cu	Ag–Au–Pd

C. Exhibiting Mass-Dependent Preferential Sputtering
(Preferentially Sputtered Component Listed First)

Al–Au	Al–Pd	Si–Ni	Si–Pt
Al–Cu	Be–Cu	Si–Pd	Si–Ti

[a]Compiled from data in References 236–254, 326, 404, and references cited therein, and the author's own unpublished work.

3. OXIDE SURFACES

The study of clean metal surfaces by AES forms a large and important field of application of the technique, in both fundamental research and technological problem solving. More often than not, however, the surfaces that an electron spectroscopist is likely to be required to study will be covered by a thin oxide film or, in some cases, may present themselves, as far as AES is concerned, as a completely oxidized surface. The extreme surface sensitivity of AES, combined with a depth-profiling facility such as inert-gas ion bombardment, makes it the ideal technique to use in the study of oxide films ranging from submonolayer coverage, to interfaces within thick oxide films, to the bulk oxides themselves.

In oxidizing gases at room temperature, the oxide layer on newly exposed metals and alloys will rapidly grow to a limiting thickness (determined by diffusion through the passivating oxide layer) to <5 nm in most cases (Fig. 6.9). This is almost entirely encompassed by the sampling depth of AES, and workers in the field of passivation are now able to obtain a detailed analysis of a particular protective film with ease. The protective layer on metals in oxidizing gases increases in thickness with increasing temperature (owing to increased diffusivity), and AES may be used to some advantage in studies of selective oxidation, surface diffusion, transport direction and depletion effects, and solid-state reactions under such circumstances. In the aqueous phase at room temperature, the oxide film may further thicken under the action of electrochemical processes such as anodic oxidation or by chemical conversion of the surface. In both atmospheric and aqueous environments, the passive oxide layer can break down, leading to extensive oxidation of the metal or alloy—a process known as corrosion. The economic significance of

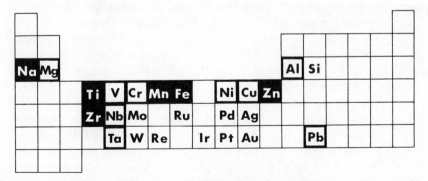

Figure 6.9.   Reactivity of clean metal surfaces with oxygen in high vacuum. ⊠ forms a thin, chemisorbed overlayer; ⊠ forms a thin oxide layer; ■ forms a thick oxide layer. (Most of the metals not represented are expected to form a thick oxide layer.)

corrosion, and the continuing need to find improved ways of reducing corrosion rates of metals in particular environments, are well recognized. The subject is a complex one, and the interested reader is referred to excellent compilations of Shreir for further information on this topic.[257] Growth of the corrosion film depends on ionic and charge-transport phenomena in the film and on phase-boundary reactions at the interfaces between the metal and the film and between the film and the environment. AES is ideally suited to measure the elemental composition of surface and interfacial layers formed through the interaction of metals and the low-atomic-number elements C, N, O, F, P, S, and Cl, the principal nonmetallic elements that are of interest in corrosion science.

One of the features of AES that make it so well suited to the study of oxides in general and metallic oxidation and corrosion phenomena in particular is its ability to differentiate between the elemental and oxidized states of many metals, that is, the availability of chemical information in the Auger spectrum. The advantages and limitations of AES for this type of application are discussed in the following section.

3.1.   Studies of Bulk Oxide Surfaces

3.1.1.   Chemical and Angular Effects on the Auger Spectrum

Changes in the energy-level diagram for an atom that result from changes in chemical environment are very minor in comparison to the changes that occur on increasing or decreasing the atomic number. Even when valence electrons, which are most drastically affected by changes in chemical environment, are involved in an Auger transition, the dominance of the core-hole energy term in the energy balance usually results in only minor variations of Auger energy, intensity, and line shape on chemical environment.

The fact that the energy levels of atoms do change in response to changes in chemical environment is of great value in XPS, where line shapes and energy resolution are generally such that chemical information is easily derived from the electron spectrum.[120] In the same way, Auger electron energies are also sensitive to the precise environment of the atom from which emission is being observed. This was first demonstrated by the pioneer of electron spectroscopy, Siegbahn, in 1958, who obtained electron spectra from elemental copper and its oxides.[258] Since this discovery of chemical shifts in the electron spectra from solid surfaces, AES has often been called on to supply chemical, as well as elemental, information on a variety of substrates. This topic is covered in more detail in Chapter 5. A useful review of chemical information in Auger spectra, particularly of oxide surfaces, is contained in Reference 259, and the problem of Auger line-shape analysis is discussed in References 39, 260, and 261. In addition to chemical information in the metal Auger features, it is also possible to obtain some information about the binding states of oxygen in oxides from the oxygen KVV spectrum. The energy and line shape of this feature have been shown to be highly dependent on the type of oxide under study, although our understanding of the reasons for these differences is still poor.[122] It has been suggested that the variation in oxygen KVV line shape may yield information concerning incorporation and nucleation in studies of metal oxidation.[262] One would intuitively expect the same information to be contained in, for example, the sulfur or carbon Auger features from sulfur and carbon-containing surfaces, but our knowledge of such systems is extremely poor at present.[263,264]

In the same way that channeling and diffraction effects can give rise to a marked anisotropy of Auger emission from single-crystal metal surfaces, a dependence of Auger electron intensity and line shape on emission angle is expected from single-crystal oxide surfaces. Preliminary angle-resolved experiments on vacuum-cleaved MgO and NiO single crystals suggests that additional effects, such as anion polarization, which is not present in metals, may significantly influence the angular dependence of Auger emission from oxides.[265] Much work is needed, however, before this aspect of AES can be understood or exploited more fully.

3.1.2. Characterization of Bulk Oxide Surfaces

Clean, bulk oxide surfaces have been studied using high-energy-resolution AES, in order to derive detailed information concerning Auger line shapes and, it is hoped, quantitative information concerning the oxide valence band. In the case of BeO,[266] Li_2O,[267] and TiO_2,[268] promising results have been obtained. For these oxides, the density of states is reflected in the CVV line shape, in the same way that the density of states of many metals is reflected in their CVV line shapes (Section 2.1). In the case of MgO[269] and Al_2O_3,[270] however, the CVV Auger line shapes are distinctly atomic-like, and contain very little valence-band information. The same is true of many fluorides.[270,271] The reason for this lies in the valence-band width, which determines the extent to which the density of states is reflected in the Auger line shape in both metals and compounds (Section 2.1).

AES is more commonly employed to determine the elemental composition and cation oxidation state in oxide surfaces. Examples are the discrimination between Cu(I) and Cu(II) oxides,[272] between Sn(II) and Sn(IV) oxides,[273] and between different iron[274] and uranium[275] oxides. AES has also been applied to the characterization of pyrolytically deposited V_2O_3 films,[276]

antimony-doped tin oxide films,[277] and mixed oxides such as cobalt fer-
rites,[278] tungsten bronzes,[279] and, of course, glass[280-286] and ceramic[285-288]
surfaces.

An excellent example of the power of AES analysis is the study of impurity
and additive segregation during the sintering of ceramic powders. For
example, oxides such as MgO are commonly mixed with Al_2O_3 to increase
the rate of sintering and to achieve higher densities in the sintered product.[289]
This was assumed to arise from a "solute-drag" mechanism, by which the
MgO additions retarded grain growth during sintering. However, AES data
showed that Ca, rather than Mg, was enriched in the intergranular Al_2O_3
fracture surfaces.[285] In another series of studies, AES has revealed a correlation between the enhancement of calcium segregation and the reduction of
flexural strength of alumina specimens during exposure to steam and steam–
CO atmospheres. The extent of calcium segregation was shown to increase
over a period of 12 days exposure.[287] Yttrium oxide has also been shown to
retard grain growth in alumina by segregating to grain boundaries.[286] The
addition of LiF or NaF to magnesia is also known to promote densification,
and AES studies of hot-pressed MgO doped with LiF showed high grain-
boundary concentrations of fluorine, which may improve material slippage
and atomic transport during hot pressing. A correlation between fluorine
segregation and sintering rate, mechanical strength, and optical properties
of MgO has also been established.[290]

Studies of chemical-vapor-deposited (CVD) silicon nitride films have
revealed an oxygen content of between 2 and 67 atom%. The AES data
indicate that oxygen incorporation into silicon nitride results in the formation of oxynitrides for $\lesssim 200$ ppm of oxygen in the CVD reactor and of silicon
oxides for higher oxygen concentrations.[288]

Hench and co-workers[280,281] have used AES and sputter-profiling to
study the corrosion of a range of different glasses. The calcium profile in a
soda–lime–silicon glass, for instance, changes markedly during exposure to
an aqueous environment at 100°C, over a period of several days. The calcium
profile shows a characteristic depleted region, which increases in thickness
with exposure time, and a rise in the "bulk" calcium concentration with
exposure, owing to the preferential leaching of sodium.[281] This calcium-rich
subsurface region is thought to impede further sodium diffusion, thereby
increasing the corrosion resistence of the glass. An important recent application of glass is for the disposal of nuclear waste materials, and corrosion
problems are obviously of extreme importance in this field. Acidic environments have been shown to selectively leach most heavy metals from glass
surfaces, resulting in a silica-rich layer of several hundred angstroms thickness. This selective leaching is followed by dissolution of the silica matrix
itself, and the glass corrosion proceeds by removal of all of the glass constituents at the same rate.[291]

Glass has also found applications in biological implants, where it may form strong and stable bonds with living bone. A composition of 45% SiO_2, 24.5% CaO, 24.5% Na_2O, and 6% P_2O_2 (by weight), called bioglass, has been found to be extremely successful in this application. The glass appears to form a stable calcium phosphate film on the surface which, in conjunction with a silica-rich subsurface layer, protects the glass from extensive *in vitro* corrosion and promotes good adhesion between the glass and bone.[292-294]

3.1.3. *Electron and Ion-Beam Effects on Oxide Surfaces*

During AES analysis of insulating surfaces in general, and oxide surfaces in particular, charge build-up may occur, which distorts the Auger signal, and electronic or thermally assisted composition changes may occur (segregation, adsorption, and desorption) which render accurate analysis difficult or impossible. All of this may occur during electron irradiation alone; the situation is further complicated by additional but similar effects due to inert-gas ion bombardment.

The problem of electron-induced artifacts has already been discussed in general terms in Chapter 3 and Section 1.2. A few sentences describing particular studies of such artifacts on oxide surfaces are relevant at this point, however, since the detrimental effects of electron irradiation are not always appreciated by electron spectroscopists.

Electron-stimulated desorption is a phenomenon well known to occur during electron bombardment of a solid, but the magnitude of desorption from glass surfaces has been studied in detail only recently.[16,284] It has been established that desorption occurs only at very high current densities, and that electron-stimulated migration (of, for example, sodium) is much more important in glass at intermediate and low current densities. Electric fields of the order of 10^4–10^5 Vcm^{-1}, and local temperature exceeding 100°C, have been calculated for this type of study.[294] It has been shown that substrate cooling to liquid nitrogen temperatures can reduce the extent of sodium and lithium migration under the electron beam.[17,294]

It is well known that a large number of oxides undergo reduction during electron bombardment, although the causes and extent of this reduction are often disputed. Examples of oxides which have reported to undergo reduction to a suboxide or elemental species are silica,[23,24] alumina,[25,26] molybdenum tri-oxide,[27] niobium pentoxide,[28] tantalum oxide,[29] tungsten oxide[29] and titanium dioxide.[30] Iron oxides have been reported as being stable to electron irradiation.[33] All of the oxides which have been observed to undergo reduction do so at only very moderate current densities, which may often be encountered in routine AES analysis (Figure 6.2). It has been claimed that native oxide films are particularly susceptible to electron-induced reduction.[14] Clearly, care must be taken in the application of AES to oxide surfaces and, where possible, standards should be run and control experiments should be carried out, if the possibility of electron-induced effects is suspected. In practice, the susceptibility of oxide surfaces to electron damage may impose severe restrictions on the spatial-resolution or detection-limit capabilities of the technique, and it will therefore be necessary to sacrifice one or both of these if damage is to be minimized (see Section 1.2).

The indiscriminate application of ion sputtering, in conjunction with AES, may also lead to erroneous conclusions in the analysis of oxide surfaces. In addition to the problem of preferential sputtering (Section 2.2.5), ion bombardment can also cause compositional changes in most oxide surfaces through the phenomenon of ion-induced reduction. Although the detailed mechanisms of this effect are still a matter of much dispute, it is well known and accepted that the great majority of oxides undergo some form of partial or total reduction under ion bombardment.[9,295] The only exceptions appear to be oxides with a high free energy of formation, such as SiO_2, Al_2O_3, SnO_2, and ZnO.[295] AES has actually been usefully applied to studies of the effects of ion damage on oxide surfaces. Such studies make important contributions to our understanding of materials behavior in environments where they are likely to undergo ion bombardment, such as inside nuclear reactors, in ion-beam lithography applications, and in outer space.[296] Studies of the extent of ion-induced reduction have also been suggested as a means of dating mineralogical samples that have been exposed to low-flux ion irradiation conditions for very long periods of time,[297] and ion-induced reduction of oxides has been suggested as a mechanism for darkening of the lunar surface.[296]

Two oxides that have been studied in some detail by AES are the pentoxides of tantalum and niobium. The former is of particular importance to surface scientists, since it has been suggested as a standard material for the characterization of ion-sources and sputter-depth profiling facilities,[298] hence its behavior under ion bombardment is of crucial importance. Although it has not been possible to identify the products of the ion-induced reduction reaction, it has been shown that the surfaces of both of these oxides are significantly depleted in oxygen following ion bombardment.[299-301] In every case, lower-energy ions were found to be more effective in oxide reduction, an effect which can be qualitatively explained on the basis that both oxygen and argon will be backscattered more effectively from niobium and tantalum ions than the reverse, and this backscattering mechanism will dominate the sputtering process at very low ion energies, since ion–surface-atom interactions increase in magnitude at lower ion energies.[299,300] Certainly, the observed strong similarity in behavior of Ta_2O_5 and Nb_2O_5 suggests that the dominating factors in ion-induced oxide reduction are the chemical and thermodynamic properties of the oxides, rather than any differences in the "elemental" sputter yields ($\gamma_{Ta} < \gamma_o \approx \gamma_{Nb}$).[301] AES has revealed extreme surface compositional changes caused by ion bombardment of TiO_2,[302] but only slight changes in iron oxides.[33,274] In oxides which undergo electron-induced reduction (such as TiO_2 and SiO_2), synergistic electron-ion beam effects are observed, owing to the fact that the sputtering yield of the electron-irradiated surface is enhanced by virtue of the fact that it has been reduced.[302,303] Ion-induced compositional changes have also been reported in a large number of oxy-salts,[304,305] although much work still needs to be carried out in this area if the precise mechanisms underlying "preferential" sputtering in complex systems is to be understood completely.

3.2. Studies of Oxidation and Corrosion

3.2.1. Atmospheric Oxidation and Passivation of Metals

The standard free energy change for the oxidation reaction

$$xM(s) + \tfrac{1}{2}O_2(g) \rightarrow M_xO(s)$$

is negative for all metals, including gold.[306] A range of experimentally derived and calculated heats of adsorption for oxygen on clean metal surfaces is presented in Table 6.6. It has been shown that there is a strong correlation between the heats of formation of oxides and the heat of adsorption of oxygen.[309] It is also apparent, from inspection of Table 6.6, that there is a correlation between the experimentally determined heat of adsorption and the saturation oxide coverage (θ_{sat}), although this latter quantity may be diffusion limited in some cases (e.g., Mo, W, Ta, and Nb).

Using the relationship

$$\Delta G_f^0 = -RT \ln Kp \qquad (6.34)$$

we can deduce the equilibrium oxygen pressure for the oxidation reaction; for copper at 300 K this is 10^{-10} torr.[306] Thus, from the thermodynamic point of view, all metals should spontaneously form an oxide at oxygen pressures of $\sim 10^{-6}$ torr or greater. Although initial chemisorption is generally characterized by an activation energy of ~ 10 kJ mol^{-1}, further interaction leading to the growth of a distinct surface oxide phase is usually associated with a higher activation energy and is therefore kinetically controlled.[306] Thick oxides may be rapidly formed, however, at high temperatures and oxygen partial pressures, or under anodization conditions (when oxide growth is assisted by an electrical field). The transformation

Table 6.6.   Experimental and Calculated Heats of Oxygen Adsorption (kJ mol^{-1})[307,308]

Metal	ΔH_{ads} (Calculated)	ΔH_{ads} (experimental)	$\theta_{sat}{}^a$
Ti	703	991	8.6
V	728		
Cr	669		
Mn	686	630	3–12
Fe	544	571	4.7
Co	477	420	2.3
Ni	481	449	2.6
Zr	824		
Nb	833	873	3.8
Mo	655	722	1.4
Tc	494		
Ru	322		
Rh	297	441	1.0
Pd	238	281	0.7
Hf	929		
Ta	937	890	3.2
W	824	815	1.3
Re	536		
Os	372		
Ir	331		
Pt	297	281	0.6

^{a}"Saturation" oxygen coverage (from Reference 308).

from a chemisorbed layer to bulk oxide has intrigued a number of investigators over the last two decades, and only recently has AES been capable of supplying answers to questions such as: At what stage does lattice penetration and oxide nucleation occur? What is the mechanism of such a process? How does the surface oxide differ from the thermodynamically stable bulk oxide? It is these practical aspects of metallic oxidation, rather than the more esoteric types of study that have been made (e.g., Reference 310), that we shall concern ourselves with here.

A large number of studies using AES have been carried out on air-oxidized ferrous alloys.[233,234,311-322] Chemical effects on the low-energy Auger transitions from the first row transition elements are such that it is possible to differentiate between metal and oxide for most of the elements encountered in stainless steels, although peak overlaps limit this application most severely.[313] It appears that AES is capable of discriminating between oxides and hydroxides on transition metal surfaces, although it is doubtful whether this information could be obtained on systems that have not been extremely well characterized.[322]

AES has shown that oxidation of 18/8 stainless steel at high temperature proceeds via preferential oxidation of chromium at low oxygen exposures, followed by surface segregation and oxidation of chromium and manganese, resulting in the complete disappearance of both iron and nickel from the steel surface.[313,323,324] At lower temperatures, chromium segregation is diffusion limited,[323] and an iron-rich oxide phase is formed.[312] Studies of simpler, two-component systems have enabled a fuller understanding of the processes of preferential oxidation and diffusion to be obtained. In the case of Fe/Ni alloys, oxidation at 500°C has been shown to result in an iron oxide (Fe_3O_4) surface layer for bulk compositions of 70% nickel or less, and in a mixed oxide layer (tentatively assigned as a spinel structure) for alloys containing more than 70% nickel.[319] This is in contrast to the "clean" alloy surface, which always exhibits a surface enrichment of nickel at equilibrium.[178,319] The spinel oxide structure observed on Fe/Ni alloys is frequently observed by AES on various stainless steels, although the constituent elements that make up the spinel may vary according to the bulk concentration and impurity levels of the steel (e.g., $MnCr_2O_4$ may be formed on manganese- and chromium-containing steels[324]). AES studies on Fe/Cr alloys have revealed the presence of a $FeCr_2O_4$ spinel oxide following oxidation at 900 K, in addition to Cr_2O_3 and Fe_3O_4 oxide phases.[318] The oxidation of Ni/Cr alloys, on the other hand, results in the growth of a thick NiO film on top of a Cr_2O_3 film, with Cr_2O_3 being formed preferentially near grain boundaries.[321]

Some measure of the variation of oxide stoichiometry with depth into the oxide layer, for depths up to several nanometers, may be obtained by considering the relative intensities of low-energy (MVV) and high-energy (LMM) Auger transitions, since these features will have different sampling depths, according to their kinetic energies. This property has been used to investigate the variation of composition with depth of native oxides on Fe–Co–Ni–P–B alloys.[320]

When used in conjunction with inert-gas ion etching, AES can provide valuable information on the variation of oxide composition with depth to depths of several microns. Although problems of interpretation may occur (owing to ion-induced artifacts, see section 3.1.3), it is possible to overcome these, to a large extent, through comparison with standard materials. A typical example of the application of sputter-depth profiling, to aerobically formed oxides on a type-304 stainless steel, is presented in Fig. 6.10. The results show that the oxide thickness increases by a factor of 10 on heating the specimen in air at 500°C for 75 min. In addition, the composition of the oxide layer changes dramatically. Whereas the room-temperature passive

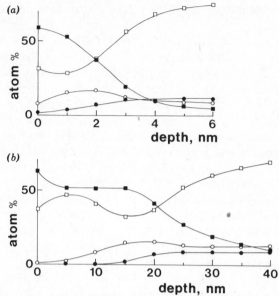

Figure 6.10.   AES sputter-depth profiles through (*a*) the natural oxide and (*b*) the oxide developed by heating in air at 500°C for 75 min on 304 stainless steel. Note the different depth scales in (*a*) and (*b*). Profiles calculated from the data in Reference 317, using standard atomic sensitivity factors (L.E. Davis, N.C. MacDonald, P.W. Palmberg, G.E. Riach, and R.E. Weber, *Handbook of Auger Electron Spectroscopy*, 2nd ed., Physical Electronics, Eden Prairie, MN, 1976): ■, oxygen; □, iron; ●, nickel; ○, chromium.

layer consisted of a mixture of iron, chromium, and nickel oxides, the heated specimen exhibited a layered structure. Near the surface, an iron-rich oxide layer, of approximately 75 Å thickness, is followed by a chromium-rich layer, and the oxide is seen to be depleted in nickel for a depth of 150 Å.[317]

An extremely detailed study of the parameters controlling the formation of protective and nonprotective oxide on 304 stainless steel has also been carried out by AES.[316] It was found that grain size, surface damage, oxygen partial pressure, and annealing treatments can all influence the oxide formed on stainless steel. The oxide formed at low oxygen partial pressures is extremely stable, and significantly improves the corrosion resistance of the steel. On large grained (40-μm) samples, a duplex oxide structure forms at pressures as low as 10^{-2} torr. The chromium concentration in the surface region may be enhanced or diminished by annealing treatments, thereby determining the probability of forming a protective Cr-rich oxide layer. The presence of carbon leads to the surface depletion of chromium during annealing at 500°C owing to the formation of a Cr-carbide species. The surface compositions as determined by AES of 304 stainless steels following various

Table 6.7. Oxidation of 304 Stainless Steel[316]

Treatment	Surface Composition (atom%) by AES						Oxide Thickness (Å)
	Fe	Cr	Ni	Mn	O	S	
Clean, annealed	60	25	9	0	0	7	
10^{-5} torr O_2, 2 hr	<1	17	0	7	72	0	
10^{-3} torr O_2, 22 min	<1	29	0	0	70	0	~2,000
Air, 15 min	<1	30	0	0	70	0	~2,000
Sputtered, cleaned, air, 15 min	34	0	2	0	63	0	~20,000

treatments are presented in Table 6.7. The mapping capabilities of AES, allowing the two-dimensional variation of composition with position on the specimen surface to be investigated, are presented in Fig. 6.11. Here, the enrichment of chromium at the grain boundaries of an oxidized 304 stainless-steel sample is easily demonstrated by comparing the secondary-electron distribution (SEM micrograph, upper picture) with the scanning Auger micrograph of chromium (lower picture). Similar experiments to those described above have been carried out on 316 stainless steel[315] and 314 stainless steel.[234]

There have been a large number of studies of the atmospheric oxidation and passivation of nonferrous metals and alloys by AES,[325-336] ranging from oxygen chemisorption on gold[331] to the growth of thick MgO films on Al/Mg alloys.[327] The substrates that have been studied cover the complete range of periodic table, from lithium $(Z=3)$[332] to plutonium $(Z=94)$.[330] In addition, there have been a number of studies of metal oxidation, in gases such as carbon monoxide[337] and dioxide,[332] nitrogen,[332] and nitric oxide[338] by AES.

Studies of the oxidation of a Nimonic 80A alloy (75Ni/19.5Cr/2.4Ti/1.3Al) at 800°C have shown that the ion-cleaned alloy has nickel, chromium, and titanium present at the surface, but oxygen exposure results in preferential oxidation of the titanium, resulting in an apparently pure TiO_2 surface phase at high exposures.[334] Similarly, the surface of 98Cu/2Be alloys (used in the fabrication of secondary-electron multiplier dynodes) has been shown by AES to be completely covered in a thick BeO layer, with no evidence of the presence of copper in the AES spectrum, following oxidation at between 750 and 870°C in an oxygen partial pressure of 5×10^{-3} torr.[329] The structure of the BeO layer, however, was found to depend on oxidation temperature, and was interpreted as arising from a diffusion-limited oxide growth mechanism. High-temperature oxidation of a Ni/14Cr/24Al/0.3Zr alloy has

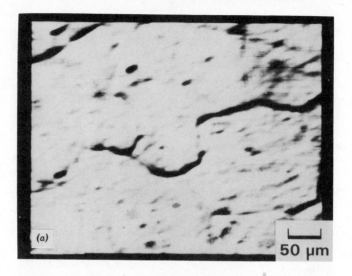

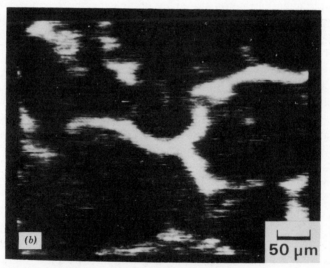

Figure 6.11. (a) Secondary-electron image and (b) chromium Auger map of 304 stainless steel oxidized in 0.05 torr O_2 at 800°C, showing chromium enrichment at grain boundaries. (Reproduced with permission from Reference 316, © North-Holland Publishing Company.)

been shown by AES to result in the formation of a 2–3-μm oxide film, with a 300-nm-thick layer of Al_2O_3 at the oxide–alloy interface.[230] The alloy just beyond the interface was shown to be depleted in aluminum but enriched in zirconium relative to the bulk alloy.

Tin–lead alloys, which form the basis of most soft solders, have applications ranging from the joining of metals to the production of electronic components. AES studies of the atmospheric oxidation of 97Pb/3Sn and 28Pb/72Sn alloys have shown that preferential oxidation of tin occurs and that this process continues until all of the surface tin is consumed, after which the lead component is oxidized.[325,326] The oxide film, at temperatures between 20 and 200°C and oxygen pressures less than 10^{-4} torr, reaches a thickness of less than 2 nm. It was found that, in the absence of oxygen, any oxidized lead phase was eventually reduced by outdiffusion and oxidation of metallic tin.[325]

Aluminum is an interesting metal for study by AES, since the chemical shift of the $L_{2,3}VV$ Auger feature, on going from metal to oxide, is 14 eV, large enough to be able to easily discriminate between these two species in the Auger spectrum. During oxidation, the oxide Auger feature (at ~ 54 eV) can be observed to increase in intensity with increasing exposure, whereas the metal feature (at ~ 68 eV) gradually diminishes in intensity with increasing exposure.[339] Once the passive oxide layer has been formed, at room temperature, there is still a contribution from the metal feature in the Auger electron spectrum, implying that either the passive layer is very thin (only one or two monolayers) or that oxidation proceeds via the formation of islands, leaving areas of bare metal between them.[339] It has been suggested that the transition from a chemisorbed state to a nucleated Al_2O_3 phase on aluminum is accompanied by the emergence of an "interatomic" Auger signal from the surface.[333] A similar argument has been used to explain the changes in Auger line shape during the oxidation of chromium surfaces.[340] A large chemical shift has also been reported for the lithium KVV feature.[332]

Chemical shifts in the Auger spectra from a number of other metals have been shown to be useful in the interpretation of oxidation mechanisms and products. Vanadium and its oxides represents one of the better examples: It has been shown that there is a linear relationship between the chemical shift of the vanadium $L_3M_{2,3}M_{2,3}$ Auger feature and the ratio of the oxygen KVV to vanadium $L_3M_{2,3}V$ peak heights in a series of bulk vanadium oxides, representing a shift of about 0.6 eV per oxidation number for this element.[341] It has been suggested that the oxidation states of first row transition metals may be derived from the experimentally measured intensity ratios of various LMM Auger features in oxides and sulfides.[342]

3.2.2. Anodization and Chemical Passivation

In an aqueous environment, the atmospherically formed oxide layer may undergo changes in structure, composition, or thickness. It may break down completely, resulting in complete dissolution of the substrate (see following section), or, under the influence of an electric field, may grow very thick. Anodic oxidation is a standard procedure for the preparation of uniform, well-characterized oxide films of known thickness from 1 nm to several hundred microns, and AES has been usefully employed in the study and characterization of anodized surfaces of aluminum and aluminum alloys,[41,228,343,344] nickel,[345,346] titanium,[347] niobium–tantalum alloys,[348] ferrous alloys,[314,349,350,352–361] and copper–nickel alloys.[351] In a number of cases, it was possible to investigate the porous nature of the anodized layer with AES,[41,345] and the effects of alloying elements on the oxide growth mechanism and morphology have been determined.[228,349,350,356–361]

A study of the anodic oxide layer formed on aluminum and the aluminum alloy 2024-T3 after anodization in phosphoric acid solution has shown that the anodization takes place more readily on pure aluminum than on the alloy.[228] AES has also shown that an enrichment of copper, caused by out-diffusion from the bulk during anodization, occurs at the oxide–alloy interface, and this may be responsible for the thinner anodized layer on the alloy.[228] In the case of the nickel, it is known that a passive oxide—NiO—is rapidly formed in air, and reaches a thickness of 9–12 Å.[345] Thick oxide films can be grown in pH 7.65 borate buffer solution by polarizing into the oxygen evolution region, but oxide growth is slow since most of the charge is consumed by O_2 evolution,[345] and the resulting oxide film is highly porous. Anodization in 0.1 N H_2SO_4 also produces a passivating anodic oxide, but only on pure nickel. In the presence of sulfur, NiO growth is inhibited and, although surface analysis indicates that OH^- ions can still be adsorbed by the surface, passivity is not attained and selective dissolution of nickel (resulting in further enrichment of sulfur) occurs. AES has shown that this inhibition of passive film formation is due to a single segregated monolayer of sulfur.[346]

Studies of titanium thin-film electrodes have shown that the surface of these specimens are covered in a mixture of oxidation states (TiO_2, Ti_2O_3, TiO), with the higher states predominating on electrochemically oxidized films.[347] Many of the conclusions of this work, however, must remain suspect, since it is known that TiO_2 reduction occurs readily under both ion and electron irradiation (see Sections 1.2 and 3.1.3). Similarly, both niobium and tantalum oxides are known to undergo reduction under ion bombardment (Section 3.1.3), hence the results of a recent AES sputter-profiling study of anodized Nb–Ta alloy films must be treated with care.[348] Such mixed-oxide films have important applications in electronics, where dielectric films such as β-Ta_2O_5 are used in the formation of integrated thin-film circuits. It has been shown that anodically grown oxides on Nb–Ta alloy films have superior electrical properties to β-Ta_2O_5 insofar as dielectric constants are concerned.[348] AES analysis has revealed a correlation between the dielectric properties of the anodic films and the thickness of the oxygen-diffused layer, and it is suggested that thicker oxygen-diffused layers result in higher donor densities.[348]

AES sputter profiling has also been carried out on anodized Fe/Ni alloy surfaces.[349] It was found that the anodic film was thicker and richer in iron in pH 6.48 solution than in pH 8.45 solution (boric acid–sodium borate), and it was concluded that a significant amount of iron in the anodic oxide was in the divalent state. It was also possible to estimate the extent of nickel enrichment at the oxide–metal interface as a function of pH and anodization potential. The results were explained in terms of a combination of (a) preferential dissolution of Fe, (b) preferential oxidation of Fe, (c) preferential anodic deposition of Fe or Ni.[349] A similar study of anodized Fe/Cr alloy revealed a significant sulfur concentration in the oxide layer for films that had been anodized in 0.5 M sulfuric acid.[350] Following mathematical treatment of the AES sputter-profiling data, it was shown that the anodic oxide depth profile was dependent on the method of anodization, exhibiting severe chronium depletion in pH 6.48 borate solution, but chromium enrichment (relative to iron) in 0.5 M H_2SO_4.[350] The inclusion of solution components in the passivating film is not necessarily detrimental, and they may contribute to its passivating properties. AES has revealed that the passivating films formed in solutions containing inhibitors such as chromate, molybdate, tungstate, or arsenate contained some Cr, Mo, W, or As, respectively.[352] The main components of films formed on iron in a calcium-containing solution were not only iron and oxygen, but also calcium, and a tentative interpretation of the data is that a mixed oxide, similar in composition to calcium ferrite ($2CaO \cdot Fe_2O_3$), is formed.[353] Such mixed-oxide films are quite protective, but the incorporation of sulfur-containing anions may be responsible for the poor ability of iron to repassivate in $Ca(OH)_2$ solutions containing S^{2-} or SO_4^{2-} ions, which points to the role of these anions in the corrosion of steels in concrete. Several AES studies of passive films on iron have failed to provide evidence for the presence of chloride (Cl^-) ions, however, despite the fact that these are thought to enter the oxide lattice quite readily.[354] AES does confirm, however, that the anodic film formed on pure iron is a duplex, consisting of γ-Fe_2O_3 in the outer part and Fe_3O_4 in the inner part.[314] There is evidence to suggest that AES can also distinguish between iron oxides and hydroxides in anodic films, by careful analysis of the low-energy $M_{2,3}VV$ Auger features.[323] According to one interpretation, an interatomic iron–oxygen Auger transition may be observed at 40 eV for the oxide,[355] but is not observed in the hydrated anodic film.[323] Interatomic metal–oxygen Auger transitions have also been reported for aluminum and chromium.[333,340]

It is well known that certain alloying elements, for example, Mo, Cr, and Si, in steels improve their corrosion resistance, and it is generally believed that this is due to a stabilizing effect of these elements in the passivating film. However, AES analysis of the anodic oxide on type-316 stainless steel failed to show any enrichment in Mo,[356] and Mo was found to be completely absent throughout the bulk of passive film formed on a molybdenum-containing steel, although it was detected on the surface, where it was assumed to arise from redeposition during passivation. [357] Similarly, no enrichment of Mo or Si is observed using AES in the anodic films formed on a selection of austenitic steels (types 304, 316, and 347).[358] The variation of surface composition with anodic polarization potential has also been studied for Fe/Co alloys and a range of chromium-containing ferritic steels.[315,359,360] In the latter case, the chromium content of the anodic oxide was found to be related to the bulk molybdenum content of the steel, even though molybdenum was not detectable in the oxide layer. From the existing data it seems that there is a substantial enrichment of chromium in the anodic films formed on both austenitic Cr–Ni and ferritic chromium steels. There appears to be a complex relationship among the bulk concentration of traditionally passivating additives, chromium enrichment in the anodic oxide, and passivity of the oxide layer on these steels. The apparent disagreement between the results of AES analyses, on the one hand (which suggest, if anything, a depletion of molybdenum in the anodic layers formed on Mo-containing steels), and the results of more-conventional analytical techniques, on the other hand (such as glow discharge spectroscopy,[359]

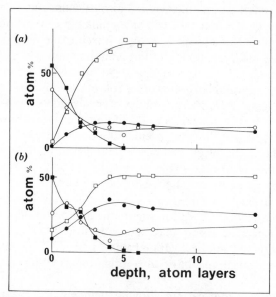

Figure 6.12.   AES sputter-depth profiles through passive oxide layers on two austenitic steels, formed after exposure to a physiological test solution: ■, oxygen; □, iron; ●, nickel; ○, chromium. Bulk steel compositions: (a) Fe, 18.2%, Cr, 10.7%; Ni, 1.3% Mo. (b) Fe, 21.8%; Cr, 24.2%; Ni, 2.7% Mo. (data from Reference 361).

γ-spectrometry,[360] and wet-chemical analysis,[314] all of which indicate Mo enrichment in the anodic film), still remains to be explained satisfactorily. A recent study of passive oxide layers formed on austenitic steels *without* anodization have reached similar conclusions regarding molybdenum concentration in the oxide,[361] with the result that the reasons for the beneficial effect of Mo additions in increasing resistance to corrosion remain unclear. Typical AES sputter profiles through passive oxide layers on austenitic steels are presented in Fig. 6.12.

3.2.3. *Corrosion and Corrosion Inhibition*

The term metallic corrosion is generally considered to embrace all interactions of a metal or alloy (solid or liquid) with its environment, irrespective of whether this is deliberate and beneficial (as in the cases of passivation and anodization discussed in the preceding sections) or advantageous or deleterious.[257] For the purposes of the present discussion, however, we will adopt the alternative definition of corrosion, that is: Corrosion is the interaction of a metal with its environment that adversely affects those properties of the metal that are to be preserved.

This definition, referred to as the deterioration definition,[257] is also applicable to nonmetallic materials such as glass, ceramics, concrete, polymers, etc., and embodies the concept that corrosion is always deleterious. The latter definition is the one generally adopted by corrosion engineers; the more-general definition is likely to find favor with corrosion scientists. The distinction between deleterious and beneficial corrosion phenomena is illustrated by comparing the relative behavior of iron and copper when exposed to industrial atmospheres. In the case of iron, a loosely bound reaction product of approximate composition $Fe_2O_3 \cdot H_2O$ (generally referred to as rust) is rapidly formed. Since it is not strongly adherent to the iron substrate, a protective oxide film is not formed, and the corrosion reaction proceeds at an approximately uniform rate until all of the iron is consumed. Copper, on the other hand, forms a strongly bound green patina, with approximate composition $CuSO_4 \cdot 3Cu(OH)_2$ (bronchantite), which is protective and isolates the metal from the atmosphere.[257] Both iron and copper, and their alloys, have been the subjects of a number of studies of corrosion phenomena using AES over the past few years.

An illustrative example of the power of AES in understanding corrosion problems is the case of corrosion in a copper pipe carrying air, in the vicinity of steel pipes carrying steam and fuel oil, buried underground without cathodic protection.[317] The copper pipe became heavily pitted in many regions, and a black region characteristic of an early stage in the corrosion process was examined by AES. Spectra from the as-received surface revealed the presence of large amounts of C, along with P, S, Cl, O, Cu, Fe, Zn, Ca, N, Mg, Al, and Si. By removal of only a thin layer of about 75 Å from the surface (by ion etching), it was apparent that the carbon, which exhibited a KVV peak shape characteristic of hydrocarbons, was only a thin overlayer, and large amounts of Cl and O were exposed. Using the spatial-resolution capabilities of AES, it was possible to demonstrate that the Cl and O were not uniformly distributed over the surface and that the high oxygen regions also contained small amounts of Ca and P as phosphate. Quantitative analysis of the spectra from high chlorine regions showed that the surface had the stoichiometry of CuCl. The conclusion, therefore, is that Cl and O are responsible for corrosion of the copper pipe.[317] In contrast, AES studies of steel pipes in the vicinity of the copper pipe, which had also undergone heavy pitting corrosion, suggested that S and O, rather than Cl, was responsible for corrosion of the steel surface. The identification of these corrosion products helped in assessing the sources of corrosion, and led to the necessary steps to protect the piping from further corrosion.[317]

The widespread use of copper and its alloys in corrosive environments has resulted in a large number of AES studies of the corrosion behavior of this metal. In a study of a range of copper alloys (Cu–Mn, Cu–Ti, Cu–Sn, and Cu–Si), AES has shown that the corrosion films formed after immersion in ammoniacal Cu(II) solutions range in composition from Cu_2O containing a little of the alloying element M (as its oxide) to M_xO containing a little oxidized copper. It did not prove possible to correlate the compositions and growth kinetics of the films with the susceptibility of the alloys to stress-corrosion cracking in the same electrolyte.[362] It was found, however, that the kinetics of dissolution of the component oxides into the ammoniacal environment (generally referred to as Mattsson's solution), and the kinetics of ion transport through the corroded film lattice, determined the resultant corrosion film composition.[362] In the case of Cu–4wt% Mn, the corrosion film was found to contain appreciable amounts of metallic copper.

The susceptibility of copper alloys to stress corrosion cracking (SCC) in ammoniacal environments has been recognized for a number of years, and the possible roles in SCC of the black tarnish films formed on brass in such solutions have been widely discussed,[363] but remain controversial. AES analysis of the tarnish formed on 64/36 brass after immersion in Mattsson's solution gives a composition Cu_2O containing 4–5 atom% Zn(II).[328] The ease of formation of this tarnish has been shown to be related to the susceptibility to SCC of the alloy, and the AES results provide experimental verification of the hypothesis that the tarnish is largely due to the formation of copper oxides.[364] The corrosion film composition is therefore very different from the atmospheric oxide composition, which shows a slight enrichment of zinc, relative to the bulk alloy composition.[328]

The SCC behavior of aluminum alloys has also been investigated by AES.[365] In this case, it was possible to show that the SCC behavior of the alloy AA 7075 was related to the presence of magnesium, copper, and zinc at the grain boundaries of the alloy, and that the concentrations of these elements were determined by sample treatment and solution temperature. AES has also been used to characterize corrosion inhibition treatments on AA 3003 aluminum alloys.[366]

Most AES studies of corrosion have concentrated on iron and steel surfaces, owing to their importance in construction and engineering applications. The phenomenon of intergranular corrosion of low-alloy steels in aqueous solutions of nitrates, carbonates, and hydroxides has been reported by several authors, and the electrochemical aspects of the problem seem to be well understood.[367] It is only since the application of AES to the problem, however, that it has proved possible to determine the exact microchemical behavior of iron surfaces in such environments.[368] AES has shown, for instance, that grain boundaries saturated with segregated N or S are not susceptible to intergranular corrosion, whereas C, probably present as fine cementite particles, can cause corrosion to be initiated. The difference in behavior between nitrate solutions, in which iron readily undergoes intergranular corrosion, and chromate solutions, in which iron is not corroded, is explained in terms of differences in the passivity of the cementite grain in these environments.[368] In another series of experiments, it has been established that intergranular attack in iron–phosphorus alloys immersed in nitrate solutions increases with increasing phosphorus content, owing to the inability of high-phosphorus alloys to form a passivation layer in this environment. Studies by AES showed that there is a correlation between the intergranular corrosion behavior of these alloys and the grain-boundary concentration of phosphorus.[369] The importance of establishing the microchemical behavior of specimens under corrosion conditions and the role played by AES in such studies have recently been emphasized in papers by Hondros and Lea,[235,370] who report on studies of SCC of mild steel in nitrate environments. A model has been proposed, based on the equilibrium oxidation potentials of alloying and impurity elements in steels, which predicts SCC behavior.[235]

A particular type of SCC, so-called sulfide stress cracking (SSC),[317] common to many commercially available steels, has been investigated by AES. In this case, the technique was used to determine the extent of impurity segregation in S- and Mn-containing steels, and has enabled a correlation between the minimum stress at failure in H_2S solutions at room temperature with the grain-boundary composition to be made.[371] AES studies of SCC phenomena in a stainless-steel pressure vessel exposed to high-sulfur and high-hydrogen atmospheres showed that there was a predominance of S in the advancing regions of the crack front, and a presence of P ahead of the crack front in the newly exposed intergranular region obtained by completing the fracture.[317] The AES results thereby demonstrate the aggressive role of S in the corrosion mechanism, and also show that SCC activity is promoted by the presence of P at the grain boundaries. The application of AES to the study of fracture surfaces is a large field in itself and is discussed in Section 5.1.

Intergranular and crevice corrosion phenomena in stainless steels exposed to H_2SO_4, HNO_3, HCl/HNO_3, HNO_3/dichromate, $NaCl/HCl$, and $NaCl/NaOH$ solutions have been studied with AES by a number of workers.[372–375] As a result of these studies, it has been demonstrated that sulfur segregation in nonsensitized steels permits intergranular corrosion in nitric–dichromate solutions, whereas chromium depletion is responsible for intergranular corrosion of the same steels in H_2SO_4–$CuSO_4$ solutions.[372] In an independent study of type-316 stainless steel, a direct relationship between chromium surface enrichment and resistance to crevice corrosion initiation was demonstrated and shown to be unaffected by the presence of nonmetallic inclusions.[373] A further study of type-304 and -316 stainless steels revealed that the enriched chromium layer was formed by selective dissolution of iron during the early stages of corrosion. The extent of chromium enrichment was shown to increase with the acidity of the crevice solution, and copper and molybdenum enrichment were bound to occur owing to the formation of Cu and Mo sulfides.[374,375] AES studies of the crevice corrosion behavior of thermally oxidized ferritic steels have revealed a correlation between the chromium content of the atmospherically grown film and the resistance of the steel to crevice corrosion, with the most resistant oxide films having thicknesses <5 nm.[232]

4. FILMS AND COATINGS

Up to this point, the main power of AES as an analytical technique has been due to its ability to perform elemental analyses of surface layers on a microscopic scale. Although we have introduced composition-depth profiling as a means of obtaining information on the depth distribution of species within a substrate, it is not until AES is applied to the characterization and investigation of films and coatings that it really comes into its own as a universal analytical tool for the three-dimensional characterization of materials.

Coatings are used in an increasingly varied number of applications to achieve appropriate or enhanced physical (optical, electrical, thermal, or hardness) and chemical (corrosion, oxidation) properties. In addition, the ever-widening trend toward microminiaturization involving more-complex multilayered structures in various areas of engineering and technology has created a pressing need for fast, reliable, quantitative approaches to the problem of characterization and elemental profiling, with depth and lateral resolution approaching atomic dimensions. The problem is compounded by the fact that the specifications and tolerances for many such coatings and multilayered devices, particularly in areas of high technology, are making increasing demands on the materials scientist to perfect better ways of preparing the devices.

A number of reviews of the various techniques that are available for the characterization and elemental analysis of films and coatings have appeared in recent years.[50,376] Of all of the techniques considered, one stands out above all the rest in its ability to perform rapid, quantitative three-dimensional materials characterization on almost any solid substrate—Auger electron spectroscopy.

4.1. Characterization of Contamination Layers and Adsorbates

AES has been used widely to study the adsorption and subsequent reactions of gases with solid surfaces, and the subject has already been discussed in Chapter 5 and Section 3.2.1 of this chapter. For the present, we are more concerned with the application of AES to the characterization of thin films of adsorbates, generally referred to as contamination, which may affect materials behavior and therefore form an important part of materials characterization. Indeed, one of the most popular areas of application of AES in materials analysis is the identification and characterization of contamination layers on specimen surfaces. The importance of understanding the sources of contamination and of being able to detect and control its presence is emphasized by the recent publication of a comprehensive text dedicated to the subject.[377] Representative examples of the application of AES to the study of different types of contamination on a wide variety of surfaces, taken from the literature,[378-394] are discussed below.

AES is used routinely to monitor contamination levels on steel surfaces, and particular studies include an assessment of surface-cleaning methods on type-316 stainless steel,[378] the effects of low-power-discharge cleaning on the level of contamination of type-304L stainless steel,[379] the removal of surface carbon from a Fe–Cr–Ni steel by exposure to a hydrogen discharge,[380] the identification of C, S, Cl, Ar, N, and O contamination on a commercially produced stainless-steel surface,[384] and the characterization of surface contaminants on industrially produced steel sheet.[382] The level of surface contamination of steel used in the construction of vacuum equipment is obviously of crucial importance, and much work has been devoted to investigating the efficacy of cleaning techniques by AES.[378,379] Attempts have also been made to assess the efficiency of various solvent-cleaning processes in the cleaning of deliberately contaminated 440C bearing steel surfaces by monitoring the level of contamination with AES.[383] AES has also been successfully applied to the identification of contamination layers on industrially produced copper[385] and tinplate.[386] In the case of tinplate, it has been possible to identify the causes of two characteristic defects in commercially produced tinplate-black staining (which is due to heavy carbon contamination in the base metal) and "wood grain" deposits (caused by intermetallic alloying and dewetting of the tin).[386] Contaminants on copper[380] and copper-base alloys,[229] magnesium,[381] titanium dioxide,[395] and lithium[332] surfaces have also been investigated by AES. In the case of air-exposed lithium, it was shown that the black-spot contamination sometimes observed on these surfaces was due to the formation of a nitride or oxy-nitride species.[332] Independent studies, however, have concluded that lithium does not react with nitrogen, either in the vacuum system or in atmosphere, and that reaction of the clean metal surface with typical atmospheric gases leads only to the formation of oxides, hydroxides, and carbonates.[396] Certainly, in the light of more-recent work, the suggestion in Reference 332 that lithium forms a carbide species on exposure to carbon dioxide now appears erroneous.[397]

Other areas in which contamination problems have benefited from AES analysis include lead-glasses,[384] mirror surfaces,[389] and aluminium-coated phosphors.[384] Interfacial contamination in thin films and coatings has also received attention,[381] but is dealt with in detail in subsequent sections of the chapter. The presence of contaminant layers on electrical contacts leading to high contact resistance,[381,387] and on adherend surfaces, leading to poor bondability,[388,392] has also been investigated by AES. One aspect of contaminant characterization by AES that should always be borne in mind is the potential perturbing effects of the primary-electron beam. It has been shown, for instance, that exposure of a clean tungsten surface to an electron beam in a CO atmosphere leads to a rapid build-up of carbon.[391] Similar problems can be caused by ion etching, which has been shown to cause carbide formation on organically contaminated silicon surfaces.[390] Clearly, while AES can provide invaluable information on the composition and nature of contamination layers on surfaces, it must always be applied with care, and the results treated with caution.[397]

4.2. Composition-Depth Profiling in Thin Films

Methods for determining the depth distribution of elements or species within a specimen may be grouped under two headings: Non-destructive and destructive composition-depth profiling.

Two methods of non-destructive composition-depth profiling are available to the Auger spectroscopist: the variation of electron inelastic mean free path with electron energy and the variation of sampling depth with emission angle. Both methods are described in detail below.

In AES the simplest way of obtaining information on the depth distribution of elements, within several nanometers of the surface, is to compare the relative intensities of widely different Auger transitions of the same element. Elements that lend themselves to study in this way are, for example, silicon and aluminum (with Auger transitions in both the low-energy, 50–100 eV, and high-energy, > 1000 eV, regimes) and the first row transition metals.[320] Owing to problems associated with the matrix dependence of inelastic mean free path,[35] electron backscattering,[128] and possible variations in line shape for the low-energy transitions (see Section 2), such a determination will usually lead to only an approximate figure for variation in elemental concentration with depth.[320] An alternative means of nondestructive profiling, again limited to several nanometers of the specimen surface, is to utilize the effective variation in sampling depth with Auger electron emission angle. This method has been discussed in the literature,[4] but several fundamental difficulties arise in its application to "real" substrates (such as the variation of primary-beam attenuation with sample angle, the complex variation of take-off angle with sample angle for a CMA analyzer,[4] and the intrinsic surface roughness of technological specimens), hence its use is restricted to well-characterized, ideal specimens and spectrometers. Both of the above methods of nondestructive composition-depth profiling have been used with considerable success in surface analytical studies using XPS, however, where fewer interpretational problems arise.[50]

Whenever more-accurate information on the depth distribution of elements within a specimen is required or whenever such information is required to depths exceeding 1 or 2 nm, it is necessary to resort to a destructive composition-depth profiling technique. Two methods are in common use in AES: sputter-depth profiling and mechanical sectioning. Of these, only the former is applicable to thin-film (< 1 μm) analysis. Mechanical sectioning is of value only in the analysis of thick films and coatings ($\gg 1$ μm), and will therefore be discussed in Section 4.4.

Sputter-depth profiling has already been introduced (Chapter 3), and the dependence of depth resolution on sputter-profiling conditions has been discussed in detail in Section 1.3.2 of the present chapter. Excellent reviews and publications on ion sputtering[45,46,398–405] and various aspects of its application to surface analysis by AES (such as depth resolution,[37–53] preferential sputtering in alloys,[236–256] and ion-induced reduction of oxides and decomposition of complex salts[295–305]) are available. From these publications, it is possible to draw the following general conclusions, as regards the application of sputter-depth profiling to materials characterization by AES:

1. Sputter yields increase with incident ion energy. This fact, combined with the experimental generalization that total ion currents of commercially available ion sources generally increase with ion energy, means that ion-etch rates increase with increasing ion energy.

2. Although the variation of sputter yields with atomic number is reasonably well understood from a theoretical point of view, our understanding of the matrix dependence of sputter yields, particularly in chemically complex or dilute substrates, is fairly poor. Overcoming ion-induced artifacts (preferential sputtering, decomposition, and reduction) and converting ion dose to a depth scale are therefore best attempted on an empirical basis.

3. Many of the deleterious ion-induced artifacts that concern Auger spectroscopists (loss of depth resolution due to atomic mixing and surface roughness) generally become more important at higher incident ion energies. Some artifacts, however (such as ion-induced reduction of oxides), have been shown to become more important at lower ion energies. Ion energies in practice range from 1 to 5 keV.

4. It may be necessary to sacrifice ion-current density (thereby reducing ion-etch rates) in order to maintain a uniform ion current over the area of analysis. Typically, the ion-etch crater should be Gaussian or rectangular in profile, with a width considerably greater than the electron-beam diameter. Etch rates of 1–10 nm/min are common in AES.

5. Owing to the increasing likelihood of surface-roughening effects at high ion doses and the long periods of time necessary to produce composition depth profiles to depths exceeding several microns with most conventional ion sources, the provision of profiles in the depth range 1 μm–1 mm is best accomplished by means of mechanical sectioning (see Section 4.4).

For many practical applications, composition-depth profiles can be obtained in a very short space of time by multiplexing the spectrometer. Using this approach, the Auger features (energy range) of interest in the spectrum are first decided upon, and the spectrometer is programmed to monitor only those regions of the spectrum during a sequential sputter profile. Although mechanical/electrical methods of multiplexing are available, the added flexibility of a dedicated minicomputer allows complex routines to be programmed into the spectrometer prior to execution. The resultant data are then displayed either in their raw form or in some processed form (such as peak-to-peak height versus ion-etch time), as required. The advantages of such multiplexing procedures are obvious, the major disadvantages are:

1. The inability, except in large and expensive computer-based multiplexing systems, to retrieve the original Auger line shape for subsequent data processing. Detailed line-shape analysis in AES sputter-depth profiling has been shown to be rewarding on a number of occasions.[406,407]

2. The possibility of overlooking the presence of unexpected elements, which have characteristic Auger features in regions of the spectrum that are not being monitored.

3. The possibility, if ion-etching is continued during Auger analysis (i.e., spectrum acquisition during continuous sputtering, rather than a sequential mode of operation), of the contribution of ion-induced Auger electron emission to the spectrum (123 and Section 2.1.4).

4.3. *In Situ* Preparation and Characterization of Thin Metal Films

In recent years, diffusion phenomena in thin films have attracted attention because of the demand for reliable interconnections for complex solid-state microelectronic circuits. The metal systems used are required to be compatible with diverse fabrication processes, such as heat treatments and mechanical stressing, and at the same time possess the proper electrical characteristics, such as ohmicity or Schottky characteristics. High reliability requirements, as well, demand a second or multilayer metallization system as an intermediary between the active semiconductor substrate and the final connector. In addition, there has been an ever-present, if less demanding, need for an understanding of thin-film phenomena in less technologically advanced fields such as adhesion, corrosion, contamination, and wear. Although many of the films and coatings covered by these topics properly belong in the subsequent sections on thick films, their introduction is relevant at this point, if only because much of our fundamental knowledge and understanding of their behavior has come from studies of model systems using thin films. The great majority of such studies have been made on evaporated films, the evaporation often being carried out *in situ*, such that the need for exposure to air (and any subsequent contamination or oxidation of the film) is removed.

AES provides apparently simple procedures for studying the growth modes of vacuum-deposited thin films. Mistakes of interpretation can occur, however, if certain factors are not carefully considered:

1. Possible changes in the sticking probability at certain coverages.

2. Dissolution of the deposited material.

3. Anomalous variations in the parameters determining the emission and attenuation of Auger electrons.

4. The possibility of growth modes in which multiple layers form simultaneously.
5. Kinetic effects due to agglomeration.
6. The effect of adsorbed impurities on the growth mode.

AES has actually been used, with some success, in studies of these various factors and their relative importance in selected systems.[109,110,408–416] In addition to the obvious fundamental studies of epitaxy and thin-film growth mode,[412–414,416] their influence on Auger intensities and line shapes,[410,412,415] and the influence of overlayer/substrate interactions,[109,110] AES has been shown to be a powerful tool in the characterization of impurity effects[409] and, in particular, segregation interdiffusion and interfacial reactions in thin films,[408,417,429] when used in conjunction with ion etching, to provide a composition-depth profile through the film and film–substrate interface. Studies of iron diffusion into nickel films, for instance, have revealed similar activation energies for bulk and grain-boundary diffusion in this system.[417] Independent studies of chromium diffusion into platinum thin films by AES without the use of sputter-depth profiling have yielded a very similar value for the activation energy for diffusion in this system,[418] whereas depth profiles of silver–gold thin films have revealed a marked difference in bulk and grain-boundary activation energies for diffusion.[419] Interfacial diffusion in thin films has also been studied by AES sputter-depth profiling in the case of platinum on gold,[408] aluminium on silicon,[420] gold on nickel,[421] and tin on molybdenum on copper.[422] Interdiffusion in such systems is important because of the frequent use of metallizations (often Pt–Au, Al–Si, Au–Ni, and Pt–Cr) in semiconductor technology, and much of the work in this area has been generated through a need to improve our understanding of metal–metal, metal–insulator, and metal–semiconductor interfaces. NiCr films, in particular, have received much attention, and it has been possible to correlate film properties such as contact resistance and temperature coefficient of resistance with the AES sputter-depth profiles.[423,424] Films deliberately contaminated with oxygen were shown to contain separate metallic nickel and chromium oxide phases.[423] Gross changes in the depth profile of an as-deposited NiCr film may also be produced by subsequent treatments; annealing at 450°C has been shown to result in preferential chromium oxidation with the consequent migration of nickel toward the film–substrate interface.[425] Other interface reactions, such as reduction of silica and alumina substrates (resulting in the formation of elemental silicon and aluminum) by electron-beam-evaporated titanium films[426] and platinum silicide formation at the interface of CVD deposited platinum films on silicon,[427] have also been revealed by AES sputter-depth profiling measurements. As an aside, the deleterious effects of electron-beam irradiation during AES profiles of Au/Ag films have also been noted, and high local heating effects (up to 700°C) have been postulated in an attempt to explain the experimentally observed dependence of the profile on primary-beam power.[428]

A comparison of the interface widths produced in ion-plated and vacuum-evaporated thin films, investigated by AES sputter-depth profiling, has shown that the interface width in an overlayer–substrate system exhibiting good solid solubility (copper on nickel) is independent of the method of deposition, whereas the interface width in an overlayer–substrate system exhibiting poor solid solubility (silver on nickel) is significantly greater in ion-plated films than in vacuum-evaporated films.[429] These results are used to explain the excellent adhesion properties of ion-plated films in terms of a graded interface produced between film and substrate due to ion implantation, atomic mixing, and ion-induced diffusion.

There are numerous other examples of AES sputter-depth profiling in the literature, and the technique has contributed a great deal to our understanding of thin-film phenomena, interface reactions, diffusion, epitaxy, and adhesion. The technique has limitations, however, in the fact that it produces gross physical changes in solid surfaces at high ion doses, leading to the formation of macroscopically rough surfaces (cone formation) and a consequent loss of depth resolution. At low ion doses (shallow depths), the depth resolution is generally limited by the Auger electron escape depth and atomic mixing (cascade and knock-on effects) within the top 2–5 nm of the surface. At high ion doses, however, various effects such as surface diffusion, the angular and material dependence of the sputter yield, and selective sputtering caused by protection of areas of the surface by low sputter-yield materials, leads to the formation of cones on many materials. As the direction of the cone axis and ion beam are observed to be the same,[376] it is possible to minimize this effect by either varying the ion-target incidence angle during analysis or by using two ion guns symmetrically inclined about the surface normal.[430] In general, however, the process of sputter etching leads to prohibitive loss of depth resolution, and is slow and wasteful of instrument time for film thicknesses greater than 1–2 μm. At these depths, recourse must be taken to mechanical-sectioning techniques such as angle lapping and ball cratering.

4.4. Composition-Depth Profiling in Thick Films

Technological applications of thick films and coatings range from corrosion-protective chromate coatings on aluminum[431] to anodic electrodeposition of paint for coil-coating galvanized steel,[432] and from dc-bias RF-sputtered TiC films on steel substrates[433] to phosphate–fluoride conversion coatings on titanium alloy substrates.[434] Surface modifications to depths exceeding 1 μm are also easily accomplished by anodization (Section 3.2.2.), ion implantation,[435] or laser surface melting.[436,437] All of these types and aspects of thick-film coatings have been investigated by AES.[431−437] In many cases, composition depth profiling through thick films may be accomplished satisfactorily by ion sputtering in conjunction with AES analysis. Such analyses of chromate-conversion coatings on aluminum, for instance, have revealed that although aluminum dissolution into the conversion bath increases with increasing immersion time, the oxide coating (characterized as a layer of hydrated chromium oxide over a layer of aluminum chromium oxide) rapidly reaches a limiting thickness of 70 nm after only 30 sec immersion.[431]

Auger sputter-depth profiles of galvanized steel surfaces following surface treatments with proprietary chrome rinse pretreatment products have revealed major differences in oxide thickness and elemental composition of the surface films depending on the type of treatment,[432] the deposition rate and adhesion of electrodeposited coatings during product manufacture.[432]

Sputter profiles to much greater depths, such as through 600-nm RF sputtering of TiC films on steel[433] and several hundred nanometers of ion-implanted steel surfaces,[435] have also been accomplished with some degree of success. Problems due to preferential sputtering and ion-induced topography become severe at such depths, however, and conclusions arising from such studies must be treated with caution.

In a comprehensive study of the relative merits of sputter-depth profiling and mechanical sectioning, Lea and Seah have been able to delineate certain preferential zones of applicability of sectioning and sputtering for general cases, taking into account the original surface roughness, the roughness of the mechanically sectional (lapped) surface, and the depth of analysis.[44] They conclude, for example, that mechanical sectioning is preferable to sputtering profiling for analysis depths exceeding ~200 nm if surface finishes of 10 nm or better can be achieved by lapping, and is preferable for *all* depths if the initial surface roughness is significantly in excess of the surface roughness that can be achieved by lapping.[44] This will be the case for many samples of technological origin, in particular, corrosion films and poorly applied coatings, when AES sputter-depth profiles are of limited use if optimum depth resolution is required.

There are two generally accepted methods of mechanical sectioning, both involving lapping or polishing to achieve the final surface finish and both of which have only recently been applied to depth profiling with AES.[44,57,438−441] In angle lapping (or taper sectioning) a taper is polished in the surface and the angle-lapped region exposes the entire depth to be

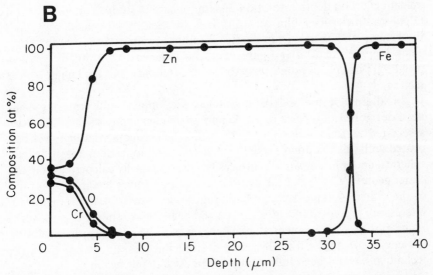

Figure 6.13. Typical ball-cratered composition-depth profile through a 33–μm electrode-posited zinc coating on mild steel. (**A**) Scanning electron micrograph of the crater generated by a 30-mm ball. Total crater diameter 2.41 mm. (**B**) Composition-depth profile through the coating, generated by performing point analyses down the side of the crater wall. (Reproduced with permission from Reference 57, © Wiley-Heyden Ltd.)

analyzed. For depths in the range 1–10 μm it is necessary to produce angles α in the range 0.1–1°. The technique has the advantage of providing a precise measurement of the depth scale and a well-defined depth resolution. On the assumption of a perfectly flat finish, the depth ΔZ analyzed using an electron beam of diameter b is simply given by

$$\Delta Z = b \tan \alpha \qquad (6.35)$$

Although the technique is useful, the provision of an angle less than 1° is difficult and tedious; in addition, it is limited to flat surfaces. An excellent example of the use of the technique and the powerful way that AES analyses can complement EPMA analyses of the same specimen are provided by a detailed study of enamel–metal interfaces in an attempt to elucidate the cause of poor enamel–metal adhesion.[440] In these studies, the polishing angle was 2°, providing a depth scale of 3.5 μm for each 100 μm across the polished surface. Using a 5-μm-diameter primary beam, it was then possible to achieve a depth resolution of 350 nm, regardless of the original coating thickness, by recording spectra at intervals of 10 μm across the taper section.[440] It was necessary to use ion sputtering in order to remove surface contamination caused by the angle lapping, but this would have an insignificant effect on depth resolution. In a more-careful study of an ideal system (a 5.4-μm-thick silver film on iron). Lea and Seah[44] have achieved depth resolutions of 150 nm using sputter profiling, compared to 105 nm using angle lapping. The latter figure could be reduced to 80 nm by sputter-profiling through a point just above the interface on the angle-lapped section.[44]

An alternative method of mechanical sectioning—ball cratering—has been developed only very recently, and overcomes many of the difficulties associated with angle lapping.[57,44] In this technique a rotating steel ball coated with fine diamond paste (0.1–1 μm) is used to fashion a well-defined spherical crater in the sample surface. The depth is easily calculated from the crater geometry,[50,57] and the depth resolution improves toward the bottom of the crater. A typical crater and Auger line scans across the region of interest are presented in Fig. 6.13. Lea and Seah[44] have shown that the technique possesses the potential of giving extremely good depth resolution (40 nm in the Ag/Fe film discussed above); but, despite its potential, the technique has not been fully exploited for AES depth profiling.

5. AES APPLIED TO INTERNAL INTERFACE ANALYSIS

While the combinations of AES and sputter-depth profiling or AES and mechanical sectioning have provided a great deal of analytical information concerning interfaces in both thin and thick films and coatings, in both

fundamental research studies and technological trouble-shooting environments (see preceding sections), the analysis of truly "internal" interfaces
(those that cannot easily be exposed by either ion sputtering or mechanical
sectioning) is best accomplished by cleaving the specimen in a controlled
environment, at, or close to, the interface region of interest. A controlled
environment is necessary in order to reduce the possibility of surface contamination (or of contamination-induced segregation) following cleavage,
and generally either takes the form of an inert atmosphere (argon or nitrogen)
in a glove-box or preparation vessel (preferably vacuum-interlocked with the
spectrometer in order to eliminate the need for any atmospheric exposure
following cleavage), or, ideally, is obtained by *in situ* fracture within a UHV
chamber of the spectrometer.

AES analysis of interfaces in solid–fluid systems (e.g., a lubricated surface)
is relatively straightforward, although certain interpretational problems may
arise through the application of organic solvents to remove all traces of the
fluid phase before analysis (Section 5.3).

AES analysis of interfaces in solid–solid systems (e.g., grain boundaries)
may be accomplished by either tensile or brittle fracture (preferably carried
out in a UHV environment), or, in studies of adhesive failure, simple delamination methods may be used. AES studies of the locus and cause of
failure of adhesive joints represent a major field of application of the technique, and are discussed in Section 5.2. *In situ* fracture studies of solid specimens, however, form one of the largest fields of application of AES, for
which it is unrivaled in its analytical capabilities.[150] These are discussed in
detail below.

5.1. *In Situ* Fracture Studies

In today's sophisticated, highly competitive technology, our ability to supply
improved structural materials in a timely and affordable manner is an
important factor in determining the performance, cost, and reliability of
future high-performance machines and structures. Despite the significant
inroads made in the design, processing, manufacture, and utilization of
organic and metal matrix composite materials and oxide-based materials
(e.g., high-performance cements[442]), the vast majority of structural materials
employed in both military and industrial applications are derivatives of
ferrous, aluminum, titanium, nickel, and cobalt base alloys, and it is likely
that this situation will remain substantially unchanged in the foreseeable
future.[443]

One of the major technical issues in these alloy systems is related to their
crack-tolerant behavior. The major crack-propagation terms are summarized
in Table 6.8.

A primary prerequisite for the systematic study of cracking behavior in

Table 6.8. Definitions of Terms Relating to Materials Failure

1. *Fracture Toughness*

A generic term for measures of resistance to crack extension (often restricted to results of fracture mechanics tests).

2. *Fatigue*

Process of progressive, localized, permanent structural change occurring in a material subjected to fluctuational stresses and strains, which may culminate in cracking or complete fracture.

3. *Corrosion Fatigue*

Fatigue aggravated by corrosion (see Section 3.2.3).

4. *Creep Fatigue*

Joint action of fatigue and creep. Creep is the time-dependent increase in strain in a solid resulting from the application of force.

5. *Stress Corrosion Cracking*

A cracking process induced by the simultaneous action of a corroding species and sustained tensile stress.

6. *Sustained Load Cracking*

Time-dependent crack propagation under load in the absence of creep or corrosion induced cracking.

7. *Fretting Fatigue*

Fatigue caused or aggravated by slight oscillatory slip (rubbing) between solid surfaces held in contact by a normal force.

8. *Hydrogen Embrittlement*

Cracking or severe loss of ductility caused by the presence of hydrogen.

9. *Temper Embrittlement*

A tendency for intergranular brittle fracture, which increases with holding time at elevated temperature.

---

materials development is the availability of valid laboratory test methods and associated analytical facilities. Only since the introduction of AES has it been possible to study the elemental composition and chemistry at the loci of failure in crack-propagation studies. Metallurgical phenomena involving segregation at internal interfaces (such as weld reheat cracking, intergranular fatigue, intergranular stress corrosion cracking, and intergranular hydrogen embrittlement) are particularly amenable to study by AES, and there are a large number of representative examples in the literature.[444-479] The most researched of the grain-boundary segregation-related problems is that of temper brittleness in low-alloy steels. Heat treatments in the temperature range 350–550°C have been known to cause certain steels to lose ductility in applications below 150°C, and it was postulated as early as 1948 that this was due to impurity segregation to grain boundaries,[445] but this was not conclusively proved until the advent of AES.[222,444] Since then AES has been an indispensable technique for the study of intergranular embrittlement. There are several useful reviews of the application of analytical techniques to the study of interfacial segregation in materials.[149-151,446]

5.1.1. Experimental Methods to Study Internal Interfaces

In general, measurements of grain-boundary segregation by AES are carried out by fracturing along the grain boundary, *in situ*, in a UHV chamber attached to the spectrometer. Depending on the metallurgical problem under study and the behavior of the material, the fracture may be performed at room temperature, on a liquid-nitrogen-cooled stage,[153] on a hot tensile stage,[447] or in a hydrogen ambient.[448] In most instruments, the fracture is produced by indirect impact using a hammer and anvil arrangement. Fractures have been produced in tension, however, via an external extensometer operating through a bellows.[153] With such an arrangement, it is possible to examine more ductile specimens, which would bend under impact. Add-on UHV fracture stages are available commercially, which allow both faces of the fracture surface to be studied *in situ* without the need for air exposure.[449]

In most cases, specimens for fracture must be specially machined to certain dimensions in order to fit into the fracture stage. In addition, fracture at a specific point in the specimen is encouraged by notching or grooving prior to fracture. Problems obviously arise if a specimen that has already undergone cracking, or is unusually weak for other reasons, is to be examined. In this case great care must be exercised in sample preparation.

Once a satisfactory fracture has been obtained, suitable areas of the fracture surface are then imaged in the spectrometer after the manner of an SEM, allowing areas of interest to be selected so that point quantitative analyses can be made of the element segregation levels.

A major limitation of the fracture method for internal interface studies is the necessity of exposing the interface of interest by the fracture method employed. This limits the technique to systems in which the cohesion at the interface is markedly less than the cohesion elsewhere, so that separation at the interface can be obtained. (Studies of adhesive, or interphase, interfaces as opposed to cohesive, or internal, interfaces are discussed in the following section, but similar conditions apply.) This factor imposes an important constraint on the evolution of surface-analysis techniques suitable for the examination of internal interfaces. Consequently, the high-spatial-resolution capabilities of AES are a distinct advantage over other forms of surface analysis such as XPS in fracture studies. Thus, in early fracture studies,[223,444,451] when relatively large electron-beam sizes were employed, a large proportion of the fracture was required to be intergranular in order to make meaningful estimations of grain-boundary compositions. With the advent of the high-resolution SAM, however, small areas of a fracture surface, that are intergranular in nature, may be selected for analysis. An ever-present danger in these types of studies, however, is the possibility of choosing a nonrepresentative region of the fracture surfaces, which may have an unusually anomalous surface composition (e.g., impurity inclusions or grain-boundary precipitates) and will therefore lead to incorrect conclusions.

Fracture surface studies are generally combined with *in situ* sputter-depth profiling in order to determine the variation of elemental composition from the free grain boundaries into the bulk. Interpretational problems can arise, however, due to the presence of preferential sputtering effects in alloy or composite materials (see Section 2.2.5) which may either mark segregation phenomena or, which is more likely, give rise to *apparent* variations in composition with depth that are, in fact, artifacts of the sputtering process.[150] Similarly, fracture studies carried out on very reactive materials (e.g., titanium) or in non-UHV conditions are prone to contamination contributions to the Auger electron spectrum, and great care must therefore be taken in both the experimental procedure and data interpretation. Contamination problems can be minimized in UHV systems if specimens are thoroughly degreased prior to insertion into the fracture stage, the entire vacuum system is baked out prior to fracture, and surfaces are analyzed as quickly as possible following fracture.

5.1.2. Theoretical Basis for Grain-Boundary Segregation

The particular application of quantitative AES to grain-boundary segregation on fracture surfaces has been treated in detail by Marcus et al.[223] and Seah and co-workers,[149-151] although the principles are identical to those pertaining to segregation studies on clean metal surfaces in UHV (see Section 2.2.3). Unlike free-surface segregation studies, however, the majority of fracture studies have been made upon technologically relevant (and therefore generally available) materials (such as standard commercial steels) as opposed to model or ideal systems (e.g., deliberately doped pure elements).

The latter, although possibly more interesting from a scientific point of view, would entail complex sample preparation techniques (see Section 2.2.3.) and may not be directly relevant to technological problems. In addition, fracture surface studies do not lend themselves to the UHV sample preparation techniques (notably surface cleaning prior to segregation) that are crucial in order to be able to carry out meaningful fundamental segregation experiments. Hence, although experiments to study single-component segregation to free surfaces *in vacuo* are easily designed and implemented (Section 2.2.3), multicomponent segregation in grain-boundary studies is the general rule. This observation, in fact, underscores the value of free-surface segregation experiments in attempting to understand the complex site-competition relationships, which undoubtedly determine multicomponent segregation at grain boundaries.

Because of their fundamental similarity, grain-boundary and free-surface segregation phenomena have generally been described by the same theories.[144-151] In general, however, the kinetics and extent of equilibrium segregation are accentuated at the free surface, owing to the increased degree of atomic freedom permitted by the solid–vacuum interface, hence care must be experienced when comparing quantitative AES data from these phenomena. Most of the necessary theoretical background has already been covered, therefore, in the preceding discussion on theoretical aspects of alloy surface compositions (Section 2.2.2). Seah and Hondros[153] have shown that the grain-boundary enrichment ratio (β), defined by the equation

$$\beta = \frac{X^S}{X^S(\text{sat}) \cdot X^B} \tag{6.36}$$

where X^S is the grain-boundary concentration of segregant X, $X^S(\text{sat})$ is the saturation value of X^S, and X^B is the bulk concentration of X, correlates well with the atomic solid solubility of X in many systems. Thus, solutes with low atomic solid solubility in the bulk of the matrix (solvent) (e.g., S or C in Fe) will tend to segregate to grain boundaries more readily than those with high atomic solid solubility (e.g., Cr or Ni in Fe). It must be emphasized, however, that the experimental data for the systems studied show only a very approximate correlation, and notable exceptions to this rule-of-thumb approach exist (e.g., β has been found to be much higher for Cr in γ-Fe than for Si in α-Fe, whereas the solid solubilities would indicate the reverse[153]). Similarly, comparison of experimentally measured and theoretical values of β [using equations such as (6.14) and (6.29)], although exhibiting fair agreement over a wide range of solute–solvent pairs, shows very poor agreement in certain cases (e.g., β for P in Fe is underestimated by almost an order of magnitude, while that for Cu in Au is overestimated to a similar extent[151]). Obviously, present theoretical descriptions of grain-boundary segregation may be used only on a semiquantitative basis, and care must be exercised when using them in a predictive sense.

5.1.3. AES Study of Fracture Surfaces

Temperature embrittlement in steels forms by far the largest single area of AES fracture studies. In one of the earliest applications of AES, Marcus and Palmberg[467] showed that the intergranular fracture surface of an embrittled 3.4Ni/1.9Cr/0.03Sb steel contained more amounts of Sᴀ, P, and Ni, compared to the bulk (transgranular fracture surface), and that this enriched grain-boundary layer had a thickness of the order of one monolayer by AES sputter-depth profiling. Similar studies on steels containing 300–600 ppm Sb, tempered and embrittled by heating at 750°C for 24 hours, clearly indicated antimony and nickel segregation at the grain boundaries. In addition, it was found that the segregation of these elements was synergistic and limited to certain grain-boundary facets. Phosphorus segregation, on the other hand, was found to be uniform over the fracture surface. The results were interpreted in terms of preferential segregation to some grain boundaries, and may be used to explain why austenite boundaries are embrittled preferentially to ferrite–ferrite boundaries during normal tempering of steels.[468,475] Temper embrittlement in SAE 3140 steel (1.26Ni/0.015P) has been shown to be due to a synergistic grain-boundary segregation effect involving both nickel and phosphorus. For this system, a complete thermodynamic, predictive model has been derived from the experimental data.[476] Site-specific segregation has been observed in temperature embrittlement of a 2.25 Cr/1Mn steel, where Sn, Sb, Cu, and S segregation to grain cavities, and not to the flat areas of grain boundaries, was observed. Phosphorus, on the other hand, was observed to segregate to noncavitated boundaries, and was found to be responsible for embrittlement. In addition, although cavity growth was caused by Sn, Sb, Cu, and S segregation, it was thought that cavity nucleation was initiated by P.[453] Phosphorus segregation at high temperatures in high (0.023%) and low (0.009%) phosphorus AISI 52100 steels has been shown to be related to the bulk concentration, in agreement with McLean's theory. At low temperatures, however, nonequilibrium segregation was observed.[457] Complex behavior in Mo-containing 2.25Cr/ 1 Ni low-alloy steel has been reported, involving Mo–P interactions in the bulk of low-phosphorus steels (when P segregation is retarded) and at the grain boundaries of high-phosphorus steels (when the embrittling effect of P is counteracted by Mo). In the latter case, grain-boundary segregation of Mo was found to be enhanced by the presence of P at the grain boundaries (cf. synergistic Ni–P interaction in 3140 steel[476]), but aging resulted in embrittlement due to the removal of grain-boundary Mo through the formation of carbides.[458] In a study of P and C grain-boundary segregation in Fe–P alloys, it was found that up to 0.01% bulk C content resulted in a suppression of intergranular fracture. Above this concentration, pearlite nodules were formed and toughness was found to decrease.[462] Sulfur and oxygen have been observed on the surface of graphite flakes in gray cast iron, but not on graphite nodules in ductile iron.[465] The interpretation here is that the magnesium and/or rare-earth additions to the ductile iron, in order to prevent flake formation, actually scavenge the surface-active impurities and prevent their segregation to grain boundaries. The action of tin or antimony additions, however, is not so clear. In this case, graphite nodule growth is retarded, but not through the scavenging of S, O, or C. AES analysis of the fracture surfaces showed the presence of small graphite nodules (~ 20 μm diameter) in antimony or tin-doped steel, with a high tin or antimony concentration at the graphite–iron boundary. It was found that the segregant was bound strongly to the iron, but not to the graphite, and the hypothesis is that the presence of tin or antimony at the Fe–C boundaries effectively prevents C diffusion, and therefore retards nodule growth.[466] Segregation at grain boundaries in vacuum melting and zone-refined iron alloys has also been observed by AES.[469] In the latter case, N was found to be the major segregant, while site competition between S and O was observed for the former. In this case, similar behavior was observed at both grain boundaries and free surfaces. The same is true of P and S segregation in high-alloy Cr, Ni steel, where temperature embrittlement has been observed to result in similar segregation to both grain boundaries and the free surface, supporting the assumption that the driving forces and mechanisms in both cases are equivalent.[471] Intergranular fracture in iron has been shown to be directly related to the grain-boundary concentration of S, but to be unrelated to O, N, or C concentration at grain boundaries.[209] Temper embrittlements of A533B and A508 steels used in reactor pressure vessels has also been studied by AES.[452]

Other AES studies of segregation in steels include observations of interphase segregation in cast irons,[477,478] correlation of rupture life and ductility with S and P segregation in 304 stainless steel,[454] B segregation in 316 stainless steel,[451] hydrogen embrittlement of austenitic stainless steels[460] and hot cracking of steel welds due to S, P, and Nb segregation.[460] Fundamental studies of grain-boundary segregation, yielding bulk and grain-boundary diffusivities, of Sn in Fe have also been made.[464]

AES fracture studies have also been conducted on aluminium and its alloys,[365,446] nickel and its alloys,[447,459,470,474] tungsten and tungsten carbide,[456,463,472] copper,[210] and molybdenum[461] as well as nonmetallic substrates such as glass,[455] ceramics,[480] and metal oxides.[376]

A number of studies have revealed the presence of discrete, particulate material at fracture surfaces in a range of different materials.[208,472-474] The degree to which surface diffusion, following fracture, affects the interpretation of such systems has also been discussed.[472,473] The problems associated with electron-induced phenomena during AES analysis of fracture surfaces have been investigated for a number of systems.[455,473,479] For sulfur on steel fracture surfaces, for instance, electron-induced enhancement of the sulfur signal has been observed,[473] whereas electron-induced diffusion of Na and K, away from the area of analysis, has been observed on fractured glass surfaces.[455] Clearly, such fracture studies must be carried out with great care if meaningful results are to be obtained.

5.2. Adhesion and Adhesive Failure

Many materials applications depend critically on bonding between dissimilar materials. The special case of interphase interfaces in macroscopically homogeneous solids has already been mentioned in the preceding sections on AES analysis of fracture surfaces. For atom-to-atom bonding across a planar interface, the thermodynamic work of adhesion is given simply by

$$W_{AB} = \gamma_A + \gamma_B - \gamma_{AB} \tag{6.37}$$

where γ_A, γ_B, and γ_{AB} are the surface free energies (in J m^{-2}) of the two phases and their interface, respectively. The results of a number of studies of W_{AB} have, in fact, shown a simple first-order correlation with γ_{AB} for a range of adhesive systems.[151] From the preceding discussion, it might be expected that interfacial segregants that reduce the interfacial free energy would lead to an increase in the adhesive strength of the interface. Although there has been no detailed study of such effects, the unique capabilities of AES are likely to lead to the resolution of this question in the very near future.[151]

The application of AES to studies of adhesion has recently been reviewed by Walls and Christie[481] and by Baun.[482] The latter has concluded that elemental characterization of adherends, especially when composition with depth is required, is best accomplished with AES. This statement is certainly supported by the growing number of applications of AES to both fundamental research[483-486] and technological problem solving[487-496] in the field of adhesion. An excellent example of the contribution of AES to our understanding of the fundamental principles and mechanisms of adhesion is the study of the variation of the coefficient of adhesion with the level of surface contamination in atomically clean systems.[483-485] A number of workers have shown that the presence of as little as one or two monolayers of oxide on the surface of clean metals is sufficient to reduce the adhesive bond strength by up to an order of magnitude.[483,484] Similar studies on the effect of submonolayer coverages of S, C, N, O, H, or P on iron adhesion have shown an overlayer-induced enhancement of adhesion up to a certain coverage (up to one-half a monolayer for C and N), with reduced adhesion at high coverages.[485] Such studies have a direct relevance to practical problems in modern technology. The enhanced adhesion effect of C and N on iron, for instance, may be related to the known effect of C and N on intergranular cohesion in steels,[485] while the effects of surface contamination on the thermocompression bonding properties of gold are widely appreciated in the microelectronics industry.[388,486] Other examples of AES applied to the study of metal–metal adhesion include braze failures in a Ni/Cu alloy,[487] a comparison of evaporated and ion-plated chromium on nickel and chromium on aluminium adhesion,[488] adhesive failure in electroplated nickel films,[489] and characterization of metal adherends prior to bonding.[481,482] Aluminium alloy adherends, in particular, have been the subject of much fruitful study by AES.[344] Metal–glass,[490] metal–enamel,[491,492] and metal–resin[493–496] adhesion have also been studied with AES. In the latter category, Gettings et al.[494–496] have shown that normally dry adhesive joints formed between mild steel and epoxy resin fracture near the metal/resin interface, while water-soaked joints fractured at the adhesive/iron oxide interface. The application of a silane-based primer was found to result in cohesive failure in the primer, in preference to interfacial failure.[494] In a more-detailed study, the same authors concluded that the environmental resistance of adhesive joints depended on the surface chemistry and roughness (specimen pretreatment) of the steels concerned,[495] and that the adhesion of silane-based primers is influenced by surface topography and morphology.[496] Organosilicon compounds, in the form of siloxane polymers, have, in fact, achieved notoriety in the field of adhesion due to their excellent release-agent capabilities. Siloxanes are often found to be responsible for poor adhesion or delamination phenomena in textile, polymer, and metal-finishing industries, and the unique capabilities of AES render it an ideal technique for problem solving in these areas.[387]

5.3. Tribology and the Solid–Fluid Interface

The science of tribology is frequently described by three words which are commonly associated with the less rigorous and more practical side of the engineering sciences: friction, lubrication, and wear. Tabor has defined tribology as the study of "How surfaces interact when they are placed together and what happens when they slide".[497] The special case of adhesion under static contact has already been dealt with in the preceding section; the application of AES to the study of sliding surfaces is considered in more detail in the present section.

Two solid surfaces in contact possibly touch each other over the entire macroscopic contact area owing to the finite roughness of either or both surfaces, and the load must be borne on an area that is initially, quite small.[498] Deformation of the contact areas can result in solid-state contact through surface films (eventually giving rise to strong adhesive bonding across the interface for clean surfaces), and the presence of gases, liquids, and solid films at the interface may markedly alter the tribological characteristics. In a series of review papers, Buckley has described the application of surface analytical techniques to tribological research,[499-501] and provides a number of examples of the application of AES in this field. Since the majority of fundamental studies in this area have been conducted on solid–solid interfaces, whereas most practical applications of the technique have been on solid–solid interfaces in the presence of a lubricating film, it is convenient to discuss the application of AES to solid–solid and solid–fluid tribological situations separately.

AES analyses of wear surfaces in unlubricated tribological situations have revealed in many cases, material transfer from one surface to the other.[500-504] This has been shown to be the case for metal on metal, metal on oxide,[500] refractory materials on metal,[502] refractory materials on oxide,[503] and metal on polymer systems.[504] In most cases, material transfer is detrimental to the mechanical properties of the system under study (friction[502] and wear[503] are both related to material transfer); however, it has been suggested that transfer in metal on polymer systems serves to minimize wear by acting as a lubricant.[504] A direct practical application of AES in a technologically relevant area of tribology is the particular case of copper commutators.[505-507] Brushes for sliding contacts in electrical machinery have generally been made out of carbon because of its physical properties, in particular, low wear rates, good conductivity, and high combustion temperatures. Notwithstanding these properties, brushes wear out and must be frequently replaced. The reasons for excessive, anomalous wear are not completely understood, but thin, silica contamination films have been proposed as a reason for high wear rates in one instance.[507] Interfacial reactions, leading to the production of a thin, graphitic-like interfacial layer, may be responsible for the observed low coefficient of friction in silicon carbide and diamond-based tribological situations.[503,508] AES has shown, for example, that silicon carbide surfaces are essentially graphitic following high-temperature heat treatment[508] or abrasion.[503] Studies of this kind are particularly important in view of the recent development of the ceramic-based internal combustion engine, which has been shown to perform very well with unlubricated silicon carbide.[509]

The definition of wear given by the Scientific Research Committee of the Organisation for Economic Cooperation and Development (OECD) as "the progressive loss of substance from the operating surface of a body occurring as a result of relative motion of the surface" clearly covers all of the above examples, but does not account for wear in electrical contacts. Even this special case of tribology is amenable to study by AES.[510]

The vast majority of solid surfaces in sliding contact are, in fact, lubricated (to a greater or lesser extent), and the tribological properties of such systems, except in extreme cases such as lubricant starvation, excessive load, or macroscopically rough surfaces will be determined by the properties of the fluid phase and the solid–fluid interface.

During the past few years, numerous workers have indicated the potential of surface-analytical techniques for studying the solid–fluid interface. The majority of these studies involved attempts to understand the fundamental wear processes operating in lubricated environments. Buckley[511,512] used AES to investigate changes in the surface compositions of sliding discs when under load in the presence of various gases. More recently, he has extended this work to studies of sliding surfaces in the presence of liquid lubricants, where dissociation and decomposition reactions may play a part in the wear process.[499] The influence of additives on the friction properties and wear behavior of tribological systems has also received much attention. Decomposition of lubricant additives at contact asperities, liberating solid films on the surface with inherently good lubricating properties, is thought to be an important factor in the wear mechanism.[513] Buckley has shown that addition of 0.5% of dimethyl cadmium to mineral oil results in a fourfold drop in the coefficient of friction of 302 stainless steel, and AES analysis of the wear surfaces, revealed complete coverage of the steel by cadmium.[499] Other studies of the reaction of lubricating oil additives with metal surfaces have concentrated on attempts to correlate the surface concentration and degree of penetration of the active elements in the wear scars with the lubricating properties of the additive.[514–517] Such studies have been carried out for sulfur- and chlorine-containing extreme pressure (EP) additives on steel surfaces[514] and for sulfur- and phosphorus-containing EP additives on steel surfaces,[515,516] and AES has proven itself as a powerful analytical technique, yielding unique and valuable information in every case.

6. SEMICONDUCTOR MATERIALS

Semiconductor devices such as transistors and rectifiers first appeared on the commercial market in the mid 1950s. Since that time the whole electronics field has been doubling in size every 5 years, and the sales of semiconductors has become a multibillion dollar industry.

A semiconductor is a material that conducts electricity less readily than conductive metal but more easily than an insulator. The electrical conductivity is governed by an applied voltage and reflects the material's impurity content and structure. By adding certain impurities at parts per billion (ppb) levels, pure silicon and germanium, for example, can be converted to p- or n-type semiconductors. AES has been applied successfully to a number of areas concerned with the surface behavior and manufacture of electrical devices. For example, many studies have been carried out on the semiconductor to oxide interface that is fundamental to the operation of MOS (metal-oxidized semiconductor) systems. This is especially true since integrated-circuit technology places more and more demands on MOS processing, and an understanding of the physics and chemistry of such interfaces is vitally important in terms of device electrical characteristics. Other studies have been concerned with substrate surface cleanliness before molecular beam epitaxy and detection of the concentration of dopants in various semiconducting materials.

Before discussing these aspects it must be emphasized that the Auger process that can actually occur in semiconductors should not be confused with semiconductor analysis by Auger techniques. This process is known as band-to-band Auger recombination and is important for narrow-gap semiconductors such as tellurium.[518] It is actually the inverse of band-to-band impact generation of free carriers. A model of the process has been described by Beattie and Landsberg.[519,520] The recombination involves two parallel processes, either electron–electron or hole–hole collisions, rather than the annihilation collision of electron–hole. Figure 6.14 illustrates Auger recombination. The two free carriers in the same band may collide. For example, in 6.14a, two carriers in the conduction band may collide. Carrier 1 may drop to the bound condition 1', while 2 may receive all the recombination energy and be promoted to an empty site, 2', in the same band. The process 6.14b depicts a collision of two carriers in the valence band. This example uses a "perfect" semiconductor, but semiconductors are usually flawed. This implies that other energy levels exist and that other processes can occur. For example, an Auger band-to-flaw effect may occur.

A strong postexcitation conductivity can be observed in semiconductors due to the Auger effect.[521] Radiative recombination has also been observed, producing line spectra for materials such as GaP, GaAs, and CdS.[522] The radiative recombination and Auger recombination processes are competing effects. The Auger effect is not important in materials where hole collisions dominate, and in the case of electron collisions, it only becomes important when the number of electrons is greater than 10^{20} cm^{-3}.[522]

We now turn to some of the work that has been carried out using AES in the semiconductor field. To give the reader an idea of the development of this area, the work carried out is generally presented in chronological order.

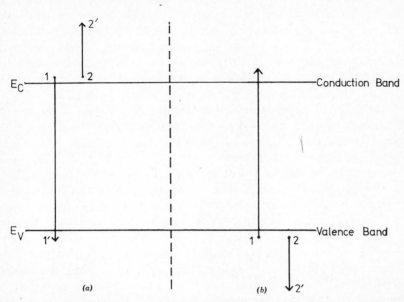

Figure 6.14. Auger recombination: (a) Two carriers in the conductive band collide with 1 dropping to the bound state 1′, while 2 may receive all the recombination energy and be promoted to an empty site in the same band; (b) collision of two carriers in the valence band.

The most prevalent of semiconductors used in the electronics industry is doped silicon. Surfaces of silicon in various forms are often encountered as dielectrics or as surface passivating layers. The material is basic to many semiconductor devices, and it has been extensively studied by AES since approximately 1970.

One of the earliest papers in this field was that of Chang,[523] who studied impurities on chemically etched n- and p-type silicon (111) and (100) surfaces by the LEED–Auger method. The samples were etched in an HNO_3, HF, HOAc, and iodine mixture, washed and then subjected to various surface experiments. The Auger spectrum was interpreted as consisting mainly of bands at 92 eV due to $L_{2,3}VV$ transitions in elemental silicon, peaks at 65 and 78 eV from SiO_2, and signals at 270 and 507 eV due to carbon and oxygen, respectively. Heating to over 1000°C was successful in removing the carbon and oxygen contaminants but peaks, due to metallic impurities (produced by diffusion) appeared at the elevated temperature. Recommendations were made for the cleaning of surfaces of silicon substrates for adsorption and epitaxial growth work.

One of the first studies of oxygen adsorption to silicon and oxidation of the element was that of Neave and Joyce.[524] The authors pointed out that AES offers considerable potential in the study of the Si–SiO$_2$ interface and, therefore, the electrical characteristic of the silicon semiconductor. By monitoring silicon and oxygen peaks it was concluded that at room temperature there is an initial rapid adsorption of oxygen followed by a rapid decrease in rate as the monolayer nears completion. At high temperatures different spectra were obtained that were considered to reflect surface oxidation.

In 1972 several papers appeared concerning AES studies of cleaning of substrates previous to deposition and/or adsorption work. For example, the production of silicon homoepitaxial films via silane pyrolysis was examined by both AES and HEED (high-energy electron diffraction[525]). In this work wafer samples were chemically treated, which included an HCl vapor-etching procedure, before examination by AES and HEED. After this pretreatment silicon was deposited by silane pyrolysis. Generally it was concluded that on "clean" silicon surfaces at about 800°C homoepitaxial growth occurs from SiH$_4$ pyrolysis and that a low-temperature substrate preparation by argon sputtering and annealing can be used to produce silicon layers free of stacking faults, growth pyramids, or crystallographic pits. Other work by Nishijuma and Murotani[526] demonstrated that silicon samples had to be heated to approximately 1200°C to remove carbon and oxygen surface contamination after chemical etching with an HNO$_3$/HF mixture. Reaction of the clean surface with various exposures of oxygen confirmed the results of previous workers with respect to the "stepwise" process of first the occurrence of chemisorption, followed by oxidation at high gas exposure. Furthermore, oxidation is enhanced at higher temperatures.

Using AES and concentration depth profiling it was shown that covering of an Si(111) surface with evaporated gold and heating at temperatures between 100 and 300°C in an oxidizing atmosphere gives an Si–Au reaction at the interface that results in the formation of an SiO$_2$ layer.[527]

Maguire and Augustus[528] discussed the importance of silicon nitride films as the diffusion mask and gate dielectric of MNS and MNOS devices and carried out an AES study of silicon oxynitride layers on the surface of silicon nitride films. Silicon substrates were treated with NH$_3$ and SiH$_4$ in a nitrogen carrier gas at 900°C. Special precautions were taken to avoid any contamination by oxygen. AES peaks at 522 and 495 eV due to oxygen were observed on both the "pure" Si and the Si$_3$N$_4$ sides of the specimen. As demonstrated by previous workers, heating over 1000°C for a few minutes was sufficient to thermally remove the oxygen from the silicon side; however, only a slight reduction in the oxygen peak was detectable on the nitride side. The authors concluded that on top of the nitride layer another layer consisting of Si$_x$O$_y$N$_2$ occurs and that this is caused by oxygen contamination of reactor gases (despite precautions taken).

Also, in 1972, some preliminary work was carried out in the detection of phosphorus in phosphorus ion-implanted silicon by AES and other techniques such as secondary-ion emission and LEED.[529,530] An implantation of a silicon (100) orientation with a 5×10^6 atoms cm^{-2} dose of phosphorus was subjected to sputtering by a 600-eV argon-ion beam in a concentration-depth-profile study. The Auger phosphorus data were used to indicate a practical detectability limit for phosphorus in silicon of $\sim 5 \times 10^{19}$ atoms cm^{-3}. The author commented that this figure is perhaps not adequate for the Auger detction of common dopants such as P, As, and B used in device fabrication (10^{17}–10^{19} atoms cm^3). Subsequently, a similar system was studied, but this time bulk-doped boron and phosphorus in silicon samples were analyzed by AES.[531] Using Irvin's curves[532] bulk resistivity measurements were related to concentration of dopants (atoms cm^{-3}) for calibration of the Auger peak ratios, P(120 eV)/Si(1619 eV) and B(179 eV)/Si(1619 eV). The specimens were continuously ion sputtered during Auger work to ensure that the surfaces were clean.

Boron and phosphorus to silicon peak ratios were plotted against dopant concentration. From these data a bulk detectability of $\sim 8 \times 10^{18}$ atoms cm^{-3} for phosphorus and boron in silicon was derived (compare $\sim 5 \times 10^{19}$ atoms cm^{-3} for concentration-depth profile study). The results were compared with other work on surface reaction of PH$_3$ with Si(111).[533]

In 1973 Chou and co-workers[534] sounded a word of caution with respect to Auger studies of small concentrations of certain species in SiO$_2$ films. This study was concerned with monitoring Si, O, and Cl Auger signals during ion etching of films of SiO$_2$ grown thermally in O$_2$ in the presence of HCl or Cl$_2$. Extended exposure to the primary-electron beam results in a mobile ionized chlorine species "piling-up" at the SiO$_2$–vacuum interface where it is desorped by impinging electrons.

Electron-beam bombardment effects were also referred to by Smith et al.[535] in an AES study of oxygen-to-silicon ratio in spin-on glass films on silicon wafers. The primary effect of a high-intensity beam is to dissociate SiO$_2$, leading potentially to desorption of oxygen and "enrichment" of silicon.

Also in 1973 the early stages of nitridation of silicon wafers (111), (311), and (100) was studied by LEED and AES.[536] Most of the experiments involved pressures of 10^{-7}–10^{-5} torr of NH$_3$ and silicon temperatures in the range of 800–1100°C. Apparently, growth of nitride on the (111) and (311) orientations is impurity induced. Subsequent growth of layers is then dominated by intralayer bonding, so that the difference in substrate symmetries has little effect.

A particularly interesting study by Chang et al.[537] was carried out on phosphorus concentration-depth profiles in doped SiO$_2$. In this work silicon single crystals were used as substrates for deposition of oxide films by reaction with PH_3 and SiH_4 at elevated temperatures. Depth profiles were obtained by either taking Auger data while simultaneously sputtering by an argon etching beam, or etching the oxide in the shape of a ramp and measuring the surface phosphorus concentration along the ramp. Elemental silicon and phosphorus Auger peaks at 107 and 120 eV, respectively, were measured by sweeping the narrow 100–130 eV energy range in order to characterize the $Si–SiO_2$ interface. As with other studies it was found that deleterious effects were caused by the incident electron beam. In this case outdiffusion of P was corrected for by taking measurements over some 12 sec.

A phosphorus concentration-depth profile (together with the Auger signal from elemental Si) was produced by ion sputtering. There appears to be a clearly P-rich layer close to the $Si–SiO_2$ interface (in the oxide film). Also adjacent to the P-rich layer is a region where there is a P-depleted zone. Experiments were carried out in an attempt to confirm that formation of the P-rich layer was not an artifact caused by either the incident-electron beam or by ion etching. It was concluded that the phosphorus-rich zone is caused by rapid reaction of phosphine at the silicon surface. When the deposition is initiated, the oxide film is very thin (< 20 Å) so that phosphine can rapidly diffuse to the silicon surface and react accordingly. The P-rich and P-depleted regions have a total thickness of ~ 100–200 Å and are formed within the first minute of deposition. These results demonstrate the power of the Auger technique in studying the interface chemistry of films and in electronic devices.

In 1975 Van Bommel et al.[538] carried out a LEED and AES study of the semiconducting material silicon carbide. Experiments with the SiC(0001) crystal surfaces show that on heat treatment these surfaces are easily covered with a layer of graphite by evaporation of silicon.

AES and concentration-depth profiling has been used in the examination of interface regions in MOS and MNOS device structures.[539] In this treatment chemical interface profiles, instrumental limitations, and SiO–Si interface morphology models were discussed. With regard to the MOS structure it was found that for the $Al–SiO_2$ interface there is a presence of an intermingled metallic Si and Al_2O_3 zone over some 200 Å. This phenomenon was attributed to the solid-state reaction

$$4Al + 3SiO_2 \rightarrow 3Si + 2Al_2O_3$$

A typical chemical-depth profile between SiO_2 and an Si substrate from thermally oxidized Si is shown in Fig. 6.15. In this diagram the various Auger transitions for different species are as follows:

	KLL		LVV
Si elemental,	1618 eV	Si elemental,	92 eV
Si in SiO_2,	1611 eV	Si in SiO_2,	78 eV
O,	502 eV		

Clearly these curves indicate that the Si atoms change their chemical state in an orderly fashion over a spatial extent of approximately 80 Å, while

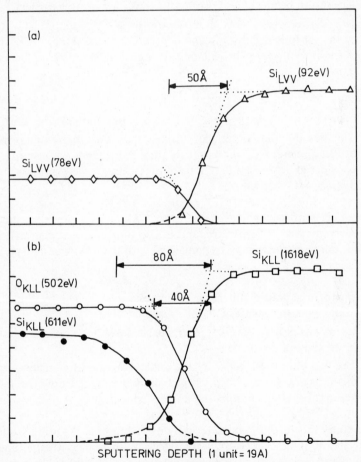

Figure 6.15.   Chemical depth profile of the interface between SiO_2 and the silicon substrate. The *LVV* transitions are shown in *a*, and the *KLL* transitions are shown in *b*. The transitions are separated and taken from real time plots of the actual Auger spectra.[539] (Reprinted with permission of American Physical Society, New York.)

the oxygen atoms stay in the same state but decrease in abundance to zero over roughly 40 Å. The Si LVV profile shows that the change from Si–O bonding to Si–Si bonding takes place within an interface layer of 50 Å.

With regard to instrumental limitations the following points are stressed (also, see previous sections in chapter):

1. The incident beam causes dissociation with subsequent desorption of oxygen.

2. Argon-ion beams may cause "knock-on" damage (see earlier).

3. In order to obtain "real" interface widths, ΔW, account must be taken of the various electron escape depths involved, viz, Si LVV versus KLL and interface "broadening" caused by the ion beam. Thus, the equation

$$\Delta W \approx \Delta W_M - \lambda - 2r_i \tag{6.38}$$

   is used, where λ is the electron escape depth and r_i is the ion range.

4. The possibility of interface "broadening" due to nonuniform ion fluxes, preferential sputtering, or inhomogeneous sample must be considered. It was felt that none of these factors was significant in the Si–SiO$_2$MO study.

The study outlined above gives strong evidence of an electronically inhomogeneous SiO$_2$–Si interface, and it also indicates that charge associated with inhomogeneity may be connected with the following morphology model:

1. There is a natural interface roughness, determined by the nature of the oxidation process.

2. Inclusions of Si in the SiO$_2$ matrix exist close to the interface, but not in the bulk of the oxide.

3. The two phases, Si and SiO$_2$, are joined by connective regions where Si atoms are coordinated by both Si and O atoms. Dangling Si bonds and nonbridging O atoms are associated with the connective regions.

This study was further refined recently by Helms et al.[540] using "state-of-the-art" thermally grown Si–SiO$_2$ interfaces. With improved experimental technique and consideration of electron escape depth and ion "broadening" processes, the width of the interface is now set at about 20 Å. Interestingly, it is thought that this width is probably due to an undulating interface, the period of which is less than 100 μm (Fig. 6.16).

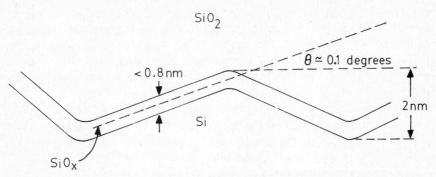

Figure 6.16.   Model for a Si–SiO$_2$ interface.[540] (Reprinted with permission of Pergamon Press, New York.)

Two other important silicon interfaces examined recently were the hydrogenated amorphous Si system[541] and the silicon–transition-metal silicide interface.[542] The former work was stimulated by interest in Schottky-barrier amorphous silicon solar cells. As with other devices the presence of a thin oxide layer is critical to the operation of the device. Experimentally, a-Si:H films were deposited from a dc cathode glow discharge in silane onto polished niobium at a temperature of about 350°C. After cleaning and annealing, AES work on oxygen adsorption was carried out. The measurements show that with exposure to oxygen the work function first increases due to the formation of a dipole layer on the surface, and then decreases because of penetration of oxygen below the surface.

In the second study thin-film specimens of silicides were prepared by depositing films of the appropriate metals onto clean Si(100) substrates. Characteristic Si $L_{2,3}VV$ Auger spectra for the silicides were compared with spectra obtained of thin silicide layers at the metal–Si interface. As with many other studies damage induced by a sputtering beam was encountered.

7.   SUMMARY AND CONCLUSIONS

From humble beginnings just over a decade ago, AES has emerged as one of the most powerful analytical techniques for the characterization of solid materials, offering practical limits to depth and spatial resolution that render it invaluable and unrivaled in applications to metals and oxides, thin films and coatings, internal interfaces, and fracture surfaces. It has found use not only in fundamental scientific research on such surfaces (where the ever-increasing volume of scientific papers testify to its popularity), but also, and perhaps most importantly from the point of view of the materials technologist or engineer, in problem solving, trouble shooting, and materials development in industrial environments. So profound has its impact been that it has become a standard analytical technique, ranking alongside the electron microprobe or infrared spectrophotometer, in many areas of industry and technology.

The present text cannot pretend to have completely covered the vast territory that is encompassed by applications of AES in materials analysis, but it is hoped that the very diversity of applications and variety of materials that have been discussed, and the numerous relevant references, some of which provide excellent starting points in the form of reviews for further literature searching in particular areas, will enable interested readers to make even better use of Auger electron spectroscopy in the years to come.

REFERENCES

1. L. A. Harris, *J. Appl. Phys.*, **39**, 1428 (1968).
2. D. T. Hawkins, *Auger Electron Spectroscopy—A Bibliography: 1925–1975*, Plenum, New York, 1977.
3. See, for example, D. M. Hercules, *Anal. Chem.*, **48**, 294R (1976); P. F. Kane and G. B. Larrabee, *Anal. Chem.*, **49**, 221R (1977); *Anal. Chem.*, **51**, 308R (1979); G. B. Larrabee and T. J. Shaeffner, *Anal. Chem.*, **53**, 163R (1981); N. H. Turner and R. J. Colton, *Anal. Chem.*, **54**, 293R (1982).
4. P. Braun, F. Rüdenauer, and F. P. Viehböck, *Adv. Electron. Electron Phys.*, **57**, 231 (1981).
5. Vacuum Generators Co., East Grinstead, U.K.: HB100 data sheet.
6. C. C. Chang, *J. Vac. Sci. Technol.*, **18**, 276 (1981); M. Prutton. R. Browning, M. M. El Gomati, and D. Peacock, *Vacuum*, **32**, 351 (1982); M. M. El Gomati and M. Prutton, *Surf. Sci.*, **72**, 485 (1978).
7. Physical Electronics PHI Data Sheet 1059, 7-80, Eden Prairie, Minnesota, 1980.
8. A. van Oostram, *Surf. Sci.*, **89**, 615 (1979).
9. See Chapter 3 of this volume. For an introduction to various experimental and theoretical aspects of ion and electron induced damage, see R. Kelly, *Surf. Sci.*, **90**, 280 (1979).
10. C. G. Pantano and T. E. Madey, *Appl. Surf. Sci.*, **7**, 115 (1981).
11. L. Prittaway, *Brit. J. Appl. Phys.*, **15**, 967 (1964).
12. D. Lichtman, *Mater. Sci. Eng.*, **53**, 73 (1982).
13. P. H. Holloway, T. E. Madey, C. T. Campbell, R. R. Rye, and J. E. Houston, *Surf. Sci.*, **88**, 121 (1979).
14. J. P. Coad, M. Gettings, and J. C. Riviere, *Discuss. Faraday Soc.*, **60**, 269 (1975).
15. A. G. Knapp and J. R. Hughes, Proceedings of the 7th International Vacuum Congress and 3rd International Conference on Solid Surfaces, Vienna 1977, p. 2161.

16. F. Ohuchi, M. Ogino, P. H. Holloway, and C. G. Pantano, *Surf. Interface Anal.*, **2**, 85 (1980).

17. C. G. Pantano, D. B. Dove, and G. Y. Onoda, *J. Vac. Sci. Technol.*, **13**, 414 (1976).

18. R. G. Musket, *Surf. Sci.*, **74**, 423 (1978).

19. H. Poppa and A. G. Elliot, *Surf. Sci.*, **24**, 149 (1971).

20. H. Tokutaka, M. Prutton, I. G. Higginbotham, and T. E. Gallon, *Surf. Sci.*, **21**, 233 (1970).

21. S. Bouquet, J. Bergner, J. Lettericy, and J. P. Langeron, *J. Elec. Spectrosc. Rel. Phenom.*, **26**, 247 (1982).

22. D. E. Clark, C. G. Pantano, and L. L. Hench, *Corrosion of Glass*, Books for Industry, New York, 1979, p. 60.

23. C. C. Chang, *Surf. Sci.*, **25**, 53 (1971).

24. S. Thomas, *J. Appl. Phys.*, **45**, 161 (1974).

25. A. Van Oostrom, *Surf. Sci.*, **89**, 615 (1979).

26. T. Smith, *Surf. Sci.*, **55**, 601 (1976).

27. T. T. Lin and D. Lichtman, *J. Vac. Sci. Technol.*, **15**, 1689 (1978).

28. T. T. Lin and D. Lichtman, *J. Mater. Sci.*, **14**, 455 (1979).

29. T. T. Lin and D. Lichtman, *J. Appl. Phys.*, **50**, 1298 (1979).

30. S. Thomas, *Surf. Sci.*, **55**, 754 (1976).

31. C. T. H. Stoddart, R. L. Moss, and D. Pope, *Surf. Sci.*, **53**, 241 (1975).

32. G. T. Burstein, *Mat. Sci. Eng.*, **42**, 207 (1980).

33. D. Buczek and S. Sastri, *J. Vac. Sci. Technol.*, **17**, 201 (1980).

34. P. H. Holloway, *Adv. Electron. Electron Phys.*, **54**, 241 (1980).

35. M. P. Seah and W. A. Dench, *Surf. Interface Anal.*, **1**, 1 (1979).

36. M. P. Seah, *Surf. Sci.*, **32**, 703 (1972).

37. S.Hofmann, *Appl.Phys.*, **13**, 205 (1977).

38. P. S. Ho and J. E. Lewis, *Surf. Sci.* **55**, 335 (1976).

39. M. P. Seah and C. Lea, *Thin Solid Films*, **81**, 257 (1981).

40. R. Smith and J. M. Walls, *Surf. Sci.*, **80**, 557 (1979).

41. T. S. Sun, D. K. McNamara, J. S. Ahearn, J. M. Chen, B. Ditchek, and J. D. Venables, *Appl. Surf. Sci.*, **5**, 406 (1980).

42. S. Duncan, R. Smith, D. E. Sykes, and J. M. Walls, *Surf. Interface Anal.*, **5**, 71 (1983).

43. C. Lea and M. P. Seah, *Thin Solid Films*, **75**, 67 (1981).

44. P. Sigmund and A. Gras-Marti, *Nuclear Instr. Methods*, **168**, 389 (1980).

45. H. H. Andersen, *Appl. Phys.* **18**, 131 (1979).

46. S. Hofmann, *Appl. Phys.*, **9**, 59 (1976).

47. R. Shimizu, *Appl. Phys.*, **18**, 425 (1979).

48. M. P. Seah, J. M. Sanz, and S. Hofmann, *Thin Solid Films*, **81**, 239 (1981).

49. H. J. Mathieu, D. E. McClure, and D. Landolt, *Thin Solid Films*, **38**, 281 (1976).

50. J. M. Walls, *Thin Solid Films*, **80**, 213 (1981).

51. S. Hofmann, J. Erlewein, and A. Zalar, *Thin Solid Films*, **43**, 275 (1977); H. W. Werner, *Surf. Interface Anal.*, **4**, 1 (1982).

52. J. S. Johannessen, W. E. Spicer, and Y. E. Strausser, *J. Appl. Phys.*, **47**, 3028 (1976).

53. P. Laty, D. Seethanen, and F. Degreve, *Surf. Sci.*, **85**, 353 (1979).
54. J. P. Chubb, J. Billingham, D. D. Hall, and J. M. Walls, *Met. Technol.*, **7**, 293 (1980).
55. P. C. Baker and N. J. Brown, *Opt. Eng.*, **17**, 595 (1978).
56. M. L. Tarng and D. G. Fisher, *J. Vac. Sci. Technol.*, **15**, 50 (1978).
57. J. M. Walls, D. D. Hall, and D. E. Sykes, *Surf. Interface Anal.*, **1**, 204 (1979).
58. J. C. Fuggle in D. Briggs (Ed.), Handbook of X-ray and Ultraviolet Photoelectron Spectroscopy, Heyden, London, 1977, and references therein.
59. D. R. Jennison, *J. Vac. Sci. Technol.*, **17**, 172 (1980).
60. P. Weightman, *Rep. Prog. Phys.*, **45**, 753 (1982).
61. G. F. Amelio and E. J. Scheibner, *Surf. Sci.*, **11**, 242 (1968).
62. J. L. Lander, *Phys. Rev.*, **91**, 1382 (1953).
63. P. J. Feibelman, E. J. McGuire, and K. C. Pandey, *Phys. Rev. Lett.*, **36**, 1154 (1976).
64. D. R. Jennison, H. H. Madden, and D. M. Zehner, *Phys. Rev. B*, **21**, 430 (1980).
65. M. Cini, *Solid State Commun.*, **24**, 681 (1977).
66. G. A. Sawatzky, *Phys. Rev. Lett.*, **39**, 504 (1977).
67. S. P. Kowalczyk, L. Ley, F. R. McFelly, R. A. Pollak, and D. A. Shirley, *Phys. Rev. B*, **9**, 381 (1974).
68. P. Weightman and P. T. Andrews, *J. Phys. C: Solid State Phys.*, **12**, 943 (1979).
69. G. Treglia, M. C. Desjonqueres, F. Ducastelle, and D. Spanjaard, *J. Phys. C: Solid State Phys.*, **14**, 4347 (1981).
70. M. Cini, *Surf. Sci.*, **87**, 483 (1979).
71. P. W. Palmberg and T. N. Rhodin, *J. Appl. Phys.*, **39**, 2425 (1968).
72. P. W. Palmberg, *Anal. Chem.*, **45**, 549A (1973).
73. H. E. Bishop and J. C. Riviere, *J. Appl. Phys.*, **40**, 1740 (1969).
74. A. Jablonski, *Surf. Interface Anal.*, **1**, 122 (1979).
75. G. C. Allen, P. M. Tucker, and R. K. Wild, *Surf. Sci.*, **68**, 649 (1977).
76. P. J. Feibleman, E. J. McGuire, and K. C. Pandey, *Phys. Rev. B*, **15**, 2202 (1977).
77. D. R. Jennison, *Phys. Rev. B*, **18**, 6865 (1978).
78. R. Lasser and J. C. Fuggle, *Phys. Rev. B*, **22**, 2637 (1980).
79. K. J. Rawlings, B. J. Hopkins, and S. D. Foulias, *J. Elec. Spectrosc. Rel. Phenom.*, **18**, 213 (1980).
80. J. A. D. Matthew, F. P. Netzer, and E. Bertel, *J. Elec. Spectrosc. Rel. Phenom.*, **20**, 1 (1980).
81. D. Chadwick and A. B. Christie, *J. Chem. Soc. Faraday II*, **76**, 267 (1980).
82. J. F. McGilp, P. Weightman, and E. J. McGuire, *J. Phys. C: Solid State Phys.*, **10**, 3445 (1977).
83. J. Vayrynen, *J. Elec. Spectrosc. Rel. Phenom.*, **22**, 27 (1981).
84. J. Vayrynen and S. Aksela, *J. Elec. Spectrosc. Rel. Phenom.*, **23**, 119 (1981).
85. L. Hilaire, P. Legare, Y. Hall, and G. Maine, *Solid State Commun.*, **32**, 157 (1979).
86. E. D. Roberts, P. Weightman, and C. E. Johnson, *J. Phys. C: Solid State Phys.*, **8**, 1301 (1975).
87. C. D. Roberts, P. Weightman, and C. E. Johnson, *J. Phys. C: Solid State Phys.*, **8**, 2336 (1975).
88. E. D. Roberts, P. Weightman, and C. E. Johnson, *J. Phys. C: Solid State Phys.*, **8**, L301 (1975).

89. P. Weightman, *J. Phys. C: Solid State Phys.*, **9**, 1117 (1976).

90. J. F. McGilp and P. Weightman, *J. Phys. C: Solid State Phys.*, **11**, 643 (1978).

91. P. T. Andrews and P. Weightman, *J. Phys. C: Solid State Phys.*, **11**, L559 (1978).

92. P. Weightman and P. T. Andres, *J. Phys. C: Solid State Phys.*, **12**, 943 (1979).

93. A. C. Parry-Jones, P. Weightman, and P. T. Andrews, *J. Phys. C: Solid State Phys.*, **12**, 1587 (1979).

94. P. T. Andrews, T. Collins, and P. Weightman, *J. Phys. C: Solid State Phys.*, **14**, L957 (1981).

95. M. Pessa, A. Vuoristo, M. Vulli, S. Aksela, J. Vayrynen, T. Rantala, and H. Aksela, *Phys. Rev. B*, **20**, 3115 (1979).

96. S. Aksela, R. Kumpula, H. Aksela, J. Vayrynen, R. M. Nieminen, and M. Puska, *Phys. Rev. B*, **23**, 4362 (1981).

97. H. H. Madden, D. M. Zehner, and J. R. Noonan, *Phys. Rev. B*, **17**, 3074 (1978).

98. T. Jack and C. J. Powell, *Phys. Rev. Lett.*, **46**, 953 (1981).

99. O. Gunnarsson, K. Schönhammer, J. C. Fuggle, and R. Lässer, *Phys. Rev. B*, **23**, 4350 (1981).

100. I. R. Holton, P. Weightman, and P. T. Andrews, *J. Elec. Spectrosc. Rel. Phenom.*, **21**, 219 (1980).

101. J. A. Tagle, V. Martinez Saez, J. M. Rojo, and M. Salmeron, *Surf. Sci.*, **79**, 77 (1978).

102. R. H. Brockman and G. J. Russell, *Phys. Rev. B*, **22**, 1302 (1980).

103. R. F. Reilman, A. Msezane, and S. T. Manson, *J. Elec. Spectrosc. Rev. Phenom.* **8**, 389 (1976).

104. T. Matsudaira and M. Onchi, *Surf. Sci.*, **72**, 53 (1978).

105. T. Matsudaira and M. Onchi, *Surf. Sci.*, **74**, 684 (1978).

106. R. Hertlein, R. Weissman, and K. Müller, *Surf. Sci.*, **77**, 118 (1978).

107. A. F. Armitage, D. P. Woodruff, and P. D. Johnson, *Surf. Sci.*, **100**, L483 (1980).

108. H. Hilferink, E. Long, and K. Heinz, *Surf. Sci.*, **93**, 398 (1980).

109. G. E. Rhead, M. G. Barthes, and C. Argile, *Thin Solid Films*, **82**, 201 (1981).

110. G. E. Rhead, C. Argile, and M. G. Barthes, *Surf. Interface Anal.*, **3**, 165 (1981).

111. L. McDonnell and D. P. Woodruff, *Vacuum*, **22**, 477 (1972).

112. B. W. Holland, L. McDonnell, and D. P. Woodruff, *Solid State Commun.*, **11**, 991 (1972).

113. T. Matsudaira, M. Watanabe, and M. Onchi, *Jpn. J. Appl. Phys. Suppl.* **2**, Pt. 2, 181 (1974).

114. S. P. Weeks and A. Liebsch, *Surf. Sci.*, **62**, 197 (1977).

115. L. McDonnell, D. P. Woodruff, and B. W. Holland, *Surf. Sci.*, **51**, 249 (1975).

116. D. Aberdam, R. Baudoing, E. Blanc, and C. Gaubert, *Surf. Sci.* **71**, 279 (1978).

117. J. M. Plociennik, A. Barbet, and L. Mathey, *Surf. Sci.*, **102**, 282 (1981).

118. T. Koshikawa, T. Von Dem Hagen, and E. Bauer, *Surf. Sci.*, **109**, 301 (1981).

119. S. J. White, D. P. Woodruff, and L. McDonnell, *Surf. Sci.*, **72**, 77 (1978).

120. T. A. Carlson, *Photoelectron and Auger Spectroscopy*, Plenum, New York, 1975.

121. C. D. Wagner, in D. Briggs (Ed.), *Handbook of X-ray and Ultraviolet Photoelectron Spectroscopy*, Heyden, London, 1977.

122. J. C. Fuggle, in C. R. Brundle and A. D. Baker (Eds.), *Electron Spectroscopy:*

Theory, Techniques and Applications, Academic Press, New York, 1981, Vol. 4.

123. For excellent reviews of ion-induced Auger emission, see R. A. Baragiola, *Rad. Eff.*, **61**, 47 (1982); F. P. Larkins, *Appl. Surf. Sci.*, **13**, 4 (1982).

124. M. P. Hooker and J. T. Grant, *Surf. Sci.*, **51**, 328 (1975).

125. R. A. Baragiola, E. V. Alonso, J. Ferron, and A. Oliva-Florio, *Surf. Sci.*, **90**, 240 (1979); J. J. Vrakking and A. Kroes, *Surf. Sci.*, **84**, 153 (1979).

126. I. D. Ward and I. J. Blattner, *Surf. Interface Anal.*, **3**, 184 (1981).

127. M. Keenlyside, F. H. Stott, and G. C. Wood, *Vacuum*, **31**, 631 (1981); O. K. T. Wu and E. M. Butter, *J. Vac. Sci. Technol.*, **20**, 453 (1982).

128. P. M. Hall and J. M. Morabito, *Surf. Sci.*, **83**, 391 (1979).

129. W. Reuter, in G. Shinoda, K. Kohra, and T. Ichinokawa (Eds.), Proceedings of the 6th International Conference on X-ray Optics and Microanalysis, University of Tokyo Press, 1972, p. 121.

130. J. Ferrange, *Acta Metall'*, **19**, 743 (1971).

131. D. Buckley, NASA TN D-6359 (1971).

132. S. H. Overbury and G. A. Somorjai, *Surf. Sci.*, **55**, 209 (1976).

133. C. D. Hartshough, A. Kock, J. Mailder, and T. Sigmon, *J. Vac. Sci. Technol.*, **17**, 392 (1980).

134. R. Anton, *Thin Solid Films*, **81**, 53 (1981).

135. P. T. Dawson and S. A. Petrone, *J. Vac. Sci. Technol.*, **18**, 529 (1981).

136. P. J. K. Paterson, H. K. Wagenfeld, W. Matthews, and P. W. Wright, *Le Vide, Les Couches Minces*, **201**, 1287 (1980).

137. T. Sekine and A. Mogami, *Le Vide, Les Couches Minces*, **201**, 1283 (1980).

138. P. Weightman, P. T. Andrews, and A. C. Parry-Jones, *J. Phys. C: Solid State Phys.*, **12**, 3635 (1979).

139. P. Weightman and P. T. Andrews, *J. Phys. C: Solid State Phys.*, **13**, 3529 (1980).

140. P. Weightman and P. T. Andrews, *J. Phys. C: Solid State Phys.*, **13**, L815 (1980).

141. P. Weightman and P. T. Andrews, *J. Phys. C: Solid State Phys.*, **13**, L821 (1980).

142. J.-M. Mariot, C. F. Hauge, and G. Dufour, *Phys. Rev. B*, **23**, 3146 (1981).

143. G. G. Kleiman, V. S. Sundaram, and J. D. Rogers, *J. Vac. Sci. Technol.*, **18**, 585 (1981).

144. M. J. Kelley and V. Ponec, *Progr. Surf. Sci.*, **11**(3) (1981).

145. A. Jablonski, *Adv. Colloid Interface Sci.*, **8**, 213 (1977).

146. A. Crueq, L. Degols, G. Lienard, and A. Frennet, *Surf. Sci.*, **80**, 78 (1979).

147. W. M. H. Sachtler and R. A. Van Santen, *Appl. Surf. Sci.*, **3**, 121 (1979).

148. F. F. Abraham and C. R. Brundle, *J. Vac. Sci. Technol.*, **18**, 506 (1981).

149. E. D. Hondros and M. P. Seah, *Internat. Met. Rev.*, **222**, 262 (1977).

150. M. P. Seah, *J. Vac. Sci. Technol.*, **17**, 16 (1980).

151. M. P. Seah, *Surf. Sci.*, **80**, 8 (1979).

152. D. McLean, *Grain Boundaries in Metals*, Oxford University Press, London, 1957.

153. M. P. Seah and E. Hondros, *Proc. Roy. Soc. A*, **335**, 191 (1973).

154. S. Brunnauer, P. H. Emmett, and E. Teller, *J. Am. Chem. Soc.*, **60**, 309 (1938).

155. E. D. Hondros and M. P. Seah, *Metall. Trans. A*, **8**, 1363 (1977).

156. M. Guttmann, *Surf. Sci.*, **53**, 213 (1975).

157. J. J. Burton and E. S. Machlin, *Phys. Rev. Lett.*, **37**, 1433 (1976).

158. P. Wynblatt and R. C. Ku, *Surf. Sci.*, **65**, 511 (1977).

159. P. Wynblatt and R. C. Ku, in W. C. Johnson and J. M. Blakeley (Eds.), *Interfacial Segregation*, American Society for Metals, Metals Park, Ohio, 1979, p. 115.

160. M. Guttmann and D. McLean, in W. C. Johnson and J. M. Blakeley (Eds.), Interfacial Segregation, American Society of Metals, Metals Park, Ohio, 1979, p. 261.

161. F. L. Williams, *Surf. Sci.*, **45**, 377 (1974).

162. G. A. Somorjai and S. H. Overbury, *Surf. Sci.*, **55**, 209 (1976).

163. B. J. Wood and H. Wise, *Surf. Sci.* **52**, 151 (1975).

164. R. C. Weast (Ed.), CRC Handbook of Chemistry and Physics CRC, Florida, 1981.

165. J. M. David and S. C. Fain, *Surf. Sci.*, **52**, 161 (1975).

166. A. S. Isa, R. W. Joyner, and M. W. Roberts, *J. Chem. Soc. Faraday Trans. I*, **72**, 540 (1976).

167. S. Thomas, *Appl. Phys. Lett.*, **24**, 1 (1974).

168. S. H. Overbury and G. A. Somorjai, *J. Chem. Phys.*, **66**, 3181 (1977).

169. A. Jablonski, S. H. Overbury, and G. A. Somorjai, *Surf. Sci.*, **65**, 578 (1977).

170. K. Watanabe, M. Hashiba, and T. Yamashita, *Surf. Sci.*, **61**, 483 (1976); F. J. Kuijers and V. Ponec, *Surf. Sci.*, **68**, 296 (1977).

171. J. Ferrange, NASA TN D-6982 (1973).

172. C. Leygraf, G. Hultquist, S. Ekland, and J. C. Erikkson, *Surf. Sci.*, **46**, 157 (1974).

173. K. Wandelt and G. Erth, *J. Phys. F*, **6**, 1607 (1976).

174. K. Goto, K. Ishikawa, R. G. Wolfe, and J. T. Grant, *Appl. Surf. Sci.*, **3**, 211 (1979).

175. R. S. Polizzotti and J. J. Burton, *J. Vac. Sci. Technol.*, **14**, 347 (1977).

176. F. J. Kuijers and V. Ponec, *Appl. Surf. Sci.*, **2**, 43 (1978).

177. K. Wandelt and G. Erth, *Z. Naturforsch. Teil A*, **31**, 205 (1976).

178. J. C. Riviere, *J. Less-Common Met.*, **38**, 193 (1972).

179. J. A. Schwartz, R. S. Polizzotti, and J. J. Burton, *J. Vac. Sci. Technol.*, **14**, 457 (1977).

180. R. Bouwman, L. H. Toneman, and A. A. Holscher, *Surf. Sci.*, **35**, 8 (1973).

181. H. Bishop and J. C. Riviere, *Acta Metall.*, **18**, 813 (1970).

182. H. P. Benzel and H. B. Aaron, *Scr. Metall.*, **5**, 1057 (1971).

183. G. J. Dooley, *J. Vac. Sci. Technol.*, **9**, 145 (1972).

184. T. Smith, *J. Electrochem. Soc.*, **119**, 1398 (1972).

185. H. H. Farrell, H. S. Isaacs, and M. Strongin, *Surf. Sci.*, **38**, 31 (1973).

186. C. A. Shell and J. C. Riviere, *Surf. Sci.*, **40**, 149 (1973).

187. L. C. Isett and J. M. Blakeley, *Surf. Sci.*, **47**, 645 (1975).

188. H. J. Grabke, *Scr. Metall.*, **9**, 1181 (1975).

189. P. W. Palmberg and H. L. Marcus, *Trans. ASM*, **62**, 1016 (1969).

190. A. Joshi and D. F. Stein, *Metall. Trans.*, **1**, 2543 (1970).

191. J. R. Rellick, C. J. McMahon, H. L. Marcus, and P. W. Palmberg, *Metall. Trans.*, **2**, 1492 (1971).

192. B. D. Powell and H. Mykura, *Acta Metall.*, **21**, 1151 (1973).

193. R. P. Simpson, G. J. Dooley, and T. W. Haas, *Metall. Trans.*, **5**, 585 (1974).

194. W. C. Johnson and D. F. Stein, *Metall. Trans.*, **5**, 549 (1974).

195. J. C. Bertolini, J. Massardier, P. Delichere, T. Tardy, and B. Imelik, *Surf. Sci.*, **119**, 95 (1982).

196. G. P. Schwartz, *Surf. Sci.*, **76**, 113 (1978).

197. D. A. Mervyn, R. J. Baird, and P. Wynblatt, *Surf. Sci.*, **82**, 79 (1979).

198. E. Raffat and D. Aberdam, *Scr. Metall.*, **27**, 1381 (1979).
199. V. Ponec, *Surf. Sci.*, **80**, 352 (1979).
200. J. Erlewein and S. Hofmann, *Surf. Sci.*, **68**, 71 (1977).
201. R. H. Stulen and R. Bartasz, *J. Vac. Sci. Technol.*, **16**, 940 (1979).
202. F. Garbassi and G. Parravano, *Surf. Sci.*, **71**, 42 (1978).
203. C. Berry, D. Majumdar, and Y. W. Chung, *Surf. Sci.*, **94**, 293 (1980).
204. T. W. Haas, J. T. Grant, and M. P. Hooker, *Appl. Surf. Sci.*, **2**, 433 (1979).
205. F. Pavlyak, I. V. Perczel, and J. Giber, *Surf. Interface Anal.*, **1**, 139 (1979).
206. B. Singh, R. W. Vook, and E. A. Knabbe, *J. Vac. Sci. Technol.*, **17**, 29 (1980).
207. A. Lanere, M. Guttman, P. Dumoulin, and C. Roques-Carmes, *Acta Metall.*, **30**, 685 (1982).
208. M. Ohring, P. W. Taubenlat, W. E. Smith, and R. S. Oswald, *Mod. Dev. Powder Metall.*, **13**, 289 (1981).
209. S. M. Bruemmer, M. T. Thomas, and D. R. Baer, *Metall. Trans. A.*, **12**, 1621 (1981).
210. W. Losch and J. Kirschner, in D. Francois (Ed.), Adv. Fract. Rev., Proc. 5th Intern. Conf. Pergamon, Oxford, England, 1982, Vol. 4, p. 1621.
211. P. H. Dumoulin and M. Guttmann, *Mat. Sci. Eng.*, **42**, 249 (1980).
212. J. Ferrante, *Acta Metall.*, **19**, 943 (1971).
213. L. Marchut and C. J. McMahon, in O. Buck, J. K. Tien, and H. L. Marcus (Eds.), *Electron and Position Spectroscopies in Material Science Engineering*, Academic Press, New York, 1979, p. 183.
214. P. R. Webber, C. E. Rojas, P. J. Dobson, and D. Chadwick, *Surf. Sci.*, **105**, 20 (1981).
215. D. G. Swartzfager, S. B. Ziemecki, and M. J. Kelley, *J. Vac. Sci. Technol.*, **19**, 185 (1981).
216. Y. Fujinaga, Surf. Sci., **84**, 1 (1979).
217. A. C. Yea, W. R. Graham, and G. R. Belton, *Metall. Trans. A*, **9**, 31 (1978).
218. M. P. Seah and C. Lea, *Phil. Mag.*, **31**, 627 (1974).
219. T. S. Sun, J. M. Chen, R. K. Viswanadham, and J. A. S. Green, *J. Vac. Sci. Technol.*, **16**, 668 (1979).
220. D. Finello and H. L. Marcus, in O. Buck, J. K. Tien, and H. L. Marcus (Eds.), *Electron and Position Spectroscopies in Material Science Engineering*, Academic Press, New York, 1979, p. 121.
221. P. M. Hall and J. M. Morabito, *CRC Crit. Rev. Sol. St. Mat. Sci.*, 53 (1978).
222. P. W. Palmberg and H. L. Marcus, *ASM Trans. Quart.*, **62**, 1016 (1969).
223. H. L. Marcus, L. H. Hacket, and P. W. Palmberg, ASTM STP 499, 90 (1972).
224. H. J. Grabke, W. Paulistschke, G. Tauber, and H. Viefhaus, *Surf. Sci.*, **63**, 377 (1977).
225. R. Bouwman, G. J. M. Lippits, and W. M. H. Sachtler, *J. Catal.*, **25**, 300 (1972).
226. R. Gomez, S. F. Fuentes, F. J. del Valle, A. Campero, and J. M. Ferreira, *J. Catal.*, **38**, 47 (1975).
227. A. E. Zagli, J. L. Falconer, and C. A. Keena, *J. Catal.*, **56**, 465 (1979).
228. J. S. Solomon and N. T. McDevitt, *Thin Solid Films*, **84**, 155 (1981).
229. P. E. Nilsson-Jatko and S. E. Karlsson, *Mater. Sci. Eng.*, **42**, 345 (1980).
230. L. A. Larson, M. Prutton, H. Poppa, and J. Smialek, *J. Vac. Sci. Technol.*, **20**, 1403 (1982).

231. K. L. Smith and L. D. Schmidt, *J. Vac. Sci. Technol.*, **20**, 364 (1982).

232. G. Hultquist and C. Leygraf, *Corros, Sci.*, **21**, 401 (1981).

233. G. Hultquist and C. Leygraf, *Corros. Sci.*, **22**, 331 (1982).

234. O. Vander Biest, G. Van Birgelen, and G. Kemeny, in P. Braderoo and V. E. Cosslett (Eds.), Proceedings of the 7th European Congress on Electron Microscopy. 1980, Vol. 3, p. 226.

235. C. Lea and E. D. Hondros, *Proc. Roy. Soc. London A*, **377**, 477 (1981).

236. G. Betz, *Surf. Sci.*, **92**, 283 (1980), and references therein.

237. E. Furman, *J. Mater. Sci.*, **17**, 575 (1982).

238. N. Q. Lam, G. K. Leaf, and H. Wiedersich, *J. Nucl. Mater.*, **88**, 289 (1980).

239. P. S. Ho, J. E. Lewis, H. S. Wildman, and J. K. Howard, *Surf. Sci.*, **57**, 393 (1976).

240. N. Saeki and R. Shimizu, *Surf. Sci.*, **71**, 479 (1978).

241. T. Koshikawa, K. Goto, N. Saeki, R. Shimizu, and E. Sugata, *Surf. Sci.*, **79**, 461 (1979).

242. L. E. Rehn and H. Wiedersich, *Thin Solid Films*, **73**, 139 (1980).

243. M. Shikata and R. Shimizu, *Surf. Sci.*, **97**, L363 (1980).

244. W. Y. Lee, *J. Vac. Sci. Technol.*, **16**, 774 (1979).

245. G. Betz, J. Morton, and P. Brawn, *Nucl. Instrum. Methods*, **168**, 541 (1980).

246. R. P. Frankenthal and D. J. Siconolfi, *Surf. Sci.*, **104**, 205 (1981).

247. R. R. Olson and G. K. Wehner, *J. Vac. Sci. Technol.*, **16**, 672 (1979).

248. H. G. Tompkins, *J. Vac. Sci. Technol.*, **16**, 778 (1979).

249. P. H. Holloway and R. S. Bhattacharya, *J. Vac. Sci. Technol.*, **20**, 444 (1982).

250. M. Yabumoto, K. Watanabe, and T. Yamashina, *Surf. Sci.*, **77**, 615 (1978).

251. G. Betz, J. Dudonis, and P. Braun, *Surf. Sci.*, **104**, L185 (1981).

252. H. J. Mathieu and D. Landolt, *Appl. Surf. Sci.*, **3**, 348 (1979).

253. P. S. Ho, J. E. Lewis, and W. K. Chu, *Surf. Sci.*, **85**, 19 (1979).

254. M. P. Seah, *Thin Solid Films*, **81**, 279 (1981).

255. P. Sigmund, *Phys. Rev.*, **184**, 383 (1969).

256. R. Kelly, *Nucl. Instr. Methods*, **149**, 553 (1978).

257. L. L. Shreir (Ed.), *Corrosion*, Newnes-Butterworths, London, 1976, Vols. 1 and 2.

258. C. Nordling, E. Sokolowski and K. Siegbahn, *Ard. Fys.*, **13**, 483 (1958).

259. H. H. Madden, *J. Vac. Sci. Technol.*, **18**, 677 (1981).

260. D. E. Ramaker, J. S. Murday, and N. H. Turner, *J. Elec. Spectrosc. Rel. Phenom.*, **17**, 45 (1979).

261. B. I. Dunlap, F. L. Hutson, and D. E. Ramaker, *J. Vac. Sci. Technol.*, **18**, 556 (1981).

262. P. Légaré, G. Maire, B. Carrière, and J. P. Deville, *Surf. Sci.*, **68**, 348 (1977).

263. W. Losch, *J. Vac. Sci. Technol.*, **16**, 865 (1979).

264. M. A. Smith, *J. Vac. Sci. Technol.*, **16**, 462 (1979).

265. J. D. Place and M. Prutton, *Surf. Sci.*, **82**, 315 (1979).

266. E. J. LeJeune and R. D. Dixon, *J. Appl. Phys.*, **43**, 1998 (1972).

267. H. H. Madden and J. E. Houston, *J. Vac. Sci. Technol.*, **14**, 412 (1977).

268. M. L. Knotek and J. E. Houston, *Phys. Rev. B*, **15**, 4580 (1977).

269. V. M. Bermudez and V. H. Ritz, *Surf. Sci.*, **82**, L601 (1979).

270. P. H. Citrin, J. E. Rowe, and S. B. Christman, *Phys. Rev. B*, **14**, 2642 (1976).

271. T. E. Gallon and J. A. D. Matthew, *Phys. Status Solidi*, **41**, 343 (1970).

272. C. Benndorf, H. Caus, B. Egbert, H. Seidel, and F. Thieme, *J. Elec. Spectrosc.*

Rel. Phenom., **19**, 77 (1980).
273. S. K. Ken, S. Sen, and C. L. Bauer, *Thin Solid Films*, **82**, 157 (1981).
274. G. Gergely and H. J. Muessig, *Phys. Status Solidi A*, **69**, K69 (1982); D. F. Mitchell, G. I. Sproule, and M. J. Graham, *J. Vac. Sci. Technol.*, **18**, 690 (1981).
275. W. P. Ellis, *Surf. Sci.*, **61**, 37 (1976).
276. O. B. Ajayi, A. A. Anani, and A. O. Obabuski, *Thin Solid Films*, **82**, 151 (1981).
277. G. B. Hoflund, D. F. Cox, and H. A. Laitinen, *Thin Solid Films*, **83**, 261 (1981).
278. F. Garbassi, G. Petrini, L. Pozzi, G. Benedek, and G. Parravano, *Surf. Sci.*, **68**, 286 (1977).
279. M. A. Langell and S. L. Bernasek, *J. Vac. Sci. Technol.*, **17**, 1287 (1980).
280. M. F. Dilmore, D. E. Clark, and L. L. Hench, *Amer. Ceram. Soc. Bull.*, **57**, 1040 (1978).
281. D. E. Clark, M. F. Dilmore, E. C. Ethridge, and L. L. Hench, *J. Amer. Ceram. Soc.*, **59**, 37 (1976).
282. A. E. Clark, C. G. Pantano, and L. L. Hench, *J. Amer. Ceram. Soc.*, **59**, 37 (1976).
283. R. A. Chappel and C. T. H. Stoddart, *Phys. Chem. Glasses*, **15**, 130 (1974).
284. F. Ohuchi, M. Ogino, P. H. Holloway, and C. G. Pantano, *J. Vac. Sci. Technol.*, **16**, 527 (1979).
285. W. C. Johnson and D. F. Stein, *J. Amer. Ceram. Soc.*, **58**, 485 (1975).
286. P. Nanni, C. T. H. Stoddart, and E. D. Hondros, *Mater. Chem.*, **1**, 297 (1976).
287. S. Sinharay, L. L. Levenson, and D. E. Day, *J. Vac. Sci. Technol.*, **16**, 503 (1979) and references therein.
288. H. H. Madden and P. H. Holloway, *J. Vac. Sci. Technol.*, **16**, 618 (1979).
289. R. L. Coble, *J. Amer. Ceram. Soc.*, **41**, 55 (1958).
290. W. C. Johnson, *Metall. Trans. A*, **8**, 1413 (1977).
291. D. E. Clark, L. Lue Yen-Bower, and L. L. Hench, in J. Mendel (Ed.), Proceedings of the International Symposium on Ceramic Nuclear Waste Management, Cincinnati, Batelle-Pacific Northwest, Richland, WA, 1979.
292. L. L. Hench and H. A. Paschall, *J. Biomed. Mater. Res.*, **5**, 49 (1974).
293. L. L. Hench, *J. Non-Cryst. Solids*, **19**, 27 (1975).
294. F. Ohuchi, D. E. Clark and L. L. Hench, *J. Amer. Ceram. Soc.*, **62**, 500 (1979).
295. See, for example, K. S. Kin, W. E. Baitinger, J. W. Amy, and N. Winograd, *J. Elec. Spectrosc. Rel. Phenom.*, **5**, 351 (1974); R. Kelly, *Nucl. Instr. Methods*, **149**, 553 (1978); S. Storp and R. Holm, *J. Elec. Spectrosc. Rel. Phenom.*, **16**, 183 (1979); R. Holm and S. Storp, *Appl. Phys.*, **12**, 101 (1977); R. Kelly, *Surf. Sci.*, **100**, 85 (1980).
296. L. I. Yin, S. Ghose, and I. Adler, *J. Geophys. Res.*, **77**, 1360 (1972).
297. H. M. Naquib and R. Kelly, *Rad. Eff.*, **25**, 1 (1975).
298. M. P. Seah and C. P. Hunt, *Surf. Interface Anal.* **5**, 33 (1983).
299. P. C. Karulkar, *J. Vac. Sci. Technol.*, **18**, 169 (1981).
300. P. H. Holloway and G. C. Nelson, *J. Vac. Sci. Technol.*, **16**, 793 (1979).
301. J. M. Sanz and S. Hofmann, 8th International Vacuum Congress, Cannes, France, 1980.
302. S. Thomas, *Surf. Sci.*, **55**, 754 (1976).
303. J. Ahn, C. R. Perleberg, D. L. Wilcox, J. W. Cohurn, and H. F. Winters, *J. Appl. Phys.*, **46**, 4581 (1975).

304. A. B. Christie, I. Sutherland, and J. M. Walls, *Vacuum*, **31**, 513 (1981).

305. A. B. Christie, I. Sutherland, J. Lee, and J. M. Walls, *Appl. Surf. Sci.* (in press).

306. M. W. Roberts and C. S. McKee, *Chemistry of the Metal-Gas Interface*, Clarendon Press, Oxford, 1978.

307. E. Miyazaki and I. Yasumori, *Surf. Sci.*, **55**, 747 (1976).

308. D. Brennon, D. O. Hayward, and B. M. W. Trapnell, *Proc. Roy. Soc. London A*, **265**, 81 (1960).

309. G. C. Bond, *Heterogeneous Catalysis: Principles and Applications*, Clarendon Press, Oxford, 1974.

310. N. R. Avery, *Surf. Sci.*, **61**, 391 (1976).

311. C. T. H. Stoddart and E. D. Hondros, *Nat. Phys. Sci.*, **237**, 90 (1972).

312. G. Betz, G. K. Wehner, L. E. Toth, and A. Joshi, *J. Appl. Phys.*, **45**, 5312 (1974).

313. G. C. Allen and R. K. Wild, *J. Elec. Spectrosc. Rel. Phenom.*, **5**, 409 (1974).

314. M. Janik-Czahor, *Corrosion*, **35**, 360 (1979).

315. G. Hulquist and C. Leygraf, *J. Vac. Sci. Technol.*, **17**, 85 (1980).

316. D. R. Baer, *Appl. Surf. Sci.*, **7**, 69 (1981).

317. A. Joshi, *Reviews on Coatings and Corrosion*, Freund, Tel-Aviv, Israel, 1979, Vol. 3; 51.

318. C. Leygraf and G. Hultquist, *Surf. Sci.*, **61**, 69 (1976).

319. K. Wandelt and G. Erth, *Surf. Sci.*, **55**, 403 (1976).

320. D. R. Baer, D. A. Petersen, L. R. Pederson, and M. T. Thomas, *J. Vac. Sci. Technol.*, **20**, 957 (1982).

321. S. Suresh and R. O. Ritchie, *Mater. Sci. Eng.*, **51**, 61 (1981).

322. J. C. Riviere, *Les Mem. Sci. de la Rev. de Metall.*, **76**, 759 (1979).

323. R. W. Revie, B. G. Baker, and J. O'M. Bockris, *J. Electrochem. Soc.*, **122**, 1460 (1975).

324. G. E. McGuire and P. H. Holloway, in C. R. Brundle and A. D. Baker (Eds.), *Electron Spectroscopy: Theory, Techniques and Applications*, Academic Press, New York, 1981, Vol. 4, p. 1.

325. R. P. Frankenthal and D. J. Siconolfi, *Corros. Sci.*, **21**, 479 (1981).

326. R. P. Frankenthal and D. J. Siconolfi, *J. Vac. Sci. Technol.*, **17**, 1315 (1980).

327. B. Goldstein and J. Dresner, *Surf. Sci.*, **71**, 15 (1978).

328. G. T. Burstein and T. P. Hoar, *Corros. Sci.*, **18**, 75 (1978).

329. Y. Fuji, F. Kanematsu, T. Koshikawa, and E. Sugata, *J. Vac. Sci. Technol.*, **17**, 1221 (1980).

330. D. T. Larson, *J. Vac. Sci. Technol.*, **17**, 55 (1980).

331. M. E. Schrader, *Surf. Sci.*, **78**, L227 (1978).

332. D. J. David, M. H. Froning, T. N. Wittberg, and W. E. Moddeman, *Appl. Surf. Sci.*, **7**, 185 (1981).

333. K. C. I. Kobayashi, Y. Shiraki, and Y. Katayama, *Surf. Sci.*, **77**, 449 (1978).

334. R. K. Wild, *Corros. Sci.*, **13**, 105 (1973).

335. P. J. Bassett and T. E. Gallon, *J. Elec. Spectrosc. Rel. Phenom.*, **2**, 101 (1973).

336. H. Windawi and J. R. Katzer, *J. Vac. Sci. Technol.*, **16**, 497 (1979).

337. M. P. Hooker and J. T. Grant, *Surf. Sci.*, **55**, 741 (1976).

338. Y. Takasu, H. Shimizu, S. Maru, and Y. Matsuda, *Surf. Sci.*, **61**, 279 (1976).

339. W. Färber and P. Braun, *Mikrochimica Acta*, **6**, 391 (1975).
340. D. Dolle, M. Alnot, J. J. Ehrhardt, and A. Cassuto, *J. Elec. Spectrosc. Rel. Phenom.*, **17**, 299 (1979).
341. F. J. Szalkowski and G. A. Somorjai, *J. Chem. Phys.*, **56**, 6097 (1972).
342. C. R. R. Rao, D. D. Sarma, and M. S. Hodge, *Proc. Roy. Soc. London A*, **370**, 269 (1980).
343. T. Smith, *Surf. Sci.*, **55**, 601 (1976).
344. J. S. Solomon and D. E. Hanlin, *Appl. Surf. Sci.*, **4**, 307 (1980).
345. B. McDougall, D. R. Mitchell, and M. J. Graham, *Corrosion*, **38**, 85 (1982).
346. P. Marcus, J. Oudar, and I. Olefjord, *Mater. Sci. Eng.*, **42**, 191 (1980).
247. N. R. Armstrong and R. K. Quinn, *Surf. Sci.*, **67**, 451 (1977).
348. K. Sasaki and T. Umezawa, *Thin Solid Films*, **74**, 83 (1980).
349. M. Seo and N. Sato, *Corros. Sci.*, **18**, 577 (1978).
350. M. Seo and N. Sato, *Surf. Sci.*, **86**, 601 (1979).
351. G. E. McGuire, A. L. Bacarella, J. C. Griess, R. E. Clausing, and L. D. Hulett, *J. Electrochem. Soc.*, **125**, 1801 (1978).
352. J. B. Lumsden, Z. Szklarska-Smialowska, and R. W. Staehle, *Corrosion*, **34**, 169 (1977).
353. H. Viefhaus and M. Janik-Czachov, *Werkst. Korros.*, **28**, 219 (1977).
354. Z. Szklarska-Snialowska, H. Viefhaus, and M. Janik-Czachov, *Corros. Sci.* **16**, 649 (1976); M. da Cunha Belo, B. Rondot, F. Pons, and J. P. Langeron, *J. Electrochem. Soc.*, **124**, 1317 (1977).
355. G. B. Erth and K. Wandelt, *Surf. Sci.*, **50**, 479 (1975).
356. J. B. Lumsden and R. W. Staehle, *Scr. Met.*, **6**, 1205 (1972).
357. P. Marcus and J. C. Charbonnier, *J. Microsc. Spectrosc. Electron.*, **6**, 329 (1981).
358. Reference 15 in Reference 314.
359. R. Berneron, J. C. Charbonnier, R. Namdov-Irani, and J. Manenc, *Corros. Sci.*, **20**, 899 (1980).
360. C. Leygraf, G. Hultquist, I. Olefjord, B. O. Elfström, V. M. Knyazheva, A. V. Plaskeyev, and Ya. M. Kolotyrkin, *Corros. Sci.*, **19**, 343 (1979).
361. J. R. Cahoon and R. Bandy, *Corrosion*, **38**, 299 (1982).
362. G. T. Burstein and R. C. Newman, *Appl. Surf. Sci.*, **4**, 162 (1980).
363. See, for example, A. J. Forty and P. Humble, *Phil. Mag.*, **8**, 247 (1963); A. J. McEvily and A. P. Bond, *J. Electrochem. Soc.*, **112**, 131 (1965); R. P. M. Procter and M. Islam, *Corrosion*, **32**, 267 (1976).
364. E. Mattsson, *Electrochim. Acta*, **3**, 279 (1961).
365. C. R. Shastry, M. Levy, and A. Joshi, *Corros. Sci.*, **21**, 673 (1981).
366. L. J. Matienzo and K. J. Holub, *Appl. Surf. Sci.*, **9**, 47 (1981).
367. See, for example, J. Flis, *Corros. Sci.*, **15**, 553 (1975); J. Flis, *Corrosion*, **29**, 37 (1973).
368. G. Tauber and H. J. Grabke, *Corros. Sci.*, **19**, 793 (1979).
369. J. Küpper, E. Erhart, and H. J. Grabke, *Corros. Sci.*, **21**, 227 (1981).
370. E. D. Hondros and C. Lea, *Nature*, **289**, 663 (1981).
371. A. Joshi, *Corrosion*, **34**, 47 (1978).
372. A. Joshi and D. F. Stein, *Corrosion*, **28**, 321 (1972).
373. G. Hultquist and C. Leygraf, *Corrosion*, **36**, 126 (1980).
374. S. Zakipour and C. Leygraf, *Corrosion*, **37**, 21 (1981).

375. S. Zakipour and C. Leygraf, Proceedings of the 8th International Congress on Metallic Corrosion, 1981, Vol. 1, p. 181.

376. See, for example, J. W. Coburn and E. Kay, *CRC Crit. Rev. Solid State Sci.*, 561 (1974); A. Benninghoven, *Thin Solid Films*, **39**, 3 (1976); P. H. Holloway and G. E. McGuire, *Thin Solid Films*, **53**, 3 (1978); A. E. Morgan and H. W. Werner, *Phys. Scr.*, **18**, 451 (1978); S. Hofmann, *Talanta*, **26**, 665 (1979); C. C. Chang, *J. Vac. Sci. Technol.*, **18**, 276 (1981); H. W. Werner, *Mater. Sci. Eng.*, **42**, 1 (1980).

377. K. L. Mittal (Ed.), *Surface Contamination: Genesis, Detection and Control*, Plenum, New York, 1979, Vols. 1 and 2.

378. A. G. Mathewson, *Vacuum*, **24**, 595 (1974).

379. Y. Gomay, R. E. Clausing, R. J. Colchin, L. C. Emerson, L. Heatherly, W. Namkung, and J. E. Simpkins, *J. Vac. Sci. Technol.*, **16**, 918 (1979).

380. R. Bouwman, J. B. van Mechelen, and A. A. Holscher, in Reference 377, Vol. 1, p. 287.

381. P. A. Lindfors, in Reference 377, Vol. 2, p. 587.

382. V. Leroy, *Mater. Sci. Eng.*, **42**, 289 (1980).

383. M. K. Bernett and H. Rayner, *ASLE Trans.*, **25**, 55 (1982).

384. R. D. Luhtuke, C. P. Gopalaraman, and V. K. Rohatgi, *Corr. Sci.*, **50**, 1055 (1981).

385. L. Mirkova, S. Rashkov, and K. Nanev, *Surface Technol.*, **15**, 181 (1982).

386. J. S. Johannsen, A. P. Grande, and T. Notevarp, *Mater. Sci. Eng.*, **42**, 321 (1980).

387. C. A. Haque and A. K. Spiegler, *Appl. Surf. Sci.*, **4**, 214 (1980).

388. G. E. McGuire, J. V. Jones, and H. J. Dowell, *Thin Solid Films*, **45**, 59 (1977).

389. L. R. Pederson and M. T. Thomas, *Sol. Ener. Mater.*, **3**, 151 (1980).

390. E. Kny, *J. Vac. Sci. Technol.*, **17**, 658 (1980).

391. M. Housley and D. A. King, *Surf. Sci.*, **62**, 81 (1977).

392. G. C. Nelson and P. H. Holloway, *Surface Analysis Techniques for Metallurgical Applications*, ASTM STP 596, 1976, p. 68.

393. J. M. Andrews and J. M. Morabito, *Thin Solid Films*, **37**, 357 (1976).

394. M. M. Bhasin, *J. Catal.*, **38**, 218 (1975).

395. J. L. Pena, M. H. Farias, and F. Sanchez-Sinencio, *J. Electrochem. Soc.*, **129**, 94 (1982).

396. P. A. Lindfors and K. M. Black, *Microbeam Anal.*, **16**, 303 (1981).

397. A. B. Christie, *Appl. Surf. Sci.*, **10**, 571 (1982).

398. H. Oechsner, *Appl. Phys.*, **8**, 185 (1975).

399. P. Kelly, *Surf. Sci.*, **100**, 135 (1980).

400. P. Sigmund, *J. Vac. Sci. Technol.*, **17**, 396 (1980).

401. G. Carter and D. G. Armour, *Thin Solid Films*, **80**, 13 (1981).

402. O. Auciello, *J. Vac. Sci. Technol.*, **39**, 841 (1981).

403. S. S. Makh, R. Smith, and J. M. Walls, *J. Mater. Sci.*, **17**, 1689 (1982).

404. J. W. Coburn, *Thin Solid Films*, **64**, 371 (1979).

405. R. Behrisch (Ed.), *Sputtering by Particle Bombardment I: Physical Sputtering of Single-Element Solids*, Springer-Verlag, New York, 1981.

406. Y. E. Strausser, D. Franklin, and P. Courtney, *Thin Solid Films*, **84**, 145 (1981).

407. S. W. Gaarenstroom, *J. Vac. Sci. Technol.*, **20**, 458 (1982).

408. C. C. Chang and G. Quintana, *Thin Solid Films*, **31**, 365 (1976).

409. C. A. Evans and R. J. Blattner, *Thin Solid Films*, **53**, 29 (1978).

410. Y. Namba and R. W. Vook, *Thin Solid Films*, **82**, 165 (1981).
411. G. Broden and H. P. Bonzel, *Surf. Sci.*, **84**, 106 (1979).
412. S. S. Chao, R. W. Vook, and Y. Namba, *J. Vac. Sci. Technol.*, **18**, 695 (1981).
413. C. Argile and G. E. Rhead, *Thin Solid Films*, **67**, 299 (1980).
414. A. Sepulveda and G. E. Rhead, *Surf. Sci.*, **66**, 436 (1977).
415. T. G. Andersson, *Thin Solid Films*, **83**, L147 (1981).
416. C. Argile and G. E. Rhead, *Surf. Sci.*, **78**, 115 (1978).
417. T. J. Chuang and K. Wandelt, *Surf. Sci.*, **81**, 355 (1979).
418. S. Danyluk, G. E. McGuire, K. M. Koliwad, and M. G. Yang, *Thin Solid Films*, **25**, 483 (1975).
419. S. K. Sen, P. M. Kluge-Weiss, and C. L. Bauer, *Thin Solid Films*, **82**, 299 (1981).
420. H. C. Card and K. E. Singer, *Thin Solid Films*, **28**, 265 (1975).
421. T. E. Brady and C. T. Singer, *Thin Solid Films*, **28**, 265 (1975).
422. R. Chopra, M. Ohring, and R. S. Oswald, *Thin Solid Films*, **86**, 43 (1981).
423. A. Hanusovsky, M. Koltai, and I. Trifonov, *Thin Solid Films*, **85**, 335 (1981).
424. S. Hofmann and A. Zalar, *Thin Solid Films*, **39**, 219 (1976).
425. Physical Electronics Applications Note No. 7601 (1976), Eden Prairie, Minnesota, U.S.A.
426. C. P. Lofton and W. E. Schwartz, *Thin Solid Films*, **52**, 271 (1978).
427. J. M. Morabito and M. J. Rand, *Thin Solid Films*, **22**, 293 (1974).
428. S. Hofmann and A. Zalar, *Thin Solid Films*, **56**, 337 (1979).
429. J. M. Walls, D. D. Hall, D. G. Teer, and B. L. Delcea, *Thin Solid Films*, **54**, 303 (1978).
430. D. E. Sykes, D. D. Hall, R. E. Thurstans, and J. M. Walls, *Appl. Surf. Sci.*, **5**, 103 (1980).
431. H. A. Katzman, G. M. Malouf, R. Bauer, and G. W. Stupian, *Appl. Surf. Sci.*, **2**, 416 (1979).
432. J. V. Standish and F. J. Boerio, *J. Coatings Technol.*, **52**, 29 (1980).
433. J. E. Greene and J. L. Zilko, *Surf. Sci.*, **72**, 109 (1978).
434. A. A. Roche, J. S. Solomon, and W. L. Baun, *Appl. Surf. Sci.*, **7**, 83 (1981).
435. I. L. Singer and J. S. Murday, *J. Vac. Sci. Technol.*, **17**, 327 (1980).
436. C. W. Draper, *J. Mater. Sci.*, **16**, 2774 (1981).
437. C. W. Draper and S. P. Sharma, *Thin Solid Films*, **84**, 333 (1981).
438. M. L. Tarng and D. G. Fisher, *J. Vac. Sci. Technol.*, **15**, 50 (1978).
439. J. P. Chubb, J. Billingham, D. D. Hall, and J. M. Walls, *Met. Technol.*, **7**, 293 (1980).
440. S. Baumgarth, W. Warnecke, and R. Holm, *Surf. Interface Anal.*, **3**, 188 (1981).
441. V. Thompson, H. E. Hintermann, and L. Chollet, *Surf. Technol.*, **8**, 421 (1979).
442. J. D. Birchall, A. J. Howard, and K. Kendall, *Chem. Britain*, **18**, 860 (1982).
443. L. R. Hettche and B. B. Rath, in J. J. Burke and V. Weiss (Eds.), *Surface Treatments for Improved Performance and Properties*, Plenum Press, London, 1982, pp. 143–171.
444. D. F. Stein, A. Joshi, and R. P. Laforce, *Trans. ASM*, **62**, 776 (1969).
445. D. McLean and L. Northcott, *J. Iran Steel Inst.*, **158**, 169 (1948).
446. A. Joshi, *Interfacial Segregation*, ASM, Metals Park, Ohio, 1979, pp. 39–109.
447. J. M. Walsh, K. P. Frunz, and N. P. Anderson, in N. S. McIntyre (Ed.), *Quantitative*

Surface Analysis of Materials ASTM STP 643, ASTM, Philadelphia, 1978, pp. 72–82.

448. R. P. Wei and G. W. Simmons, *Scr. Metall.*, **10**, 153 (1976).
449. Vacuum Generators Scientific, East Grinstead, U.K.
450. A. Joshi and D. F. Stein, ASTM STP 499 (1972), p. 59.
451. R. K. Wild, *Mater. Sci. Eng.*, **42**, 265 (1980).
452. S. G. Druce and B. C. Edwards, *Nucl. Technol.*, **55**, 487 (1981).
453. U. Franzoni, H. Goretzki, and S. Sturkse, *Scr. Metall.*, **15**, 743 (1981).
454. C. L. White, R. A. Padgett, and R. W. Swinderman, *Scr. Metall.*, **15**, 777 (1981).
455. P. Cahmption and J. P. Lacharme, *Verres Refract.*, **35**, 65 (1981).
456. R. K. Viswanadham, T. S. Sun, E. F. Drake, and J. A. Peck, *J. Mater. Sci.*, **16**, 1029 (1981).
457. T. Ando and G. Krauss, *Metall. Trans. A*, **12**, 1283 (1981).
458. P. Dumoulin, M. Guttmann, M. Foucault, M. Palmier, M. Wayman, and M. Biscondi, *Met. Sci.*, **14**, 1 (1980).
459. J. Vitart-Barbier, A. Larere, C. Roques-Carmes, and G. Saindrenan, *Mem. Etud. Sci. Rev. Metall.*, **78**, 359 (1981).
460. H. Hanninen and E. Minni, *Tutkimuksia-Valt. Tek. Tutkimuskerkus*, 32 (1981).
461. T. Noda, M. Okada, and N. Tominaga, *J. Nucl. Mater.*, **101**, 354 (1981).
462. K. Abido, S. Suzuki, and H. Kimura, *Trans. Jpn. Inst. Met.*, **23**, 43 (1982).
463. C. Kim, W. L. Phillips, and R. J. Weimer, ASTM STP 733, 314 (1981).
464. J. Bernardini, P. Gas, E. D. Hondros, and M. P. Seah, *Proc. Roy. Soc. London A*, **379**, 159 (1982).
465. W. C. Johnson and H. B. Smanth, *Metall. Trans. A*, **8**, 553 (1977).
466. W. C. Johnson and B. V. Kovacs, *Metall. Trans. A*, **9**, 219 (1978).
467. H. L. Marcus and P. W. Palmberg, *Trans. Met. Sco. AIME*, **245**, 1664 (1969).
468. A. Joshi, *Scr. Met.*, **9**, 251 (1975).
469. M. T. Thomas, D. R. Baer, R. H. Jones, and S. M. Bruemmer, *J. Vac. Sci. Technol.*, **17**, 25 (1980).
470. E. Kay, W. Slolz, R. Stickler, and H. Goretzki, *J. Vac. Sci. Technol.*, **17**, 1208 (1980).
471. W. Losch, *Acta Metall.*, **27**, 567 (1979).
472. M. Menyhard, *Surf. Interface Anal.*, **1**, 175 (1979).
473. M. Menyhard, paper presented at 3rd European Conference on Surface Science, Cannes, France (1980).
474. G. C. Allen and R. K. Wild, *Appl. Surf. Sci.*, **8**, 278 (1981).
475. A. Joshi, *Scr. Met.*, **7**, 735 (1975).
476. M. P. Seah, *Acta Met.*, **25**, 345 (1977).
477. W. C. Johnson and H. B. Smartt, *Met. Trans. A*, **8**, 553 (1977).
478. W. C. Johnson and H. B. Smartt, *Scr. Met.*, **9**, 1295 (1975).
479. J. P. Coad, J. C. Riviere, M. Guttmann, and P. R. Krake, *Acta Met.*, **25**, 161 (1977).
480. H. L. Marcus and M. E. Fine, *J. Amer. Cer. Soc.*, **55**, 568 (1972).
481. J. M. Walls and A. B. Christie, in D. M. Brewis (Ed.), *Surface Analysis and Pretreatment of Plastics and Metals*, Applied Science, London, 1982.
482. W. L. Baun, *Appl. Surf. Sci.*, **4**, 291 (1980).
483. D. H. Buckley, *J. Vac. Sci. Technol.*, **13**, 88 (1976).

484. M. D. Pashley and D. Tabor, *Vacuum*, **31**, 619 (1981).
485. W. Hartweck and H. J. Brabke, *Surf. Sci.*, **89**, 174 (1979).
486. P. H. Holloway, *Gold Bull.* **3**, 99 (1979).
487. D. T. Larson and F. W. Korbitz, *J. Vac. Sci. Technol.*, **12**, 721 (1975).
488. C. W. B. Martinson, P. J. Nordlander, and S. E. Karlsson, *Vacuum*, **27**, 119 (1977).
489. J. R. Waldrop and H. L. Marcus, *J. Test. Eval.*, **1**, 194 (1973).
490. A. E. Yaniv, D. Katz, I. E. Klein, and J. Sharon, *Glass Technol.*, **22**, 231 (1981).
491. W. Warnecke, S. Baumgarth, and R. Holm, *Mikrochim. Acta*, **9**, 281 (1981).
492. D. E. Clark, C. G. Pantano, and G. Y. Onada, *J. Amer. Ceram. Soc.*, **58**, 336 (1975).
493. T. Smith and D. H. Kaelble, in R. L. Patrick (Ed.), *Treatise on Adhesives and Adhesion*, Dekker, New York, 1981, Vol. 5, p. 139.
494. M. Gettings, F. S. Baker, and A. J. Kinlock, *J. Appl. Polym. Sci.*, **21**, 2375 (1977).
495. M. Gettings and A. J. Kinlock, *Surf. Interface Anal.*, **1**, 165 (1979).
496. M. Gettings and A. J. Kinlock, *Surf. Interface Anal.*, **1**, 189 (1979).
497. D. Tabor, Proceedings of the NASA Symposium on Interdisciplinary Approach to Friction and Wear, San Antonio (1967).
498. W. E. Jamison, *J. Vac. Sci. Technol.*, **13**, 76 (1976).
499. D. H. Buckley, *Thin Solid Films*, **53**, 271 (1978).
500. D. H. Buckley, *Wear*, **46**, 19 (1978).
501. D. H. Buckley, *Progr. Surf. Sci.*, **12**, 1 (1982).
502. K. Miyoshi and D. H. Buckley, *Wear*, **77**, 253 (1982).
503. D. H. Buckley, NASA Technical Memo 82753 E1069 (1981).
504. S. V. Pepper, *J. Appl. Phys.*, **45**, 2947 (1974).
505. B. Singh and R. W. Vook, *Electr. Contacts*, **26**, 53 (1980).
506. B. Singh, B. H. Hwang and R. W. Vook, *Vacuum*, **32**, 23 (1982).
507. D. E. Savage, M. G. Lagally, and M. E. Schroeder, *Appl. Surf. Sci.*, **7**, 142 (1981).
508. K. Miyishi and D. H. Buckley, *Appl. Surf. Sci.*, **10**, 357 (1982).
509. S. Timoney and G. Flynn, private communication.
510. K. S. Majumder and Y. E. Strausser, *Thin Solid Films*, **53**, 63 (1978).
511. D. H. Buckley, *ASLE Trans.*, **17**, 36 (1974).
512. D. H. Buckley, *ASLE Trans.*, **17**, 206 (1974).
513. C. N. Rowe and J. J. Dickert, *ASLE Trans.*, **10**, 85 (1967).
514. J. J. McCarroll, R. W. Mould, H. B. Silver, and M. L. Sims, Proceedings of the 7th International Vacuum Congress, Vienna, (1977).
515. M. R. Phillips, M. Dewey, D. D. Hall, T. F. J. Quinn, and H. N. Southworth, *Vacuum*, **26**, 451 (1976).
516. R. Schumacher, E. Gregner, A. Schmidt, H. J. Mathieu, and D. Landolt, *Tribology Int.*, **13**, 311 (1980).
517. R. J. Bird and G. D. Galvin, *Nature (London)*, **254**, 130 (1975).
518. A. B. Grebene, *J. Appl. Phys.*, **39**, 4866 (1968).
519. A. R. Beattie and P. T. Landsberg, *Proc. Roy. Soc. London A*, **249**, 16 (1959).
520. P. T. Landsberg and A. R. Beattie, *J. Phys. Chem. Solids*, **8**, 73 (1959).
521. J. D. Cuthbert, *J. Appl. Phys.*, **42**, 747 (1971).
522. P. T. Landsberg and M. J. Adams, *Proc. Roy. Soc. London A*, **334**, 523 (1973).
523. C. C. Chang, *Surf. Sci.*, **23**, 283 (1970).

524. B. A. Joyce and J. H. Neave, *Surf. Sci.*, **27**, 499 (1971).
525. R. C. Henderson and R. F. Helm, *Surf. Sci.*, **30**, 310 (1972).
526. M. Nishijima and T. Murotani, *Surf. Sci.*, **32**, 459 (1972).
527. T. Narusawa, S. Komiya and A. Hirabi, *Appl. Phys. Lett.*, **21**, 272 (1972).
528. H. G. Maguire and P. D. Augustus, *J. Electrochem. Soc.*, **119**, 791 (1972).
529. T. E. Gallon, M. Prutton, and L. Wray, *J. Vac. Sci. Technol.*, **9**, 911 (1972).
530. J. M. Morabito and J. C. Tsai, *Surf. Sci.*, **33**, 422 (1972).
531. J. H. Thomas and J. M. Morabito, *Surf. Sci.*, **41**, 629 (1974).
532. J. C. Irvin, *Bell System Tech. J.*, **41**, 387 (1962).
533. A. J. Van Bommel and J. E. Crombeen, *Surf. Sci.*, **36**, 773 (1973).
534. N. J. Chou, C. M. Osburn, Y. J. Van der Meulen, and R. Hammer, *Appl. Phys. Lett.*, **22**, 380 (1973).
535. J. N. Smith, S. Thomas, and K. Ritchie, *J. Electrochem. Sic.*, **121**, 827 (1974).
536. R. Heckingbottom and P. R. Wood, *Surf. Sci.*, **36**, 594 (1973).
537. C. C. Chang, A. C. Adams, G. Quintana, and T. T. Sheng, *J. Appl. Phys.*, **54**, 252 (1974).
538. A. J. Van Bommel, J. E. Crombeen, and A. Van Tooren, *Surf. Sci.*, **48**, 463 (1975).
539. J. S. Johannessen, W. E. Spicer, and Y. E. Strausser, *J. Vac. Sci. Technol.*, **13**, 849 (1976).
540. C. R. Helms, W. E. Spicer, and N. M. Johnson, *Solid State Commun.*, **25**, 673 (1978).
541. B. Goldstein and D. J. Szostak, *Appl. Phys. Lett.*, **33**, 85 (1978).
542. J. A. Roth and C. R. Crowell, *J. Vac. Sci. Technol.*, **15**, 1317 (1978).

Varian Chart of Auger Electron Energies (See Appendix 2.) Reprinted with permission of Varian Associates, Palo Alto, California. Chart prepared by Yale E. Strausser and John J. Uebbing, 1971.

CONCISE COMPARISON OF METHODS FOR SURFACE ANALYSIS

1. INTRODUCTION AND SURVEY

Over the past decade, there has been considerable research and development in other techniques designed chiefly for surface analysis. As is evident from this text, progress in these methods has been stimulated by the move to thin-film technology and understanding of such areas as catalysis, corrosion, adhesion, surface diffusion, nucleation, failure and device operation, etc. With regard to the practical consequences of these areas, it is instructive to consider the definition of a surface; according to Honig[1] –"the outermost atomic layer bounding a solid;" and Lichtman[2] –"a surface is that part of the bulk from one to ten atomic layers from the surface monolayer." Although these statements are quite unequivocal, the perceptions and requirements of the solid-state physicist interested in a submonolayer of an adsorbate on a single crystal in a 10^{-11} torr vacuum are expected to be a little different from the polymer chemist interested in a deposited Langmuir–Blodgett film. In other words, the physics and chemistry of one particular situation can in fact be completely ascribed to the properties of layers within the bounds of the definitions above, but there are cases where perhaps a looser definition of a "surface" is justified. Keeping these points in mind, it behooves us to consider the type of information that characterizes a surface. According to Duke[3] –"to specify the microscopic 'structure' of a solid surface, one must provide four types of information: the chemical identity of atomic species in the uppermost few layers; the positions of these species; their vibrational amplitudes; and the ground state properties and excitation spectrum of the valence fluid in which they are immersed." These criteria have been expressed in point form, with respect to the ideal properties a technique must possess for surface examination:[4]

1. Capable of providing information in the monolayer range.
2. Detection of elements.
3. Identification of chemical compounds including organic species.

4. Elucidation of surface "geometry" and topography.
5. Capability of isotope separation.
6. High sensitivity.
7. High spatial resolution.
8. Application to a wide range of samples, for example, to rough and nonconducting surfaces.
9. No discrimination against any component of the surface.
10. No influence of the analysis process on surface composition and structure.

It goes without saying that no single technique satisfies all these criteria, nor is it feasible to expect that one method will answer all questions for a particular sample. In other words, complementarity is a common feature in the examination of surfaces.

The methods discussed in this section conform loosely to the common theme of study and analysis of information carried by emitted photons, ions, and electrons after impact of these entities as probes or application of electric fields. These are summarized in Table 7.1 after an expansion of the classifications offered by Hofmann,[5] although it is arguable that several methods do not fall into the pure surface-analysis field, at least according to the definitions presented above. Before comparing the information generated by these techniques, we discuss the salient features of a number of the more-important methods, although it should be recognized that the significance of a particular method is definitely influenced by the viewpoint of the practitioner.

The basis of SIMS is the bombardment of solids by 1–30 keV ions, which results in the ejection of substrate species in positively and negatively charged atomic and molecular particles (and neutrals).[6,7] The charged moieties are mass analyzed by high-sensitivity mass spectrometry. The penetration of the primary-ion particle into the solid and the transfer of its energy via a series of collisions to the lattice atoms initiates collisional cascades among other lattice atoms. Within the appropriate escape depth, secondary particles will be ejected from the surface in a neutral or charged state. The method is used in both dynamic and static modes, where in the latter a relatively large target area (~ 0.1 cm^2) is bombarded by a primary beam of small current density ($\sim < 10^{-9}$ A cm^{-2}). This results in a low sputtering speed and a relatively long average lifetime of the monolayer.

In RBS, the surface is bombarded with H$^+$ or He$^+$ ions in approximately the 300 keV to 3 meV energy range. A small fraction (about 10^{-6}) of the incoming particles undergo a Rutherford collision and are backscattered. The resulting spectrum is a convolution of a mass scale (established by the

Table 7.1. Some Surface-Analysis Methods Classified According to Excitation Mode and Emitted Species (after Hofmann)[a]

Excitation or Probe	Exit Species and Information Carrier		
	Photons	Electrons	Ions (neutrals)
Photons	X-ray fluorescence (XRF) Laser optical-emission spectroscopy (LOES) Light (Raman) scattering spectroscopy (LS) Fourier-transform infrared spectroscopy (FTIR) Ellipsometry (E) Evanescent wave spectrofluorimetry (EWS)	X-ray photoelectron spectroscopy (XPS) Ultraviolet photoelectron spectroscopy (UPS) Auger electron spectroscopy–x-ray excitation (XAES)	Photodesorption (PD)
Electrons	Electron microprobe (EMP) Appearance potential spectroscopy (APS) Cathodoluminescence (CD) Scanning-electron-microscopy x-ray detection (XSEM)	Auger electron spectroscopy (AES) Scanning electron microscopy (SEM) Low-energy electron diffraction (LEED) Electron-impact energy-loss spectroscopy (EELS)	Electron stimulated desorption (ESD)
Ions	Ion-induced x-ray spectroscopy (IIX) Proton-induced x-ray spectroscopy (PIX) Surface composition by analysis of neutral species and ion-impact radiation (SCANIIR) Glow-discharge optical spectroscopy (GDOS)	Ion-neutralization spectroscopy (INS) Ion-induced Auger electron spectroscopy (IAES)	Secondary-ion mass spectrometry (SIMS) Ion-scattering spectroscopy (ISS) Rutherford backscatter spectroscopy (RBS)
Electric Field		Field electron microscopy (FEM)	Atom-probe field-ion microscopy (APFIM)

[a]Table is reducible according to parent technique, for example, AES, IAES, XAES, but all methods specified to delineate clearly probe and exit species identification.

backscattering process) and depth scale (formed by energy loss in the sample before and after the backscattering event). In ISS, low-energy ions (0.5–2 keV) bombard the surface to provide energy spectra characteristic of the masses of the scattering centers. Besides the use of the lower-energy exciting beam, ISS differs from RBS in that the scattering angle is commonly 90° instead of 180°. Since these low energy ions penetrate to only a few atomic layers, and elastic scattering occurs only for the outermost atomic layer, the spectrum provides qualitative and quantitative information only on this layer.

With SEM electrons from a filament are accelerated and focused by magnetic lenses to produce a fine electron beam, which can be focused on the sample surface.[8] There are scanning coils before the final lens, and the spot is moved across the surface in the form of a square raster. The current passing through the scanning coils is passed through the deflection coils of a cathode ray tube. Secondary electrons strike the collector and are used to moderate the brightness. Since the time associated with the emission and collection of the electrons is much smaller than that of the scanning process, there is a good correspondence between the number of secondary electrons from a point on the surface and the brightness of the corresponding point on the CRT streen. Many modes of operation of SEM are used according to the collection of different signals produced by the interaction of the electron beam with the sample. These include emissive (secondary electrons), reflective (backscattered electrons), transmission, cathode luminescence as well as x-ray and AES processes. In most of these techniques SEM provides only topographical information.

The generic term LEED applies to the experiment where electrons of low energy, ~ 5–500 eV, are impacted on a surface at an angle to the normal of about 5° with consequent study of the scattering process.[9] The interaction of such electrons with the valence electrons of the solid can result in elastic (ELEED) or inelastic (ILEED) scattering. As with AES and XPS, the surface sensitivity of the method rests on the fact that electrons in this low-energy range rapidly lose energy by inelastic collisions and, therefore, those which escape with little or no energy loss must have originated in the surface layers of the solid. The ELEED method is wisely used for the study of atom positions with respect to surface crystallography. As mentioned earlier EELS is concerned with discrete energy losses suffered by a monoenergetic impinging electron beam and is often used to study the vibrational frequencies of surface-adsorbed molecules.[10]

The introduction of Fourier-transform infrared spectroscopy has added a very powerful tool to the array of physical methods available for the analytical chemist.[11,12] The heart of an FTIR spectrometer is the Michelson-type interferometer where the transform from interferogram to frequency-domain spectrum is carried out by computer. The advantages of the method

(Felgett-scan speed, Jacquinot-sensitivity, and Conne-laser calibration) have led to significant advances in infrared spectroscopy of thin films and techniques where high scan speed is essential such as the GC–IR combination. With respect to surface analysis, attenuated total reflection (ATR) and reflection–absorption (RA) methods are used in the subtraction mode. Here, spectra corresponding to the solid substrate and gas are subtracted from the surface system to obtain that of the surface species of interest.

2. COMPARISON

All the techniques require that samples be studied in a high vacuum. Since electrons or ions are used as emitting probes of surface structure, these particles require a long mean free path (> 10 cm) to be able to exit the sample and then reach the detector without suffering a gas-phase collision. Furthermore, for the surface to be operated free of contamination ultra high vacuum (UHV) conditions (10^{-11}–10^{-10} torr) are mandatory. However, AES, XPS, SIMS, and SEM are often operated at HV ($\sim 10^{-8}$ torr) where the contamination that occurs is recognized and tolerated in order to facilitate the speed of analysis (and more inexpensive pumping needs). These techniques when operated in the "dirty" mode can be contrasted with LEED, EELS, and UPS, where the surface must be free of adsorbed gases and contaminants. In UPS, the introduction of the necessary He gas is often avoided by use of an aluminum window, which unfortunately reduces photon intensity. Interestingly, the high electric field in the FIM has been used to keep the sample surface free from contamination.

The demands placed by interpretation of the diffraction pattern from LEED and the micrograph from FIM with respect to atomic positions means that these two methods are virtually only applicable to single-crystal surface studies. Also, with LEED and EELS the requirement for a contaminant-free surface often results in heating/sputtering cycles, particularly in the case of the former method. This means that LEED crystals are frequently stored in the analysis chamber for long periods of time. Depending on the type of study AES, XPS, and SIMS require little or no sample preparation. The sample, in thin film, powder, crystal, or other solid form, is mounted onto a specimen holder, which is usually in electrical contact with the spectrometer to reduce sample changing. In SEM, in order to prevent the build up of charge, samples have to be coated with a thin conducting layer, or the microscope is operated at low potentials. Finally, in contrast to LEED, the techniques AES, XPS, SIMS, SEM, and EELS can be applied to rough and/or disordered surfaces.

As mentioned previously, it is desirable that the technique not cause significant alteration of the sample in the course of the analysis (aside, of course, from deliberate attempts at depth profiling by surface removal). In general, most of the techniques are damaging to the surface, but the extent of destructiveness varies with the method and type of sample. There is no doubt that certain substrates are affected by the relatively high-energy beams in AES, XPS, and SEM. As discussed earlier, in AES the incident electron beam can cause significant damage, compared to x-rays, because of the higher energy absorbed per unit time per unit weight at the surface. The effect is most apparent when working with insulating surfaces (e.g., organic and polymer samples), but does not usually cause concern when dealing with metallic species. With XPS, there are relatively few systems for which appreciable radiation damage occurs during the time taken to record a spectrum. However, the view might be expressed that such damage is more prevalent than is realized, and when it does occur without recognition, serious errors in spectral assignment can be made. Sample consumption accompanies the analytical excitation to a definite degree in both ISS and SIMS. If low ion-current densities and energies are employed, both techniques permit the analysis of a monolayer of sample while consuming a fraction of that layer. In static SIMS ($<10^{-9}$ A cm^{-2}), about 1% of a monolayer or less is sputtered during the analysis. LEED is also destructive to the surface, but to a lesser extent than in AES, SEM, or XPS, and EELS is the least damaging method because the excitation source is very low in energy.

At this point it should be mentioned that RBS does not require sample consumption in order to perform a depth profiling analysis because all the depths over the region that the probing beam penetrates are sampled simultaneously. Radiation damage is a consequence of extended exposure to the probing beam. However, the manner in which the probing ions interact with the sample minimizes sample alteration in the analytical volume itself. The majority of ion energy lost in this zone results from interaction with electrons. At the depth where the ion has lost sufficient energy to allow a significant number of collisions with target atoms, the interaction is so deep (>1 μm) that no sputtering can take place. In addition, ions backscattered in this region have insufficient energy to escape the sample. An exception is organic materials which do sustain damage due to electron irradiation. Otherwise, RBS is a nondestructive technique.

The sensitivity of surface methods for elements varies widely from technique to technique and within a particular method. Basically, AES, XPS, SIMS, RBS, and ISS can be used to carry out an elemental identification, whereas LEED, SEM (without x-ray adjunct), EELS, FTIR, and FIM are not directly employable for such analysis. Important features of SIMS are its ability to detect hydrogen and isotopes with very high mass resolution. Another significant aspect of SIMS is that it constitutes the most sensitive of all the methods, with a detection limit of up to 10^{-8} monolayer or about 10^{-14} g cm^{-3} for some elements. It is, however, very dependent on the chemical condition of the surface (matrix effects), because variations occur in the ionization coefficients of the different elements under energetic ion bombardment. They vary from 10^{-3} for clean surfaces to nearly 1 in polar compounds.[13] The large variations in relative ion yield and consequent low detection sensitivity for certain elements can be alleviated by the use of negative SIMS, since these elements that have a low positive yield generally have a high negative yield. In spite of some problems, Evans[14] points out that SIMS does not encounter significant interference difficulties until analysis is performed at levels well below the detection limits of most surface techniques.

The sensitivity of AES is dependent on the relative Auger electron efficiencies of the particular elements in the sample, on operating parameters such as the primary-beam energy and current, on data acquisition characteristics, and on the nature of the large inelastic electron background. As with XPS, the variation in sensitivity from the most sensitive to the least sensitive element is very approximately a factor of 10. One disadvantage of Auger is the competition offered by fluorescence as a decay process after creation of the initial vacancy. As we have seen Auger electron ejection is the major deexcitation route for compounds containing light elements and fluorescence emission is the major pathway for heavier atoms. AES is said to be more surface sensitive than XPS (detection limit of $\sim 10^{-10}$ g cm^{-3} or $\sim 10^{-3}$ monolayer compared to 10^{-8} cm^{-3}) because of the current of Auger electrons generated by electron excitation and because the inelastic mean free path for these electrons is about 1 nm. However, Powell[15] has argued that differences in the latter for AES and XPS are not really very great. Finally, EELS and LEED have comparable limits of detection to AES.

In RBS, the detection limit is determined by the cross section for nuclear backscattering of the probing ion by the target atom. For a given probing ion, the cross section is proportional to the square of the atomic number (Z) of the target atom. Thus, the sensitivity for lead ($Z=82$) is about 100 times greater than that for oxygen ($Z=8$). This has lead to the conclusion that detection limits of 10^{-4}–10^{-3} monolayer is possible for high-Z elements but only 10^{-1} monolayer for low-Z elements.[16] Elemental sensitivities are not readily predictable for ISS because the size of a given scattered peak depends on a number of parameters including the scattering cross section, a neutralization factor, and geometric considerations. The detection limits vary by only a factor of 5 from O to Bi, but the average detection limit of ISS is in the order of 10^{-2} monolayer.[14] The major contribution to this poor detection limit is that almost all of the incident ions undergo charge neutralization during scattering and cannot be detected by the analyzer. With both RBS and ISS mass resolution decreases dramatically as the target atom's mass greatly exceeds that of the incident particle. Taking RBS as an example, the resolution depends on the type of spectrometer used to analyze the backscattered ions. The most widely used 1–2-MeV H^+ or He^+ ions and solid-state barrier detectors permit the resolution of adjacent elements and even isotopes up to approximately mass 40, while at heavier masses the resolution degrades to about 10 mass units.[16] Owing to the nature of the spectra produced by the backscattering process, a coalescence of a mass and a depth scale,[17] ions backscattered from deep-lying high-mass elements are indistinguishable from those scattered from low-mass elements in or on a high-mass substrate. In contrast to high-energy backscattering in RBS, it is possible in ISS to resolve low-mass elements on a high-mass substrate because the contribution from substrate atoms below the surface is small.[18]

It is interesting to note that AES and RBS have been identified as complementary techniques in that for high-mass elements on low-mass substrates the excellent elemental sensitivity of AES complements the high sensitivity but poor elemental analysis of RBS.[19] Conversely, for low-mass elements on high-mass substrates, AES has much higher sensitivity than the approximate 100 monolayer sensitivity of RBS.

Molecular or valence information is available from SIMS, XPS, EELS, UPS, and FTIR. The first technique in this group is particularly valuable because of its capabilities concerning the identification of organic moieties and "inorganic" clusters. However, the interpretation of the relationship of the SIMS ions to surface structure is attracting relatively large numbers of workers and problems still remain. The chemical-shift effect in XPS is associated with the separation of photoelectron signals caused by the particular valence environment of an atom. This effect has aroused great interest because of its potential in structural studies, viz., metal oxidation states and nonequivalent atoms in organic molecules. Furthermore, the electron count rates corresponding to PE signals faithfully mirror the stoichiometric ratios of such inequivalent atoms. Unfortunately, the scope of XPS is limited by the resolution that can be achieved with use of conventional x-ray sources. Both EELS and FTIR can generate the information expected from vibrational spectroscopy. For example, the former has been used to study the nature of π and σ bonding of unsaturated hydrocarbons on a surface.[20] Finally, it has been amply demonstrated in this text that use of the Auger chemical-shift effect is rather restricted because of difficulties in interpretation of the Auger energy in terms of the three levels involved in the transition.

Many of the techniques described above can be used in conjunction with concentration-depth profiling, although it is clear that the problems delineated earlier must be kept in mind. Only RBS samples all depths simultaneously since the depth of a subsurface feature is established by the energy-loss process as the probing ion traverses the sample atomic layers before and after the backscattering event. This energy loss is well established in the published literature and by physical principles so that depth-scale assignments may be made without the need of comparative standards. It thus provides a quantitative method of thickness measurement and has been suggested as a means of calibrating ion sputtering used in other techniques.[21] Typical depth resolution attainable is of the order of 100–300 Å.

The ability to perform a lateral analysis with high spatial resolution is an important part of the complete microchemical characterization of the sample. Of the techniques considered, AES, SEM, and SIMS are widely used for this type of work since the probe beam sizes are appropriate for microanalytical examination. When ultrafocused beams of 0.1 μm diameter are used in scanning Auger microscopy, two-dimensional compositional analyses with submicron spatial resolution are obtained. Combined with CDP, the method is capable of producing three-dimensional elemental imaging of a number of constituents through multiplexing. AES combined with SEM is a similar method that can generate a surface micrograph with specific elemental analysis of localized volumes. In RBS, the probe beam sizes are of the order of 0.3–1 mm in diameter. Although the beam size can be reduced, the consequent increase in analysis time becomes prohibitive and one risks radiation damage because of the high-energy particles being used. Finally, the electron microprobe deserves mention at this point. The method makes use of 30-keV incident electrons with analysis of emitted x-rays. Although reasonable lateral resolution can be achieved (1–2 μm), the sampling depth is of the order of 10,000–20,000 Å, so the method cannot really be included in the surface category.[18]

One of the most important considerations when evaluating an analytical technique is the ability to evaluate quantitatively the data produced by that technique. Generally, this applies to the conversions of spectral intensities to atomic concentrations. Of the techniques under discussion, RBS is the most easily quantitated.[14] All of the parameters influencing relative backscattering intensities are based on well-established physical principles, thereby permitting absolute determination of relative atomic concentrations. There is no ambiguity in this determination as there is in depth-scale assignments. Conversely, all the other techniques being examined depend primarily on comparative standards to establish relative sensitivities or correction factors for quantitation of the data.

In ISS, quantitative measurements are complicated by the unpredictable dependence of the neutralization coefficient on the surface composition, geometric effects, double scattering, primary-beam contamination, and a correlation between the primary-ion energy and the information depth.[22,23]

It is generally assumed that the application of a backscattering coefficient correction to the relative intensities from a given sample provides quantitative analysis.

The large variation in the secondary-ion yields and matrix or chemical effects complicate quantitation of the SIMS technique where work with many of the elements requires a very complete set of standards. The difficulties encountered may be reduced by employing a slight modification.[24] When O^+ primary ions with high current densities are used instead of noble-gas ions, the surface will be oxidized up to a level of saturation and emit mainly ions rather than neutrals. The sensitivity of the technique is maximized in this case and fairly constant, since the chemical conditions are fixed by the oxygen saturation. Thus the stoichiometries of the surface layers can be measured with little error, but the undisturbed original surface monolayer cannot be investigated by this variation.

UPS can be largely ruled out of quantitative analysis since there is no direct identification of elements and one has superimposed broad peaks on a scattered electron background. There are also problems in the calibration of the energy scale. Mixing or coating the sample with a standard compound for calibration, as in XPS, is undesirable because the valence bands of the calibrant are usually not well separated from those of the sample, resulting in considerable overlap of the two spectra. Also, the addition of a standard may shift the Fermi level, which is the usual level in solids.

AES and XPS are more promising because core-level ionizations are atomic in nature and will not be very affected by the chemical environment. The representation of various atom populations in the sample by peak area and greater simplicity of the ionization mechanism of XPS have made this technique more amenable to quantitative measurements than AES. The latter is discussed in more detail in an earlier section. In summary, one must either theoretically estimate relative Auger transition sensitivities for different atoms or calibrate values as is done for XPS. Theoretical values are not available and calibration is difficult because it is not possible to measure areas of the peaks on a derivative plot. Using peak-to-peak heights as an estimate of intensity (as is often done) is only valid if the peak shape does not differ from compound to compound; a situation that is commonly untrue. The second difficulty is the often very destructive nature of the electron beam on the surface being examined leading to dissociation, desorption, segregation, and bulk dissolution.

A number of the points outlined above are summarized for some techniques in Table 7.2. This part of the text is not meant to be all encompassing and, of course, developments are occurring rapidly all the time, viz., the increasing use of synchrotron radiation and efforts to produce lateral resolution in XPS. Another area that is attracting growing interest at the

Table 7.2. Summary of Characteristic Features of some Methods of Surface Analysis[a]

Characteristic	AES	XPS	SIMS	ISS	LEED	EELS
Excitation beam	Electrons	X-ray photons	Ions	Ions	Electrons	Electrons
Energy (keV)	0.1–5	1–10	0.1–100	0.5–2	0.2–0.5	0.003–0.008
Diameter (μm)	25–100 →1 raster	10	10^3 →1 raster	10^3	10^3	10^3
Information depth (Å)	3–25	10–30	3–20	3–10	0–10	0–10
Monolayers	2–10	3–10	1–4	2	0–2	0–2
Detection capability						
Elements	$Z>2$	$Z>1$	All	$Z>1$	Not directly	Not directly
Elemental sensitivity range	10	10	10^4	10	—	—
Isotopes	No	No	Yes	Restricted	No	Restricted
Chemical valence	Special cases	Yes	Indirect	No	No	Yes
Organics	No	Yes	No	No	No	No
Beam damage	Small	Occasionally	Yes, dynamic; uppermost layer, static	Small	Small	No
Lateral resolution	<1 μm possible	~1.0 mm	<1 μm possible	10 μm	2.0 mm	1.0 mm
Detection limits						
"Surface" (g cm^{-2})	10^{-10}	10^{-9}	10^{-13}	10^{-10}	10^{-10}	10^{-10}
Bulk (atomic fraction)	10^{-3}	10^{-3}–10^{-2}	10^{-3}–10^{-4}	10^{-2}		
Advantages	Sensitive to low-Z elements; minimal matrix effects; high-lateral resolution	Information on chemical bonds; no beam damage	Detection of all elements and isotopes; good detection sensitivity; high lateral resolution	Outermost atomic layer analysis	Atomic structure of "ordered" surface	Direct information on interaction of adsorbate
Disadvantages	Difficult to quantitate; no H, He detection	No lateral resolution; slow profiling; no H detection	Difficult to quantitate matrix effects	Low sensitivity, poor lateral resolution; slow profiling	No elemental analysis pattern; often difficult to interpret; long analysis time	No elemental analysis; long analysis time

[a]Many characteristics to be regarded as "ball-park" figures due to influence of variable factors, type of analysis, etc. For more detail see excellent comparison reviews: S. Hofmann, *Talanta*, **26**, 665 (1979); C. A. Evans, *Anal. Chem.*, **47**, 818A (1975); R. E. Honig, *Thin Solid Films*, **31**, 89 (1976).

time of writing is the *in situ* study of liquid/solid interfaces, which are particularly important in electrode processes and biomedical engineering. For example, many technological problems are posed by the desirability of studying a "monolayer" of adsorbed protein, particularly with regard to tertiary structure and conformation. Spectroscopic ellipsometry and evanescent-wave spectrofluorometry are just two of the methods undergoing development at the present time.

REFERENCES

1. R. E. Honig, *Thin Solid Films*, **31**, 89 (1976).
2. D. Lichtman, in A. W. Czanderna (Ed.), *Methods of Surface Analysis*, Elsevier, New York, 1975.
3. C. B. Duke, NATO Course on Electron Emission Spectroscopy, University of Ghent, 1972.
4. A. Benninghoven, *Surf. Sci.*, **35**, 427 (1973).
5. S. Hofmann, *Talanta*, **26**, 665 (1979).
6. H. W. Werner, in L. Fiermans, J. Vennik, and W. Dekeyser (Eds.), *Electron and Ion Spectroscopy of Solids*, Plenum Press, New York, 1978.
7. J. A. McHugh, in A. W. Czanderna (Ed.), *Methods of Surface Analysis*, Elsevier, New York, 1975.
8. J. A. Belk, *Electron Microscopy and Microanalysis of Crystalline Materials*, Applied Science Publ., London, 1979.
9. T. M. Buck, in A. W. Czanderna (Ed.), *Methods of Surface Analysis*, Elsevier, New York, 1975.
10. H. Ibach (Ed.), Topics in Current Physics, Vol. IV, Electron Energy Loss Spectroscopy for Surface Analysis.
11. J. L. Koenig, *Acc. Chem. Res.*, **14**, 171 (1981).
12. J. R. Ferraro and L. J. Basile (Eds.), *Fourier Transform Infrared Spectroscopy Applications to Chemical Systems*, Academic Press, 1978, 1979, Vols. 1 and 2.
13. A. Benninghoven and A. Mueller, *Phys. Lett. A.*, **40**, 169 (1977).
14. C. A. Evans, *Anal. Chem.*, **47**, 818A (1975).
15. C. J. Powell, *Surf. Sci.*, **44**, 29 (1974).
16. C. A. Evans, *J. Vac. Sci. Technol.*, **12**, 144 (1975).
17. W. K. Chu, J. W. Mayer, M. A. Nicolet, T. M. Buck, G. Amsel, and F. H. Eisen, *Thin Solid Films*, **17**, 1 (1973).
18. J. W. Mayer and A. Turos, *Thin Solid Films*, **19**, 1 (1973).
19. R. G. Musket and W. Bauer, *Thin Solid Films*, **19**, 69 (1973).
20. H. Ibach and S. Lehwald, *J. Vac. Sci. Technol.*, **15**, 407 (1978).
21. R. Behrisch, B. M. U. Scherzer, and P. Staib, *Thin Solid Films*, **19**, 57 (1973).
22. E. Taglauer and W. Heiland, *Surf. Sci.*, **47**, 234 (1975).
23. H. Niehus and E. Bauer, *Surf. Sci.*, **47**, 222 (1975).
24. N. Treitz, *J. Phys. E.*, **10**, 573 (1977).

APPENDIX

1

SOME REVIEW PAPERS ON THE AUGER PROCESS AND APPLICATIONS OF AES

The following list constitutes a sampling of review papers on the principles, modes of operation and application of AES, Auger data and discussion in the context of other surface-oriented and electron spectroscopic methods.

1950–1974

E. Bauer, *Z. Metallk.*, **63**, 437 (1972).

I. Bergström and C. Nordling, in K. Siegbahn (Ed.), *The Auger Effect*, in *Alpha-, Beta-, and Gamma-Ray Spectroscopy*, North Holland, Amsterdam, 1965, p. 1523.

D. Betteridge, *Analyst*, **99**, 994 (1974).

R. Bouwman, *Ned. Tijdschr. Vacuumtech.*, **11**, 37 (1973).

C. R. Brundle, *Proc. Soc. Anal. Chem.*, **10**, 194 (1973).

C. R. Brundle, *J. Electron Spectrosc. Relat. Phenom.*, **5**, 291 (1974).

E. H. S. Burhop, *The Auger Effect and Other Radiationless Transitions*, Cambridge University Press, Cambridge, 1952.

E. H. S. Burhop and W. N. Asaad, *Advan. At. Mol. Phys.*, **8**, 163 (1972).

T. A. Carlson, *Proc. Intern. Conf. Inner Shell Ioniz. Phenomena Future Appl.*, **4**, 2274 (1973).

C. C. Chang, *Surface Sci.*, **25**, 53 (1971).

C. C. Chang, in P. J. Kane (Ed.), *Analytical Auger Electron Spectroscopy*, in *Characterization of Solid Surfaces*, Plenum Press, New York, 1974, p. 509.

G. L. Connell and Y. P. Gupta, *Mater. Res. Stand.*, **11**, 8 (1971).

J. S. Dionisio, *Ann. Phys. (Paris)*, **8**, 747 (1963).

T. E. Gallon and J. A. D. Matthew, *Rev. Phys. Technol.*, **3**, 31 (1973).

D. Haneman, *Proc. Roy. Aust. Chem. Inst.*, **38**, 45 (1971).

K. Hayakawa, S. Kawase, and H. Okano, *Hyomen*, **12**, 518 (1974).

W. C. Johnson, D. F. Stein, and A. Joshi, *Can. J. Spectry.*, **17**, 88 (1972).

S. Komiya and K. Lyo, *Shinku*, **15**, 35 (1972).

M. A. Listergarten, *Bull. Acad. Sci. USSR, Phys. Ser.*, **24**, 1050 (1960).

H. P. W. Losch. *Met. ABN (Ass. Brasil. Metais)*, **29**, 493 (1973).

Ph. Maitrepierre, *Bull. Cercle Etud. Met.*, **12**, 721 (1974).

J. A. D. Matthew, *Endeavour*, **33**, 86 (1974).

381

W. Mehlhorn, in *Proc. Intern. Conf. Phys. Electron. At. Collisions*, North-Holland, Amsterdam, 1972, p. 169.

Y. Murata, *Kagaku* (*Tokyo*), **44**, 82 (1974).

T. Murotani, K. Fujiwara, M. Otani, and M. Nishijima, *Mitsubishi Denki Giho*, **47**, 667 (1973).

K. Nakayama, *Boshoku Gijutsu*, **20**, 255 (1971).

M. Nishio, *Sumitumo Keikinzoku Giho*, **15**, 41 (1974).

H. Okada, H. Ogawa, and H. Omatu, *Hyomen*, **12**, 148 (1974).

P. W. Palmberg, in D. A. Shirley (Ed.), *Proc. Intern. Conf. Electron. Spectrosc.*, North-Holland, Amsterdam, 1972, p. 835.

J. C. Rivière, *Contemp. Phys.*, **14**, 513 (1973).

J. C. Rivière, in T. Mulvey and R. K. Webster (Eds.), *Auger Spectroscopy*, in *Mod. Phys. Tech. Mat. Technol.*, Oxford University Press, Oxford, 1974, p. 187.

K. D. Sevier, *Low Energy Electron Spectrometry*, Wiley-Interscience, New York, 1972, p. 51.

E. N. Sickafus, *J. Vac. Sci. Technol.*, **11**, 299 (1974).

E. N. Sickafus and H. Bonzel, *Progr. Surface Membrane Sci.*, **4**, 115 (1971).

G. Stupian, in J. H. Richardson (Ed.), *Auger Electron Spectrometry*, in *Systematic Materials Analysis*, Academic Press, New York, 1974, p. 57.

N. J. Taylor, *Tech. Met. Res.*, **7**, Pt. 1, 117 (1972).

J. C. Tracey, in W. Dekeyser, L. Fiermans, G. Vanderkelen and J. Vennik (Eds.), *Electron Emission Spectroscopy*, Reidel, Dordrecht, 1973, p. 295.

J. C. Tracey and J. M. Burkstrand, *Crit. Rev. Solid State Sci.*, **4**, 381 (1974).

Y. Tsuji, *Kinzoku Hyomen Gijutsu*, **24**, 111 (1973).

H. J. Ullrich, S. Däbritz, H. Schreiber, and K. Kleinstück, *Technik*, **26**, 391 (1971).

J. Vennik and L. Fiermans, *Silicates Ind.*, **39**, 1973 (1974).

J. R. Waldrop and H. L. Marcus, *J. Test. Eval.*, **1**, 194 (1973).

1975

P. Auger, *Surf. Sci.*, **48**, 1 (1975).

L. Fiermans and J. Vennik, *J. Rev. M.*, **21**, 65 (1975).

K. Hayakawa, S. Kawase, M. Ichikawa, and H. Okano, *Oyo Butsuri*, **44**, 1294 (1975).

K. Hayakawa, H. Okano, S. Kawase, M. Ichikawa, and Y. Goto, *Shokubai*, **17**, 9 (1975).

D. M. Holloway, *J. Vac. Sci. Technol.*, **12**, 392 (1975).

A. Joshi, L. E. Davis, and P. W. Palmberg, *Sci. Technol.*, **1**, 159 (1975).

Ph. Maitrepierre, *Circ. Inf. Tech., Cent. Doc. Sider*, **32**, 1749 (1975).

J. J. McCarroll, *Surf. Sci.*, **53**, 297 (1975).

W. E. Moddeman, C. R. Cothern, and J. N. Black, **18**, No. 5, 27 (1975).

R. Shimizu, T. Koshikawa, and K. Goto, *Shinku*, **18**, 415 (1975).

D. F. Stein, *J. Vac. Sci. Technol.*, **12**, 268 (1975).

H. Tokutaka, K. Nishimori, and K. Takashima, *Nippon Bustsuri Gakkaishi*, **30**, 673 (1975).

1976

D. Chattarji, *Theory of Auger Transitions*, Academic, New York, 1976.

J. E. Defrance, *Rev. Quest. Sci.*, **147**, 339 (1976).

J. C. Eriksson and C. Leygraf, *Kem. Tidskr.*, **88**, 46 (1976).

K. Goto and K. Ishikawa, *Oyo Butsuri*, **45**, 174 (1976).

E. Isoyama and T. Uchiyama, *Kinzoku Hyomen Gijutsu*, **27**, 378 (1976).

E. Isoyama and T. Uchiyama, *Kinzoku Hyomen Gijutsu*, **27**, 435 (1976).

D. Martinez-Duart and J. M. Albella, *Quim. Ind.*, **22**, 977 (1976).

W. Mehlhorn, *NATO Adv. Study Inst. Ser., Ser. B* (1976).

J. M. Morabito and P. M. Hall, *Scanning Electron Microsc.*, **9**, Pt. 1, 221 (1976).

G. Nagy, *Kem. Ujabb Eredmenyei*, **34**, 219 (1976).

M. Ono and H. Shimizu, *Oyo Butsuri*, **45**, 355 (1976).

1977

B. E. Blekher, I. A. Brytou, N. I. Komyak, and V. V. Korablev, *Zavod. Lab.*, **49**, 1069 (1977).

A. Bukaluk and R. Suida, *Postepty Fiz.*, **28**, 29 (1977).

R. Holm, *Chem Rundsch*, **30**, 6 (1977).

N. C. MacDonald and C. T. Hovland, *Pro. Annu. Conf. Microbeam Anal. Soc.*, **12**, 64A (1977).

J. M. Morabito, in R. Dobrozemsky, F. Ruedenauer, and F. P. Veihboech (Eds.), *Microelectronic Technology*, Pro. Int. Vac. Congr., 7th, **3**, 2267 1977).

T. Ohnishi, *Kagaku Sosetsu*, **16**, 161 (1977).

R. L. Park and M. DenBoer, in R. Vanselow and S. V. Tong (Eds.), *Chem. Phys. Solid Surf.*, 191 (1977).

F. J. Szalkowski, *J. Coll. Interface Sci.*, **58**, 199 (1977).

M. Thompson, Talanta, **24**, 399 (197).

S. Usami, *Zairyo Kagaku*, **14**, 96 (1977).

N. Wu, Z. Lui, and K. Chen, *Wuli*, **6**, 353 (1977).

1978

M. G. Cattania Sabbadini, V. Ragaini, and M. Tescari, *Chim. Ind.*, **60**, 8 (1978).

T. E. Gallon, *NATO Adv. Study Inst. Ser. B* (Elec. Ion Spec. Solids), 230 (1978).

O. Gijeman and G. A. Bootsma, *Ned. Tijdschr. Natuurkd. A*, **44**, 114 (1978).

P. H. Holloway, *Scanning Electron Microsc.* **1**, 361 (1978).

R. Holm, *Scanning*, **1**, 42 (1978).

K. Way, *At. Data Nucl. Data Tables* **22**, 125 (1978).

1979

D. Aberdam, *J. Microsc. Spectrosc. Electron.*, **4**, 431 (1979).

T. Aberg and G. Howat, *Res. Rep. Helsinki Univ. Technol.*, *Lab Physics*, **5** (1979).

N. A. Alford, A. Barrie, I. W. Drummond, and Q. C. Herd, *SIA Surface Interface Anal.*, **1**, 36 (1979).

I. M. Bronshtein and V. M. Stozharov, *Izv. Akad. Nauk. SSSR Ser. Fiz.*, **43**, 500 (1979).

C. C. Chang, *Annu. Rep., Conf. Electr. Insul. Dielectric Phenom.*, **45-8**, (1979).

D. Finello and H. L. Marcus, in O. Buck, J. K. Tien, and H. L. Marcus (Eds.), *Advances and Quantification of Auger Electron Spectroscopy*, Academic, New York, 1979.

S. Hofman, *Talanta*, **26**, 665 (1979).

S. Konczak, B. Parka, and K. Wojtowicz, *Zesz. Nauk. Politech, Slask.*, *Mat.-Fiz.*, **30**, 59 (1979).

J. P. Langeron, *J. Mcirosc. Spectrosc. Electron.*, **4**, 371 (1979).

G. LeGressus, D. Massignon, and R. Sopizet, *J. Microsc. Spectrosc. Electron*, **4**, 409 (1979).

H. J. Mathieu and H. D. Landolt, *Vide*, 273 (1979).

M. Onchi, *Kagaku*, **34**, 856 (1979).

W. Palczewska, in J. Fijalkowski (Ed.), *Electronic and Ionic Methods for the Analysis of Solid Surfaces*, Ossolineum, Wroclaw, Poland, 1979.

P. J. K. Paterson and P. W. Wright, *Met. Forum*, **2**, 55 (1979).

T. N. Rhodin and J. W. Gadzuk, in T. N. Rhodin and G. Ertl (Eds.), *Electron Spectroscopy and Surface Chemical Binding*, North-Holland, Amsterdam, 1979.

R. Riwan, *Vide*, 199 (1979).

H. Siegbahn, *NATO Adv. Study Inst. Ser.*, *Ser. C*, **46**, 273 (1975).

G. Svehla (Ed.), *Comprehensive Analytical Chemistry*, Elsevier, Amsterdam 1979, p. 306.

I. Szymerska, and A. Joblanski, in J. Fijalkowski (Ed.), *Auger Electron Spectroscopy*, Ossolineum, Wroclaw, Poland, 1979.

L. Zhao, *Hua Hsuch Tung Pao*, **2**, 131 (1979).

1980

H. Aksela, *Acta Univ. Ouluensis, Ser. A*, **89** (1980).

R. J. Blattner and C. A. Evans, *J. Educ. Modules Mater. Sci. Eng.*, **2**, 1 (1980).

M. L. DenBoer, *Diss. Abstr. Int. B*, **40**, 4359 (1980).

A. P. Janssen and J. A. Venables, *Conf. Ser. Inst. Phys.*, **52**, 357 (1980).

W. C. Johnson, *Microstruct. Sci.*, **8**, 49 (1980).

M. Klaua and G. Oertel, in O. Bruemmer, J. Heydenreich, and K. H. Krebs (Eds.), *Solid State Analysis with Electrons, Ions and X-rays. Auger Electron Spectroscopy*, VEB Dtsch. Verlag Wiss., Berlin, 1980.

1981

G. C. Allen and R. K. Wild, *CEGB Res.*, **11**, 12 (1981).

A. Baiker, *Chimia*, **35**, 485 (1981).

J. A. C. Broekaert, *Spectrochim. Acta, Part B*, **36B6**, 1981.

K. Furuya, *Hyomen*, **19**, 673 (1981).

W. Katz, *Microbeam Anal.*, 16th, 2871 (1981).

H. H. Madden, *J. Vac. Sci. Technol.*, **18**, 677 (1981).

G. F. McGuire and P. H. Holloway, *Electron Spectrosc., Theory, Tech. Appl.*, **4**, 1 (1981).

M. Seo, *Nippon Kinzoku Gakkai Kaiho*, **20**, 290 (1981).

M. Soma, *Shokubai*, **23**, 455 (1981).

G. R. Sparrow and I. W. Drummond, *Ind. Res. Dev.*, **23**, 112 (1981).

D. F. Stein and A. Joshi, *Annu. Rev. Mater. Sci.*, **11**, 485 (1981).

1982

J. Cazaux, D. Gramari, D. Mouze, J. Perrin, and X. Thomas, *Conf. Ser. Inst. Phys.* **61**, 425 (1982).

C. R. Helms, *J. Vac. Sci. Technol.*, **20**, 948 (1982).

N. H. Turner and R. J. Colton, *Anal. Chem.*, **54**, 293R (1982).

APPENDIX

2

VARIAN CHART OF AUGER ELECTRON ENERGIES*

The fold-out chart shows the energies of the Auger electrons that arise from the bombardment of the different elements by medium energy electrons (100 eV–10 keV). It is a composite of experimentally observed and identified Auger electron spectra and theoretically predicted Auger peak locations. An approximate indication of the relative sizes of the experimental peaks from each element for which experimental data are available is given by the size of the dot; · represents small peaks, • represents medium sized peaks; and ● represents large peaks. The atomic number of the elements is plotted along the abscissa, and the ordinate represents the energy of the Auger electron. Data are presented for elements up to atomic number 100 over the energy range from 20 eV to 3000 eV on a logarithmic scale.

The lines connecting theoretical points for different atomic numbers identify the type of transition generating those points. The lines are labeled with three subscripted letters. The letters used are the standard x-ray designations for the atomic shells, K being the innermost shell, L the next shell, M the third shell, etc. Transitions involving the valence band are sometimes designated by the letter V in the appropriate step. Subshells are designated by subscript numerals. For example, L_1, L_2, L_3 are the three subshells of the L shell.

The first letter and subscript of a three-letter Auger transition designator indicates the subshell in which the initial vacancy of the Auger transition occurs. The second letter plus subscript gives the subshell from which an electron decays to fill the initial vacancy. The third letter and subscript shows the subshell from which the Auger electron is then released. Thus, the line labeled KL_1L_2 connects all theoretically predicted peaks arising from transitions in which the initial vacancy is created in the K shell. An electron then decays from the L_1 subshell to fill that vacancy and an Auger electron is emitted from the L_2 subshell to conserve energy.

---

*Reprinted with permission of Varian Associates, Palo Alto, California. Chart prepared by Yale E. Strausser and John J. Uebbing, 1971.

387

The theoretical peak positions were calculated using the method suggested by Burhop.[1] The Auger electron energy for a particular transition in an element of atomic number Z is equal to the binding energy of the electron removed in the initial ionization minus the binding energy of the electron that decayed into that vacancy, corrected for the ionized state of the atom, minus the binding energy of the electron that escaped as the Auger electron. The correction is made by using the binding energy of the corresponding electron in the element whose atomic number is $Z+1$. Auger electron energies were calculated for all transitions that can be initiated by a primary electron with an energy of 5000 eV or less and for all elements from $Z=3$ to $Z=100$, using the binding energies given in references 2 and 3. All of these possible Auger peaks were then compared to the experimental data, where available, and those transitions which are actually observed were thus identified. These transitions are the ones shown on the chart.

The experimental Auger electron energies shown here were taken from plots of the first derivative of the energy distribution and represent the points of maximum negative slope, on the high energy side of the peak in the actual energy distribution.

Since the original compilation of these data, equipment for the measurement of Auger spectra has evolved to a level of sophistication that makes it possible to observe differences contributed by work function and chemical shift variations. As a result, variations up to three or four eV are observed in the location of many of the peaks when measured on different samples or in different systems. Transitions involving valence band electrons, as for example in the case of the LMM spectra of Mg, Al, and Si, show very large changes in the relative height of the different peaks in their spectra, as a result of chemical effects.

APPLICATION

This chart may be used in several ways. The primary application is in identifying the peaks in an Auger spectrum of an unknown material (or unknown contaminants on known base material). The best way to do this is to look along lines of constant energy, corresponding to the energy of each peak present in the spectrum, and identify each element which has a peak in that position. Elements can then be eliminated from the list of possibilities on the basis that if an element were present, all of the peaks from that element would appear in the spectrum. If this is not sufficient to eliminate all of the uncertainty, then the relative sizes of the peaks from an element can provide an additional indication of the elements present.

Other uses of the chart include identifying transitions that are observed experimentally and predicting the results of proposed measurements to ensure that they can be made unambiguously.

REFERENCES

1. E. H. S. Burhop, *The Auger Effect and Other Radiationless Transitions*, University Press, England (1952).
2. J. A. Bearden and A. F. Burr, *Rev. Mod. Phys.*, **39**, 125 (1967).
3. R. D. Hill, E. L. Church, and J. W. Mihelich, *Rev. Sci. Instr.*, **23**, 523 (1952).

INDEX